Water Science and Technology Library

Volume 86

The aim of the Water Science and Technology Library is to provide a forum for dissemination of the state-of-the-art of topics of current interest in the area of water science and technology. This is accomplished through publication of reference books and monographs, authored or edited. Occasionally also proceedings volumes are accepted for publication in the series.

Water Science and Technology Library encompasses a wide range of topics dealing with science as well as socio-economic aspects of water, environment, and ecology. Both the water quantity and quality issues are relevant and are embraced by Water Science and Technology Library. The emphasis may be on either the scientific content, or techniques of solution, or both. There is increasing emphasis these days on processes and Water Science and Technology Library is committed to promoting this emphasis by publishing books emphasizing scientific discussions of physical, chemical, and/or biological aspects of water resources. Likewise, current or emerging solution techniques receive high priority. Interdisciplinary coverage is encouraged. Case studies contributing to our knowledge of water science and technology are also embraced by the series. Innovative ideas and novel techniques are of particular interest.

Comments or suggestions for future volumes are welcomed.

Vijay P. Singh, Department of Biological and Agricultural Engineering & Zachry Department of Civil Engineering, Texas A&M University, USA Email: vsingh@tamu.edu

More information about this series at http://www.springer.com/series/6689

Martina Zelenakova
Editor

Water Management and the Environment: Case Studies

Editor
Martina Zelenakova
Technical University of Košice
Košice
Slovakia

ISSN 0921-092X ISSN 1872-4663 (electronic)
Water Science and Technology Library
ISBN 978-3-030-07705-1 ISBN 978-3-319-79014-5 (eBook)
https://doi.org/10.1007/978-3-319-79014-5

Printed on acid-free paper

This Springer imprint is published by the registered company Springer International Publishing AG part of Springer Nature
The registered company address is: Gewerbestrasse 11, 6330 Cham, Switzerland

Preface

Water is a fundamental natural resource. It is a vital component of the natural environment, but it is also a basic prerequisite for all human economic and social activities in general. Water is a form of wealth which requires protection; its usage needs to be regulated, and its supply needs to be regenerated. Water may be continuously renewed in nature, but only on the precondition that the fundamental principles of its protection are respected. Everybody carrying out some activity which may affect the state and relations of surface and underground waters has the obligation to make all necessary efforts for their preservation and protection (Fig. 1).

The first step towards effective protection of water resources is to know their size and distribution, but also their quality. Systematic investigation and evaluation of the occurrence and condition of surface and underground waters within the European countries as well as worldwide is a basic responsibility of the state, as an indispensable requirement for ensuring the preconditions for permanently sustainable development as well as for maintaining standards of public administration and information. Sustainable development of water management is based on the principle that water as a natural resource may be utilized only to that extent which

Fig. 1 Water in the environment (*Photo* Fialová)

ensures future generations sufficient usable supplies of water in the seas, rivers, lakes and reservoirs, and that reserves contained in porous environments below the surface of the land remain preserved in the same quantity and quality. It is evident that surface waters are more vulnerable than those underground in terms of their hygienic quality and safety, but also of their protection as a natural ecosystem and maintenance of their amounts. They are an important medium regarding the transport, decomposition and accumulation of pollutants, whether of natural or anthropogenic origin, which in excessive amounts represent considerable risks for all kinds of living organisms, thus also for human beings. For this reason, it is necessary to devote all the more attention to the protection of water sources. The basic requirement in this context is to optimize their monitoring, the assessment of their quality and the implementation of necessary environmental measures.

The essence of protection of the environment in general and the water environment in particular lies in establishing a system enabling acceptable development of anthropogenic activity while preserving the quality of the environment, natural resources, ecosystems and health. Application of a process of evaluation of environmental risks in the conditions of water-flow catchment areas has now become indispensable. It is necessary to engage in environmental risk assessment in response to highly-intensive anthropogenic activity in river basins. This consists primarily in agricultural and industrial production and the associated building and operating of production plants, forestry management, waste disposal solutions, infrastructure development, as well as water management installations.

The benefit of this book lies in bringing together scientists, researchers, academics and lecturers in the field of water management to share experiences and successes in addressing water management. It deals with a wide variety of water resource management issues from water quality to water quantity, considering all impacts of water management on the environment. The book presents international approaches to utilizing the latest developments in both the theoretical basis and the applicability of state-of-the-art knowledge which can be effectively used for resolving a variety of pressing problems in integrated water resource management. The main problems focused on in the book are water pollution, whether physical, chemical or biological, and hydrology issues including limnology projects considered also from the geographical and human point of view.

The editors would like to thank the authors for their constructive contributions and the publisher for providing the opportunity for this edition.

The authors are responsible for the quality of the text, the quality of the language and the compliance of the copyright rules set by Springer.

Košice, Slovakia
February 2018

Martina Zelenakova

Contents

Editor and Contributors

About the Editor

Martina Zelenakova is currently an Associate Professor at the Institute of Environmental Engineering, Faculty of Civil Engineering, the Technical University of Košice. In the context of her scientific research activities she has focused on environmental impact assessment, water management, and the assessment of environmental risks in river basins in relation to flood events, drought and sources of pollution. The results of her work have been published in books, national and international journals, and national and international conferences proceedings. Dr. Zelenakova has engaged in national and international projects, also as the principal investigator.

Contributors

Zuzana Boukalová VODNÍ ZDROJE, a.s., Prague 5, Czech Republic

Jitka Fialová Department of Landscape Management, Faculty of Forestry and Wood Technology, Mendel University, Brno, Czech Republic

Andrzej Gałaś AGH University of Science and Technology, Kraków, Poland

Slávka Gałaś AGH University of Science and Technology, Kraków, Poland

Florina Grecu Department of Geomorphology, Pedology and Geomatics, Faculty of Geography, University of Bucharest, Bucharest, Romania

Binod Das Gurung CISD, Kathmandu, Nepal

Ketil Haarstad Research Professor, Division of Environment and Natural Resources, NIBIO—Norwegian Institute of Bioeconomy Research, Oslo, Norway

Petr Hlavínek Centre AdMaS, VUT Brno, Brno, Czech Republic

Gabriela Ioana-Toroimac Faculty of Geography, University of Bucharest, Bucharest, Romania

Mariana Jakubisová Arborétum Borová Hora, Zvolen, Slovakia

Daniel Kahuda VODNÍ ZDROJE, a.s., Prague 5, Czech Republic

Piotr Klimaszyk Department of Water Protection, Faculty of Biology, Adam Mickiewicz University, Poznań, Poland

Istvan Kocsis National Administration "Romanian Waters", Cluj-Napoca, Romania

Pavla Kotásková Department of Landscape Management, Faculty of Forestry and Wood Technology, Mendel University, Brno, Czech Republic

Włodzimierz Marszelewski Faculty of Earth Sciences, Department of Hydrology and Water Management, Nicolaus Copernicus University in Toruń, Toruń, Poland

Adam M. Paruch Division of Environment and Natural Resources, NIBIO—Norwegian Institute of Bioeconomy Research, Aas, Norway

Lisa Paruch Division of Environment and Natural Resources, NIBIO—Norwegian Institute of Bioeconomy Research, Aas, Norway

Bożena Pius Faculty of Earth Sciences, Department of Hydrology and Water Management, Nicolaus Copernicus University in Toruń, Toruń, Poland

Pavlína Procházková Department of Landscape Management, Faculty of Forestry and Wood Technology, Mendel University, Brno, Czech Republic

Gheorghe Romanescu Department of Geography, Faculty of Geography and Geology, Alexandru Ioan Cuza University of Iasi, Iasi, Romania

Piotr Rzymski Department of Environmental Medicine, Poznan University of Medical Sciences, Poznań, Poland

Daniel Sabău National Administration "Romanian Waters", Cluj-Napoca, Romania

Agnieszka Stec Department of Infrastructure and Water Management, The Faculty of Civil and Environmental Engineering and Architecture, Rzeszow University of Technology, Rzeszów, Poland

Petrică Stroi National Administration "Romanian Waters", Cluj-Napoca, Romania

Răzvan Stroi National Administration "Romanian Waters", Cluj-Napoca, Romania

Daniel Słyś Department of Infrastructure and Water Management, The Faculty of Civil and Environmental Engineering and Architecture, Rzeszow University of Technology, Rzeszów, Poland

Gheorghe Şerban Faculty of Geography, Babeş-Bolyai University, Cluj-Napoca, Romania

Miloslav Šlezingr Department of Landscape Management, Faculty of Forestry and Wood Technology, Mendel University, Brno, Czech Republic

Jan Těšitel METCENAS o.p.s., Plzeň, Czech Republic

Liliana Zaharia Faculty of Geography, University of Bucharest, Bucharest, Romania

Adéla Žižlavská Centre AdMaS, VUT Brno, Brno, Czech Republic

Part I
Water and Landscape

Chapter 1
Possible Use of Water Areas by Disabled People

Pavla Kotásková, Jitka Fialová, Mariana Jakubisová, Miloslav Šlezingr and Pavlína Procházková

Abstract People with reduced mobility have specific needs regarding their access to areas within the landscape. This follows not only from the findings of a questionnaire survey, but also from the experience with designing barrier-free buildings and their surroundings, where for example in the Czech Republic, it is necessary to respect Decree no. 398/2009, on general technical requirements for barrier-free use of buildings. This chapter presents the basic parameters for designing roads in the countryside as well as recreational areas and access roads to water bodies, view points, fishing places or just water. Stable banks need to be selected for access to water bodies. The bank stability can be enhanced by means of biotechnical modifications. Examples from Slovakia and the Czech Republic show that pleasant places accessible for all can be created also in the vicinity of water bodies if a universal design is used in which persons with reduced mobility are considered.

Keywords Wheelchair users · Surface · Fishing · Swimming
Trails for disabled people · Universal design

P. Kotásková (✉) · J. Fialová · M. Šlezingr · P. Procházková
Department of Landscape Management, Faculty of Forestry and Wood Technology, Mendel University, Zemědělská 3, 613 00 Brno, Czech Republic
e-mail: pavlakot@mendelu.cz

J. Fialová
e-mail: jitka.fialova@mendelu.cz

M. Šlezingr
e-mail: miloslav.slezingr@mendelu.cz

P. Procházková
e-mail: xproch41@node.mendelu.cz

M. Jakubisová
Arborétum Borová Hora, Borovianska Cesta 2171/66, 960 53 Zvolen, Slovakia
e-mail: mariana.jakubisova@tuzvo.sk

M. Zelenakova (ed.), *Water Management and the Environment: Case Studies*, Water Science and Technology Library 86,
https://doi.org/10.1007/978-3-319-79014-5_1

1.1 Introduction

When creating a suitable barrier-free natural environment around water bodies or rivers (water areas), it is necessary to be acquainted with the requirements of people with reduced mobility as well as the parameters of their movement. People with reduced mobility are wheelchair users, persons with pushchairs and people accompanying children younger than three years, the elderly, pregnant women and persons with crutches, canes, walkers or other walking aids. People with physical disabilities form a large and very diverse group.

The situation is most difficult for wheelchair users, which is why this group was selected as the topic of this chapter.

The current trend is to discover new ways allowing people with physical disabilities to choose their own way of life and live without relying on the help of others, needing constantly to ask for aid, as much as possible. The basic prerequisite for somebody's active participation in social life is the accessibility of areas and buildings, and the ability to use them and move freely in them. This means the fulfilment of the right to freedom of movement in the broadest sense of the word (http://www.czp-msk.cz/pdf/uzitecne/ATHENA_PRIRUCKA_KOMPLET.pdf).

It is necessary to design not only paths and parking lots, but also access to water and hiking trails, meaning walking paths, bike paths or educational trails, in order to allow wheelchair users independent, safe, easy and smooth motion. Additionally, the wheelchair users need to be able to pass by other pedestrians or even bicycles in a natural environment. Routes already implemented abroad are good examples (Fig. 1.1).

However, the issue of passing by or conflicts with other pedestrians, skaters, but especially cyclists, needs to be addressed specifically. Discussions with wheelchair users have shown that this is a great problem for some of them. Some wheelchair users negatively evaluate the behaviour of cyclists on their shared trails. These trails

Fig. 1.1 Signpost directing the viewer to a rest area with a view of the scenery—adapted for wheelchair users—Scotland (*Photo* Fialová)

are thus dangerous for them. If the financial situation and the spatial arrangement allow, it is appropriate to propose separate paths for these two groups.

The issue of providing comparable conditions for recreation and its full use by people with reduced mobility has been dealt with abroad for a significantly longer time than in the Czech Republic. Some countries have created detailed methodological guides or recommendations which present the needs of these people and simple solutions to the satisfaction of their needs, so that people with reduced mobility can actively participate in recreation. The range of activities available is very varied. There are opportunities to go fishing, sailing, go into water and bathe, sit by the fire or use relaxation points. The measures that allow people with reduced mobility access to recreational activities are in fact usually very simple and are not demanding in terms of the material used. An example can be the guide by Ylva Lundell, a researcher from the Swedish University of Agricultural Sciences.

Sweden is a good example of a country providing high-quality solutions to barrier-free access, as this issue has been legally regulated there since the 1960s. An active approach to this issue is taken by the whole of society. For example, authorities, public institutions and private companies adopt policies on removing barriers. Any initiative related to this active approach is perceived very positively and is often used for promotion. Ordinary citizens are active as well, as they point out the existence of barriers and ask for their removal. The result of this society-wide activity has been the adoption of a law which considers the inaccessibility of public spaces as discrimination with all its consequences. Confirmation of this active and correct approach lies in the fact that before the regulations are adopted, there is extensive verification of the proposed measures, in the form of practical studies with the participation of people with various limitations. The result is then a proposal of an appropriate solution that takes account of the needs of all these people. Sweden also has various provisions in the building code which require modifications to existing public buildings and spaces leading to barrier-free access. The guarantor of the construction modifications, including their barrier-free solutions, is a state authority which also enforces the related legislation. The Swedish capital, Stockholm, strives to be recognized as the most accessible capital in the world. It is clear, therefore, that it takes measures in all areas of social life to make it available for people with reduced mobility and orientation, and to eliminate the prejudices within society concerning them (Antonovičová 2014).

This topic has already been looked into by researchers, such as Loučková and Fialová (2010), Jakubis and Jakubisová (2012), Jakubis (2013, 2014), Junek and Fialová (2012), Kotásková and Hrůza (2013), Jakubisová et al. (2015).

There are many among us who like a quiet corner of nature, and others who like to climb rocky peaks, descend into valleys and breathe fresh air on the banks of roaring torrents. However, there are also people who, due to their medical condition, are unable to do so and we have somehow omitted to give them the opportunity (Junek and Fialová 2012).

It is important to provide all people with impaired mobility with some compensation for their limited movement. Unless society as a whole is aware of this, they may become excluded from social life. Support for people in wheelchairs can help them engage in various activities that will keep them active and prevent their social exclusion (Vítková 2006).

The proposals presented in this chapter are based on results of the projects financed by the Internal Grant Agency of Mendel University in Brno and project financed by the Visegrad Fund—Trails for disabled people in the V4 Countries.

1.2 Convention on the Rights of Persons with Disabilities

The UN Convention on the Rights of Persons with Disabilities (the Convention) is a very important document which deals with the rights of people with disabilities. Article 30 of the Convention, entitled "Participation in cultural life, recreation, leisure and sport", states that people with disabilities have the right to take part in cultural life, sport and recreational activities and tourism equally to others. Through physical activity, they improve their physical condition and mental health, as well as endurance and courage. In order to fulfil one of the objectives of the Convention, all spaces (both architectural and outdoors) should be wheelchair accessible, including tourist trails in the countryside. In the concept of these proposals, it is also very important to remove barriers to communication, bearing in mind that they are most limiting for people with hearing disabilities. In the practical management of removing barriers to accessibility of environments, products, information and services, it is necessary to implement many systemic measures, which should be then translated into legislative regulations and control mechanisms, such as: inclusion of principles of universal accessibility into national programs; adoption of control mechanisms for compliance with accessibility for all; promotion of and education in issues of accessibility and universal design; and support for related research.

1.3 Universal Design and Disabled People

In the concept of universal design for barrier-free access, it is important to know the needs of the people and conditions of the environment in which the individual person will move (e.g. wheelchair users, mothers with strollers, visually impaired pedestrians). In its implementation, the following principles are of importance: flexibility in use, simple and intuitive use, perceptible information. We can use these seven "Principles of Universal Design" to evaluate existing designs, not only in architectural built environments but also in proposals for wheelchair access to facilities in the natural landscape (thus in fact in any environment): size and space for approach and use (appropriate size and space is provided for approach, reach, manipulation, and use regardless of user's body size, posture, or mobility), flexibility in use (the design accommodates a wide range of individual preferences and abilities), equitable use (the design is useful and marketable to people with diverse abilities), tolerance for error (the design minimizes hazards and the adverse consequences of accidental or unintended actions), simple and intuitive use (use of the design is easy to understand, regardless of the user's experience, knowledge, language skills, or current concentration level), low physical effort (the design can be used efficiently and comfortably and with a minimum of fatigue), perceptible information (the design communicates necessary information effectively to the user, regardless of ambient conditions or the user's sensory abilities). (https://www.thefreelibrary.com/Segs4Vets+making+mobility+accessiblea0179736624), North Carolina State University.

1.4 Requirements for Wheelchair Accessibility in Slovakia

Legislation on designing the Forest Transportation Network in Slovakia and associated legal standards are: STN 73 6101 Design of Roads and Motorways; STN 73 6110 Design of Local Roads; STN 73 6108 Forest Transportation Network; TP 10/2011 Technical Conditions—Design of debarrierization measures for persons with reduced mobility and orientation on roads, MDVRR SR: 2011; Decree of the Ministry for Home Affairs (MV) of the Slovak Republic No. 9/2009 Coll. implementing the Law on Road Traffic and amending and supplementing certain other laws, as amended; ResAP (2007)3: Resolution ResAP (2007)3 "Achieving full participation through Universal Design"; Decree of the Ministry for the Environment (MŽP) of the Slovak Republic No. 532/2002 Coll. specifying details of general technical requirements for construction and general technical requirements for structures utilized by persons with limited movement and orientation abilities; Law No. 317/2010 Coll. ratifying the UN Convention on the Rights of Persons with Disabilities; Communication from the Ministry for Foreign Affairs (MZV) of the Slovak Republic No. 317/2010 Convention on the Rights of Persons with Disabilities.

The research conducted focused on the preparation of materials concerning the building regulations in Slovakia. Universal design for implementation of barrier-free hiking trails (hereinafter the "BHT") in forest areas of the Slovak Republic recommends the following principles: unobstructed width of the BHT should be at least 1800 mm; it may be narrowed to a width of 900 mm only where justified, for example when terrain demands or technical equipment is installed; the longitudinal slope of a BHT section should be 1:21 (4.8%) at most and any section with such slope should not be longer than 20 m; if it is longer, it must be interrupted by a flat terrain with a relaxing bench; if the longitudinal slope of a BHT section exceeds the recommended value, it must be designed as a ramp equipped with handrails in compliance with construction regulations; the transverse slope of the BHT may be 1:50 (2%) at most; head clearance must be at least 2200 mm; the BHT surface must be even and hardened to be usable for a person with a wheelchair, walking aids, stroller and similar; if the BHT surface is made of a metal grid, the maximum size of the grid holes must be 20 mm × 20 mm; if the BHT surface is made of balks, they must be placed transversely to the direction of movement and the spaces between them may not be wider than 10 mm; if the BHT surface is made of natural stone tiles, the spaces must not be wider than 10 mm; naturally, the BHT surface must be non-slip; BHT sections where a risk of falling is imminent must have a solid sidewall with filler at a height from 100 to 1100 mm; any bridge or footbridge on the BHT with a height up to than 500 mm above the ground must have an elevated rim on both edges up to a height of 100 mm or a guide rail at a height of 300 mm; a bridge or footbridge on the BHT with a height over 500 mm above the ground must have an unobstructed width of at least 900 mm; solid sidewalls with filler must be placed on both edges of the bridge at a height from 100 to 1100 mm; rest areas must be placed along the BHT (the recommendation is one rest area after per 200 m) outside the main course of the route, with benches and area for parking wheelchairs or strollers; benches for a rest must have backrests and armrests; there must be a hardened surface beside the bench to park a wheelchair or a stroller; multi-sensory means must be used for the information and orientation systems of the BHT (e.g. relief maps, plans, labels with inscriptions); there should be wheelchair-accessible toilets along the hiking trail. Wheelchair-accessible hiking trails in open countryside are designed for locations with favourable terrain and moderate slopes. They should be formed as a closed circuit with various lengths of routes, to provide a variety of choice and suit the demands of various users. Their design is related to the current legislation. These trails are usually not designated for one type of movement or user only—they can be used by others, for example cyclists.

1.5 Requirements for Wheelchair Accessibility in the Czech Republic

Decree No. 398/2009 on general technical requirements ensuring barrier-free use of buildings stipulates the obligation to propose adjustments in order to facilitate independent movement of persons with reduced mobility.

For passage, a wheelchair user needs a path width of 900 mm. However, where the wheelchair user needs to pass by a pedestrian or where turn, a width of 1500 mm is necessary. In the case when two people in wheelchairs need to pass by each other, a width of 1800 mm must be provided. The minimum space for turning a wheelchair by 90° to 180° is a rectangle sized 1200 mm × 1500 mm. The minimum handling space required for turning the wheelchair in different directions in angles greater than 180° is a circular area with 1500 mm in diameter (Fig. 1.2).

Therefore, pavements and hiking trails must be at least 1500 mm wide, ideally 2000 mm. Obstacles on the path, such as benches, information boards and trees must be placed so that the area for walking along them is at least 1500 mm wide. In justifiable cases, these can be placed so that the walking area is narrowed to 900 mm, but only locally.

The following general requirements must be met for barrier-free use by people with reduced mobility: the maximum elevation of a step a person in a wheelchair can overcome is 20 mm; the maximum longitudinal gradient of the path must be 1:12 (8.33%), or 1:8 (12.5%) on ramps not longer than 3 m; the maximum transverse

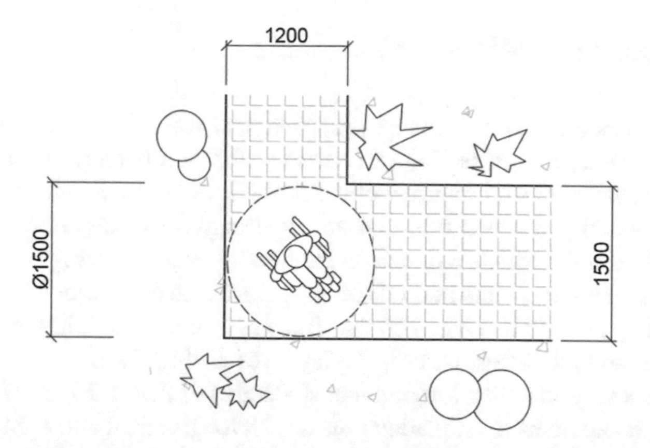

Fig. 1.2 Minimum handling space required for wheelchair turning (*Prepared by* Kotásková)

Fig. 1.3 Stairs and ramp leading to a pier which serves as a boat transport stop (*Photo* Kotásková 2015)

gradient should not exceed 1:50 (2%); fat bridge structures, the gradient must not exceed 1:40 (2.5%) with a minimum width of 900 mm. At gradients greater than 1:6, for example at the ends of a bridge, the wheelchair user might fall forward from the wheelchair. When going up, there is a real risk of the wheelchair toppling backwards. The wheelchair user cannot ride up a steep ramp without outside help (Filipiová 2002).

It is advisable to design different measures and solutions together at a single point on the path for various groups to overcome barriers. For example, to overcome a step in the terrain, people with reduced mobility might make use of a ramp with a smaller incline, cyclists might use a ramp with a steeper incline, and the others might use stairs (Fig. 1.3).

1.6 Proposing Trails and Viewpoints

Roads as well as hiking trails can be proposed near water bodies or rivers in some areas. The advantage of areas along rivers is mostly their flat terrain, so the paths in the vicinity usually have minimum inclinations. It is not necessary for the whole trail to run in the vicinity of water. It is advisable to design closed circuits or create a link to a network of trails which will enable shortening or lengthening of the trail. It is desirable to place viewpoints and relaxation points with comfortable benches along the trails. They need to be placed so that free passage of min. 1500 mm remains at any point on the trail, at best by creating lay-bys (see Fig. 1.4).

If there is a section with a longitudinal slope greater than 1:20 (5.0%) and longer than 200 m, it should have relaxation points with longitudinal and transverse slopes of not more than 1:50 (2.0%). When implementing these relaxation points, space needs to be left for a wheelchair or a stroller to be parked next to the bench or table (Fig. 1.5).

It is often not required to provide direct access to water; only a viewpoint in close proximity to the bank can be designed for the needs of the wheelchair user. The banks must be reinforced, mostly with simple vegetation adjustments.

The most important forest-type groups correspond to the structure of riparian stands based on the systematic division proposed by Mezera (1956). Non-autochthonous, non-indigenous, introduced, exotic and fruit species should be eliminated. Within riparian stands the most frequently used species are willow (*Salix*), alder (*Alnus*), elm (*Ulmus*), maple (*Acer*), ash (*Fraxinus*) and poplar (*Populus*). With respect to shrubs, the most frequent ones are in particular dogwood (*Corn*us), shrub willow (*Salix*), buckthorn (*Frangula*), hawthorn (*Crataegus*) and spindle tree (*Euonymus*). Accompanying stands can be made up of maple (*Acer*), ash (*Fraxinus*), lime (*Tilia*), elm (*Ulmus*), hornbeam (*Carpinus*), English oak (*Quercus robur*); disseminated mazzard (*Cerasus avium*), birch (*Betula*) or crane (*Sorbus*); the undergrowth may consist, for instance, of honeysuckle (*Lonicera*), hazelnut tree (*Corylus*) or privet (*Ligustrum*) (Šlezingr and Lichtneger 2011). The example of the riparian stand from the Czech Republic can be seen in the Fig. 1.6.

However, some places, particularly around water bodies, have significantly damaged banks, where erosion walls are formed and the soil is washed away. At these places, landslides are possible and so the viewpoint is dangerous.

Biotechnical revetment is one of the most suitable types of bank stabilization. Its technical element is placed in the most stressed part of the bank, and the vegetation elements are used in the less stressed parts (wave run-up zone). The elements overlap. Possible wooden stabilization structures are fences made from pole timber (Fig. 1.7), single- or multi-line woven fences, fascine and fascine-gravel cylinders, vegetated cabins, or vegetated flat bands of rubble masonry. Technical stabilization methods can also be used, e.g. stone used at the bottom of slopes, paving, stone rip-rap, prefabricated revetment, concrete or reinforced concrete retaining walls, or gabions (Ondrejka et al. 2013a).

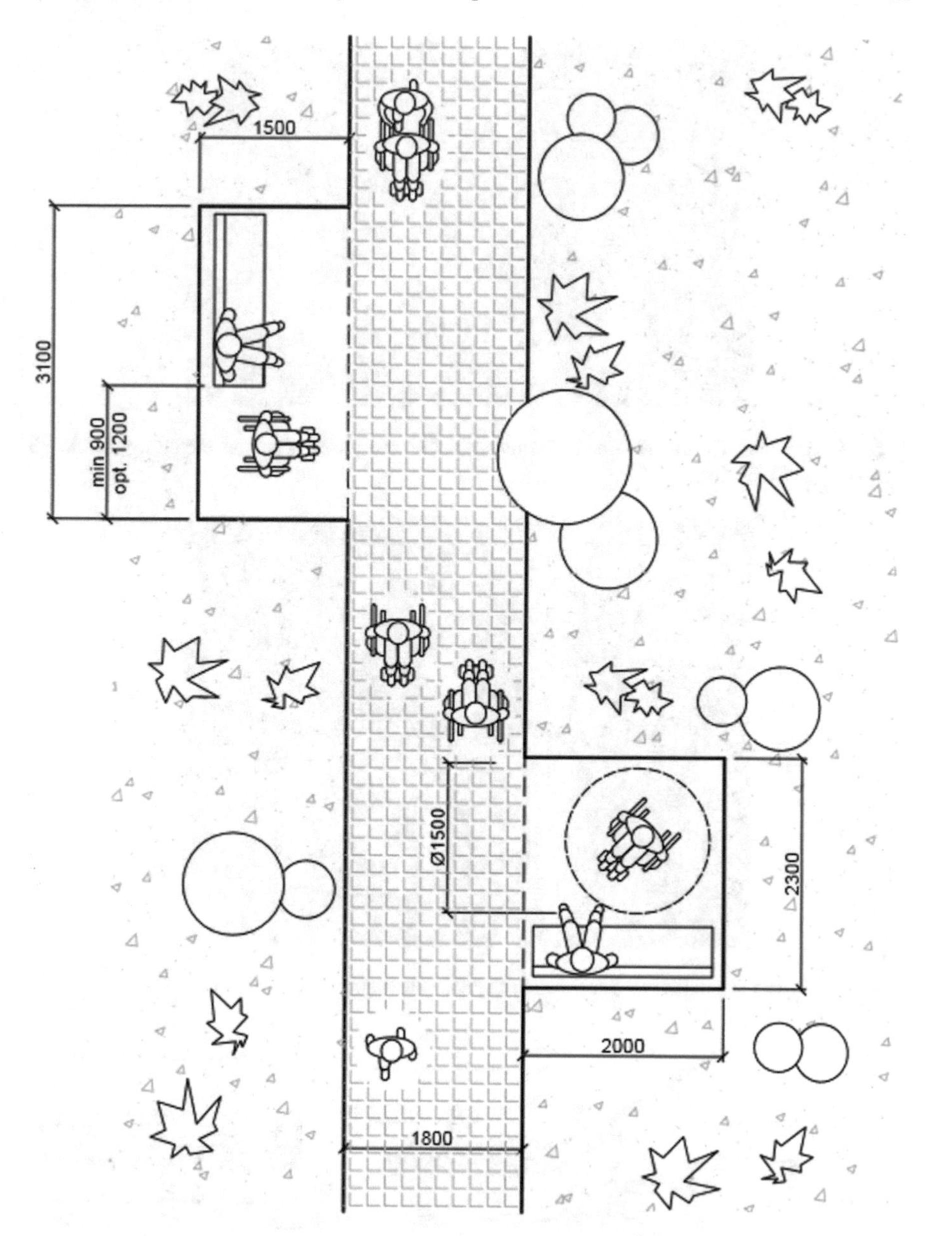

Fig. 1.4 Examples of the design of relaxation points (*Prepared by* Kotásková)

It is generally assessed that the most suitable type of revetment is a well-founded stabilization made of quarry stone of the required size together with stone rip-rap or rip-rap vegetated with willow cuttings at the bottom of the slope, and then grass cover and bank-side trees at the top of the slope (Fig. 1.8) (Šlezingr 2004).

Fig. 1.5 Example of an implemented relaxation point for every kind of person, including wheelchair users (*Photo* Kotásková)

Fig. 1.6 Natural vegetation on the riverbank (*Photo* Šlezingr)

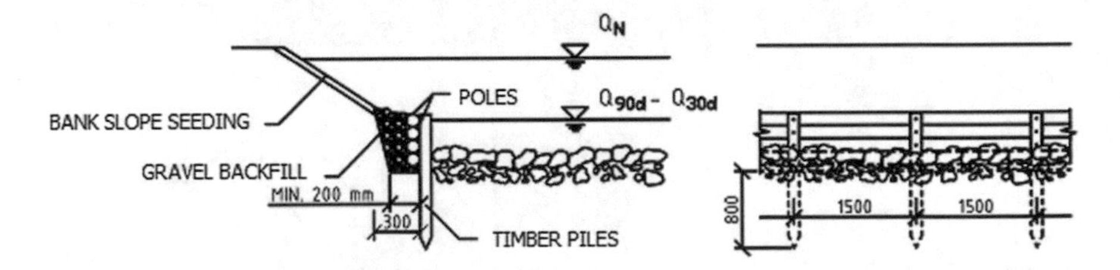

Fig. 1.7 Pole timber fences (*Prepared by* Kotásková)

The gabion is a very suitable stabilization element (Fig. 1.9) as well as other popular wire-stone structures, e.g. wire baskets filled with quarry stone. Gabions copy small uneven places in the terrain and water can flow through them. Stone is a natural material, and thus, it suits the landscape better than prefabricated concrete

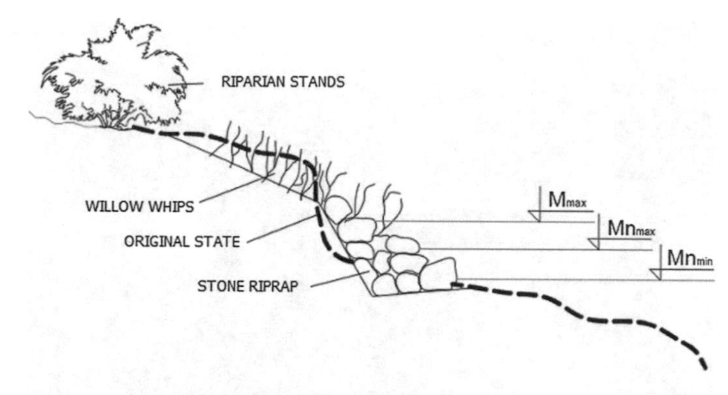

Fig. 1.8 Vegetated stone rip-rap (*Prepared by* Šlezingr)

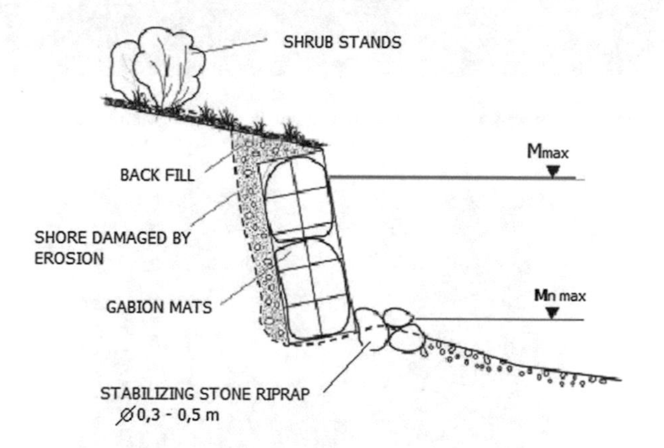

Fig. 1.9 Placement of gabion for bank revetment (*Prepared by* Šlezingr)

panels or vegetation blocks, especially when vegetation begins growing through the stone (Ondrejka et al. 2013b).

Relaxation points and viewpoints, or fishing spots, can be made using a paved area or wooden footbridges (Fig. 1.10). Additionally, they can be placed above the body of water on a bridge structure (Fig. 1.11). Even large-span wooden bridges for high loads are possible when sufficiently strong load-bearing materials are used; especially useful in this case is pasted laminated wood. Wood is able to sustain short-term overloading without adverse effects. Thanks to the development of connecting elements, it is possible to design modern and highly aesthetical platforms and bridges (Kotásková and Hrůza 2013).

Ideally, the path surface should suitably fit in the natural environment; however, it is not suitable to use unpaved surfaces or surfaces paved with a non-cemented material for wheelchair users, as these surfaces become unusable, especially after spring thaw or long-lasting rains. Therefore, the surface of a path which leads along a

Fig. 1.10 Making the surroundings of a watercourse accessible for wheelchair users—Finland (*Photo* Fialová)

Fig. 1.11 Example of making a watercourse and viewpoints accessible for wheelchair users (a section of the barrier can be opened) in Finland (*Photo* Fialová)

Fig. 1.12 Path drainage—barrier for wheelchair users on an otherwise accessible trail (*Photo* Kotásková)

stream must be paved with either close-to-natural means, bitumen or concrete, based on the character of the surroundings and the frequency of use.

The only recommendable non-cemented surface is a layer of mechanically reinforced aggregate with maximum fraction size of 32 mm.

Mechanically reinforced aggregate can be a suitable surface, provided that the finest fraction size is used when building the road and correct technological procedures are followed. This type of surface is mainly recommended for public spaces and parks due to its properties, which are in particular strength and durability. However, it should be noted that this surface holds water, and when it is used it must have proper drainage (Calkins 2009; Axelson and Chesney 1999).

Other types of non-cemented layers with maximum fraction size of 63 mm are not recommendable with regard to the width of tires and size of the wheels of non-motorized means of transport (Hrůza 2015).

Additionally, the surface can be paved with compact asphalt layers. It is also possible to use tiles with minimal joints (interlocking concrete pavers), though the material might not always fit in with the environmental character.

Asphalt surfaces seem to be the most suitable coverings, as the surface is even, the risk of slipping is reduced, movement is comfortable, and these surfaces are also highly functional and durable. The highest quality asphalt covers commonly used are coated aggregate, asphalt concrete or asphalt blanket. Pavements can be made from cast asphalt with concrete underlay. Currently, the most common cover for paths with asphalt pavement is penetration macadam. However, for this purpose it is probably the least suitable one of all the bituminous covers. In the structure of this pavement aggregate of rough fractions (0–32, or up to 64 mm) is used, which leads to a rough surface due to the more pronounced structure of the material (Juško 2015).

The path must not include any barriers such as drainage or soakaways reinforced with stone (see Fig. 1.12).

A lawn, especially if it is well maintained, is very well perceived as a natural surface. However, it must be remembered that wheelchair users need to exert much

Fig. 1.13 Inappropriate design of pavement with a timber surface (*Photo* Kotásková)

more effort on grass than when moving on a totally smooth surface, and a grass surface can also be damaged by the wheels of wheelchairs. This damage can be prevented by the use of grass stabilization plastic mats (Axelson and Chesney 1999; Calkins 2009).

Timber surfaces are suitable for paths, provided that the timber used is of high quality, strong and well-treated, and larger spaces are left in the construction because the volume of timber can change depending on the air humidity. If these requirements are not complied with, the use of this surface can cause considerable difficulties. Woodchips are not an appropriate surface for people with reduced mobility in general (Axelson and Chesney 1999).

When timber is used, its long life should be ensured by means of constructional protection. Timber should not rest directly on the ground. This causes degradation due to increased humidity and thus also moisture. Moisture in timber brings about volume changes, so it is not appropriate to fill the spaces between the timber elements with concrete, as seen in Fig. 1.13.

Wooden corduroy walkways are used in waterlogged or otherwise inaccessible areas. The walkable area is placed between longitudinal beams and consists of prisms—longitudinally cut logs or planks attached to the beams with nails. The joints between the individual elements should be at least 1 cm so that water can flow away from the surface. The wooden elements must not be placed parallel to the demanded movement as the longitudinal joints and potential unevenness of the individual wooden components could form traps for wheelchair users (Fig. 1.14). Additionally, there should be guiding rails taller than 60 mm to prevent potential falling off the path in the case of swerving (Fig. 1.15). A similar design can be used for the viewpoints and wooden piers.

No decree or regulation covers this issue, and there are no binding rules. The Czech Fishing Union made an attempt to cover this gap and published recommendations concerning the fishing spots suitable for disabled fishers on its website.

According to the recommendations, a place suitable for disabled fishers should bear the symbol of a disabled fisherman and should be included in the list registered by the Czech Fishing Union. However, the symbol does not mean that the place is reserved for disabled anglers. It only has informative character and

Fig. 1.14 Unsuitable design of a corduroy road (*Photo* Kotásková)

Fig. 1.15 Suitable surface of a corduroy trail for wheelchair users (*Photo* Fialová)

points out the suitable access, parking and fishing places for people with physical limitations (https://www.rybsvaz.cz/?page=reviry%2Fztp%2Fztp&lang=cz&typ=mpr&id_svaz=&moznosti_rybolovu_prehled=ztp#zalozka).

The places suitable for fishing are those with good access to water in a wheelchair (preferably a paved access road without mud, sand, larger bumps or other complications), where the wheelchair can stop in a suitable flat place on the bank, and the fish can be caught and taken out using a common landing net without the assistance of another person.

With regard to the recreational use of places for bathing by people with reduced mobility, it is essential that the beaches and approaches are adapted for their easy access (Fig. 1.16).

The parameters of public beaches and water access are not defined in any binding regulation in the Czech Republic. There are mainly grassy and sandy beaches in the Czech Republic. However, movement by a disabled person on a sandy or grassy

Fig. 1.16 A suitable access road near a water body: the asphalt surface is laid on a base made from mechanically reinforced aggregate which is well compacted (*Photo* Kotásková)

Fig. 1.17 An example of a lifting device to move wheelchair users into the water (*Photo* by ALTECH, spol. s r.o.; http://www.altech.cz/produkty/bazenovy-zvedak/bazenovy-zvedak-delfin/)

beach is difficult. Additionally, water access provided by means of steps with rails can only be used by some of the disabled. Another option to get out of the wheelchair into water involves using a suitable lifting device placed on a pier (wooden platform) above the water surface (see Fig. 1.17).

The device must be characterized by simple handling and easy fitting to the construction of the pier, or the bottom. The device, which is currently manufactured and used in swimming pools, does not require installation under water, power supply or motor; only pressure from standard water supply is needed. The connection of the water supply with a valve located on the jack is effected using a hose with an internal

Fig. 1.18 Board on a nature trail; positioned at a suitable height for wheelchair users—Scotland (*Photo* Fialová)

diameter of 10 mm. This device can be used in places where there is the mains water supply (0.4 MPa). The lifting device is operated using a lever. A special safety lock secures the seat until the user is comfortably seated. The movement of the seat is provided by water pressure, which releases the safety lock in the upper position of the jack. The seat is made of polypropylene and can be loaded with a weight up to 110 kg. On customer request, the pool lift comes with a clamping belt for maximum safety and comfort. The jack is commonly installed in swimming pools with a minimum depth of 1100 mm. Wheelchair users can use it for all water sports and activities, if it is located near water reservoirs or other water bodies where swimming is possible (http://www.altech.cz/produkty/bazenovy-zvedak/bazenovy-zvedak-delfin/).

Around rivers and water bodies, there can be trails designed as educational with information boards (Fig. 1.18). For people in wheelchairs, appropriate access and sufficient space in front of the information board should be provided to allow turning of the wheelchair. The information board should be of suitable height corresponding to the eye-height of people in wheelchairs. The centre of the information board should not be higher than 120 cm above the ground (Lundell et al. 2005).

Technical solutions of proposal trails in urban areas (Figs. 1.19, 1.20 and 1.21) are different in comparison with proposals for the countryside.

As an example of differences in barrier-free proposals near the river within a town, we present the EUROVEA waterfront in Bratislava (Fig. 1.22) and the waterfront in Lyon (Fig. 1.23). The difference in the examples of proposals and in approaches to creation of the waterfront with barrier-free access is evident. Why is the issue of barrier-free access or universal principles so important? Because the creation of a thoughtful environment must be characterized by tolerance for human diversity, for weaker and older people, and for all users if possible. In areas for recreation with landscape protection, it is important to create a barrier-free environment, paying respect to cultural heritage, natural attractions, places with a view of the countryside, rivers, birdwatching, wildlife, plants and trees. People with disability have either no possibility of access or limited access to tourism.

Fig. 1.19 View of Horné Lánice Park near the River Hron in Zvolen (Slovakia) with the possibility of leisure activities "for All" (*Photo* Jakubisová)

Fig. 1.20 Hiking trail near the River Hron in Zvolen (Slovakia), suitable for wheelchairs. In the background, an artificial canal built for water slalom-kayaking (*Photo* Jakubisová)

Fig. 1.21 Detail of arch bridge with wheelchair access over a tributary of the River Hron (Kováčovský potok) on hiking path in Zvolen, Slovakia (*Photo* Jakubisová)

1.7 Example of the Study—Počúvadlo (Slovakia)

Here is an example (Fig. 1.24) of other possibilities for movement by wheelchair users around bodies of water such as lakes or water reservoirs suitable for recreation and tourism in Slovakia. The natural conditions of trails for wheelchair users are specific, and proposals should be close to being formally standardized. Interventions in the

Fig. 1.22 Bratislava waterfront (*Photo* Jakubisová)

Fig. 1.23 Lyon waterfront (*Photo* Jakubisová)

countryside (e.g. in a Protected Landscape Area) are subject to strict laws and may be allowed only in exceptional cases. We therefore propose parameters and conditions different from the standards stipulated for architectural design of buildings. For these reasons, we suggest attractive places for movement in a wheelchair that do not need any major intervention or changes in the environment and are suitable for tourist and recreational activities for wheelchairs. The protected areas may still be in the "pleasure periphery", or they are destinations for sustainable tourism activities.

Information about the technical parameters (Table 1.1; Fig. 1.25) and access including the localization of these territories is very important for disabled people (Table 1.2).

Fig. 1.24 Hiking trail for disabled people near Počúvadlo reservoir in the Štiavnické vrchy Protected Landscape Area, Slovakia (UNESCO) (*Photo* Jakubisová)

Table 1.1 Basic data about Počúvadlo hiking trail for wheelchair people (Jakubisová 2014)

Summary of the characteristics of the route for wheelchair people	Data
Total length of the route	1.82 km
Maximum longitudinal slope between stations S11–S12	12.08%
The maximum cross slope on the road	3.00%
Max. height/depth of barriers on the road	0.08 m
Estimated time of the presentation on individual stops (in hours)	¼
Estimated speed of wheelchair users (km/hours)	1 km/0.6
Estimated time of the route without stopping (in hours)	1.09

Explanatory notes For overcoming obstacles for wheelchair users is important: maximum longitudinal slope does not exceed the length 150 m on the trail; cross slope of the route is max. 2%; surface unevenness not exceed the height and depth of 8 cm

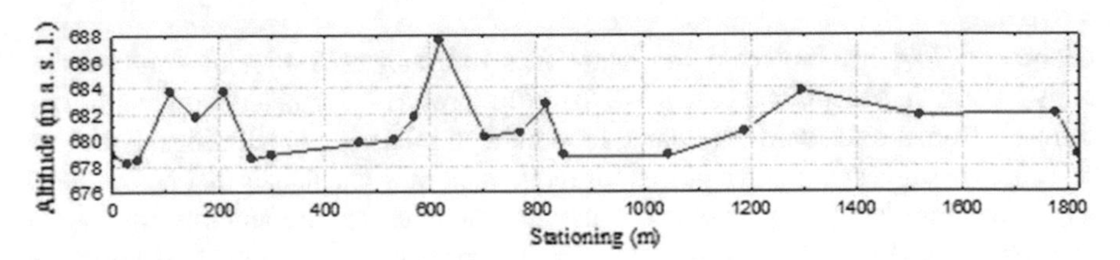

Fig. 1.25 Longitudinal profile of Počúvadlo—technical parameters of hiking trail for wheelchair users (*Source* Jakubisová 2014)

1.8 Example of the Study—Brno Dam Area (Czech Republic)

As the model area, the Brno Reservoir was chosen. It is one of the most significant water reservoirs in the Morava River drainage basin. Brno Reservoir is located in the Podkomorský forest north-west of Brno city, on the River Svratka. It is one

Table 1.2 Localization of territory Štiavnické vrchy Protected Landscape Area in Slovakia

Country	Slovakia
Geographic region	Banská Bystrica
Governing body	Štiavnické vrchy Protected Landscape Area Administration in Banská Štiavnica
Location	Central Slovakia
Coordinates	N48° 24′ 42″ E18° 52′ 21″
Established	22 September 1979
Area	77,630 ha
Water reservoir	Počúvadlo
Altitude	550 m a.s.l.
Area	12.3 ha
Coordinates	N48° 21′ 59″ E18° 50′ 05″

Fig. 1.26 Barrier-free toilets and changing rooms (*Photo* Kotásková)

of the water management structures whose history began before the Second World War. The dam was brought into operation in 1940. The main purpose for the dam construction was to ensure sufficient amounts of irrigation and drinking water for the Brno agglomeration, and the next purpose was flood protection. From the beginning, it was planned well as a source of energy and recreation. A feature unique in Europe is the public transport provided by electrically driven boats.

The water capacity is 21 million m^3, and the surface elevation is 229.08 m above sea level. The reservoir occupies an area of about 270 hectares, has a length of 10 km and extends to the town of Veverská Bitýška. The maximum reservoir depth is 23.5 m (Šlezingr 1998).

We can see barrier-free access in some parts around the Brno Reservoir. Reinforced roads, for example in Kozí Horka, and barrier-free toilets and changing rooms have been constructed (Fig. 1.26).

Another good example of barrier-free access is the construction of new barrier-free boat transport stops. Embarking and disembarking from the boat is done by means of piers. The barrier-free regulation imposes no requirements on the pier's technical

Fig. 1.27 Boat transport stop Kozí Horka—an example of good solution for the movement of the disabled (*Photo* Kotásková)

implementation; it only deals with the parameters of wheelchair-accessible ramps. These parameters should be analogically used when designing barrier-free piers, with respect to the specific local conditions. In general, the barrier-free decree emphasizes the use of appropriate surface, sloping and railing of the ramp, which should be followed even in the case of a pier. Naturally, a smooth barrier-free transition from the pier to the following road must be provided too. The Brno Reservoir has a newly implemented barrier-free boat transport stop at Kozí Horka as shown in Fig. 1.27. There was a bad access path to the old stop where people had to take uncomfortable steps.

The new solution is an extended pier with a pathway: the pier construction is fixed, and the pathway is taken away when the boating season finishes. It is a steel structure with a grid and non-slip treatment. The grid appears to be a very suitable surface. Water does not remain there and so the surface is not slippery even after rain. The pier is equipped with double-sided railings. There is a smooth transition to the path of mechanically reinforced aggregate, leading to the beach and the parking lot (see Fig. 1.28).

Other options for barrier-free access at the Brno Reservoir are currently being planned. So far the plans for possible use of the public beach with access to water using lifting equipment have not been completed.

There are several places used as public beaches around the Brno Reservoir. There is usually a grassy surface, and the beach boundaries are not precisely defined. Access to water has a natural character and is very steep in many places. Some locations have reinforced banks. In the framework of a project implemented in the previous period by the Morava River Basin Company, banks with a length of 590 m were reinforced at the sites Rakovecká zátoka and Kozí Horka, with grass stabilization concrete panels and in one place quarried stone. Currently, there are quite a lot of new access paths to the water. These paths usually use steps with railings on one or both sides. They are generally 1.5 m wide. Some of them in the middle of the site

Fig. 1.28 Smooth transition from the boat-stop pier to the path made of mechanically reinforced aggregate (*Photo* Kotásková)

Fig. 1.29 Concrete steps providing access to water (*Photo* Kotásková)

near a newly rebuilt fenced playground have a width of 3.5 m. Monolithic reinforced concrete was used for their construction as shown in Fig. 1.29.

A pier with a lifting device or other apparatus enabling wheelchair users to swim in the water is missing from the whole area of the Brno Reservoir.

There are places where disabled anglers could go fishing, but not all of them meet the conditions for their safety. The reasons are, e.g., steep slopes of the banks, access roads with bumps, or rocky or loamy–sandy banks.

1.9 Conclusion

This chapter lists the parameters that have to be respected for a universal design of the environment around water bodies or near rivers. This means designing for all, especially barrier-free access. As the examples show, it is necessary to design not only a suitable surface and road slope, but also places for relaxation or viewpoints. It is also possible to create suitable barrier-free water access points or fishing spots. Naturally, when designing such a place, parking spaces as well as barrier-free toilets at a sensible distance are a necessity.

Acknowledgements The chapter was created with support of the project Trails for disabled people in the V4 countries (International Visegrad Fund's Small Grant No. 11510242) and with financial support of the Internal Grant Agency of the Faculty of Forestry and Wood Technology, Mendel University in Brno, project No. LDF PSV 2016016 Making forest accessible in the changing social requirements and conditions and project No. LDF PSV 2016002 Minimizing losses forest and agricultural land due to erosion and abrasion processes in the landscape. We would like to thank ALTECH company for providing us with the picture used in this chapter.

References

Act No. 317/2010 Coll. ratifying the UN convention on the rights of persons with disabilities

Antonovičová M (2014) Přístupnost veřejného prostředí ve Švédsku. Smart Cities. [on line]. [cit. 2016-12-10]. Dostupné z: http://www.scmagazine.cz/casopis/04-14/pristupnost-verejneho-prostredi-ve-svedsku?locale=cs

Axelson PW, Chesney D (1999) Accessible exterior surfaces: technical article. Santa Cruz, California, p 16

Bazénový zvedák Delfín—ALTECH [online]. [cit. 2017-06-06]. Dostupné z: http://www.altech.cz/produkty/bazenovy-zvedak/bazenovy-zvedak-delfin/

Calkins M (2009) Materials for sustainable sites: a complete guide to the evaluation, selection and use of sustainable construction materials. I. Title. Wiley, Hoboken, New Jersey. ISBN 978-0-470-13455-9

Communication from the Ministry of the Foreign Affairs (MZV) of the Slovak Republic No. 317/2010 convention on the rights of persons with disabilities

Decree Nr. 398/2009 Coll, on general technical requirements for the use of buildings by persons with reduced mobility and orientation

Decree of the Ministry of Environment (MŽP) of the Slovak Republic No. 532/2002 Coll. specifying details of general technical requirements for construction and general technical requirements for structures utilized by persons with limited movement and orientation abilities

Decree of the Ministry of Interior (MV) of the Slovak Republic No. 9/2009 Coll. implementing the Act on Road Traffic and amending and supplementing certain acts, as amended

Filipiová, D. (2002) Projektujeme bez bariér. Ministerstvo práce a sociálních věcí, Praha 2002, 104 s., ISBN 80-86552-18-7

Hrůza P (2015) Využití lesní cestní sítě a směřování zpřístupňování lesa v České republice pro osoby se sníženou pohyblivostí. In Chodníky pre telesne postihnutých ľudí na vozíku v krajinách V4. 1. vyd. Zvolen: Technická univerzita vo Zvolene, 2015, s. 53–61. ISBN 978-80-228-2757-7

http://www.czp-msk.cz/pdf/uzitecne/ATHENA_PRIRUCKA_KOMPLET.pdf. Centrum pro zdravotně postižené Moravskoslezského kraje o.p.s. [online]. [cit. 2015-10-10]. Dostupné z: www.project-athena.cz, http://www.utok.cz/sites/default/files/data/USERS/u24/2014_RaOP_1st%20part.pdf, http://www.utok.cz/sites/default/files/data/USERS/u24/2013_RaOP.pdf

Jakubis M, Jakubisová M (2012) Proposal of educational-touristic polygon in Račkova valley (West Tatras) in Tatras National Park. In: Fialová J (ed) Public recreation and landscape protection—hand in hand. Conference proceedings. Mendelova univerzita v Brně, Brno, pp 58–62

Jakubis M (2013) Torrent as an important component of recreational and touristic potential of the landscape. In: Fialová J, Kubíčková H (eds) Public recreation and landscape protection—with man hand in hand. Conference proceeding. MUB Facultas Silviculturae et Technologiae Ligni,

Brno, pp 216–220. ISBN 978-80-7375-746-5. Online: http://www.utok.cz/sites/default/files/data/USERS/u24/2013_RaOP.pdf

Jakubis M (2014) Historical water reservoirs in the region of Banská Štiavnica and the possibilities of their recreational, tourist and educational utilization. In: Fialová J (ed) Public recreation and landscape—with man hand in hand, Conference Proceeding. Mendel University, Brno, pp 132–136. Online: http://www.utok.cz/sites/default/files/data/USERS/u24/2014_RaOP_1st%20part.pdf

Jakubisová M (2014) The proposal of recreational and educational trail for disabled people in wheelchair around the historical water reservoir Počúvadlo. In: Fialová J (ed) Public recreation and landscape—with man hand in hand, Conference Proceeding. Mendel University, Brno, pp 288–295. Online: http://www.utok.cz/sites/default/files/data/USERS/u24/2014_RaOP_2nd%20part.pdf

Jakubisová a kol (2015) Chodníky pre telesne postihnutých ľudí na vozíku v krajinách V4. Zborník príspevkov zo seminára s medzinárodnou účasťou, Technická univerzita vo Zvolene, Zvolen 2015, 163 s. ISBN 978-80-228-2757-7. Online: http://www.tuzvo.sk/files/3_7_OrganizacneSucasti/ABH/zbornik-vedeckych-prac_seminar-trails_abh-2015.pdf

Junek J, Fialová J (2012) Prezentace a zpřístupnění chráněných území osobám s tělesným a pohybovým omezením - Bez bariér v národních parcích TANAP a PIENAP. In: Fialová J (ed) Public recreation and landscape protection—hand in hand. 1. vyd. Brno: Mendelova univerzita v Brně, s. 63–68. ISBN 978-80-7375-611-6

Juško V (2015) Technické parametre líniových trás pre pohybovo hendikepované osoby na Slovensku. In Chodníky pre telesne postihnutých ľudí na vozíku v krajinách V4. 1. vyd. Zvolen: Technická univerzita vo Zvolene, 2015, s. 41–52. ISBN 978-80-228-2757-7

Kotásková P (2015) Technické požadavky na dřevostavbu rodinného domu bez bariér pro osoby se sníženou schopností pohybu. Stavební partner. 2015. sv. II, č. 2, s. 12–17. ISSN 1805-5958. Online: http://partnerstvi-stavebnictvi.msdk.cz/emagazin/2015-02

Kotásková P, Hrůza P (2013) Bridges and footbridges in the landscape environment for recreational and tourist use. In: Fialová J, Kubíčková H (eds) Public recreation and landscape protection—with man hand in hand. 1. vyd. Brno: Mendelova univerzita v Brně, 2013, s. 14–18. ISBN 978-80-7375-746-5

Loučková K, Fialová J (2010) The study of the nature trail equipped by the exercise elements for disabled people and seniors. [CD-ROM]. In: Colloquium of landscape management. ISBN 978-80-7375-397-9

Lundell Y, Rolison N, Wennerberg A (2005) Access to the forest for disabled people. Skogsstyrelsens förlag, Jönköping. ISSN 1100-0295

Mezera A (1956) Stanovištně typologický přehled lesních rostlinných společenstev/Mezera - Mráz – Samek. Brandýs nad Labem: Lesoprojekt. 92 p. + tables

Ondrejka Harbuľáková V, Zeleňáková M (2013a) Technical measures of riverbank stabilization in engineering practice, In: Visnik: Teoria i praktika budovnictva. No 756, pp 200–206, ISSN 0321-0499

Ondrejka Harbuľáková V, Zeleňáková M (2013b) Riverbank stabilization as flood protection measures, In: Hydrologic risks—flood and droughts, Košice: TU, pp 140–151, ISBN 978-80-553-1492-1

Přehled revírů: Revíry vhodné pro handicapované rybáře. Český rybářský svaz. Online: https://www.rybsvaz.cz/?page=reviry%2Fztp%2Fztp&lang=cz&typ=mpr&id_svaz=&moznosti_rybolovu_prehled=ztp#zalozka

Resolution ResAP (2007)3 Achieving full participation through universal design

Šlezingr M (1998) Brněnská přehrada a lidé kolem ní. VUT – FAST, Brno, 84 s. ISBN:80–214-1127-9

Šlezingr M (2004) Břehová abraze. Příspěvek k problematice zajištění stability břehů. CERM, Brno. ISBN 80-7204-342-0

Šlezingr M, Lichtneger P (2011) Selection criteria for designing of deciduous species as part of bankside trees and shrubs [CD-ROM]. In: Colloquium on landscape management. ISBN 978-80-7375-518-8

Slovakian Technical Norm 73 6101 Design of roads and motorways
Slovakian Technical Norm 73 6108 Forest transportation network
Slovakian Technical Norm 73 6110 Design of local roads
Technical Parameters 10/2011 Technical conditions—design of debarrierization measures for persons with reduced mobility and orientation on roads
The principles of universal design were developer by the center for universal design Carolina State University. Online: https://www.thefreelibrary.com/Segs4Vets+making+mobility+accessiblea0179736624
UN convention on the rights of persons with disabilities. Online: https://www.un.org/development/desa/disabilities/convention-on-the-rights-of-persons-with-disabilities.html
Vítková M (2006) Somatopedic aspects [in Czech: Somatopedické aspekty]. 2nd revised and expanded edition. Paido, Brno: 302. ISBN 80-7315-134-0

Chapter 2
Construction of Hydrotechnical Structures in Terms of Rational Management of Mineral Resources

Slávka Gałaś and Andrzej Gałaś

Abstract The following chapter discusses the use of non-renewable mineral resources occurring in dam reservoir basins. The work presents an assessment of construction of hydrotechnical objects in terms of mineral deposits management, taking into consideration construction of flood control basins with associated mineral exploitation, exploitation of minerals associated with maintaining retention of water reservoirs and limited access to mineral deposits as a result of flooding of the reservoir basin. After analysis of mineral management in several selected flood protection reservoirs in Poland: Racibórz Dolny, Świnna Poręba, Nysa and Tresna, it can be stated that construction of a hydrotechnical object with associated mineral exploitation seems to be the most rational solution considering sustainable development and protection of non-renewable resources. The concept of spatial–technical solution based on combination of a step-by-step construction of a reservoir with successive exploitation of natural aggregates deposited in the basin of the Racibórz Dolny dry polder, will increase the capacity of the reservoir, reduce the costs of reclamation of the post-mining areas and allow further use of agricultural and forest land on its shore area. The mineral resources management should be carried out with the emphasis on the most complete extraction of minerals, which is essential in the case of planned flood protection reservoirs, as their basins often contain minerals that can be used as building materials.

Keywords Reservoirs · Construction · Mineral resources · Exploitation
Rational management

S. Gałaś (✉) · A. Gałaś
AGH University of Science and Technology, Kraków, Poland
e-mail: sgalas@geol.agh.edu.pl

A. Gałaś
e-mail: pollux@geol.agh.edu.pl

M. Zelenakova (ed.), *Water Management and the Environment: Case Studies*, Water Science and Technology Library 86,
https://doi.org/10.1007/978-3-319-79014-5_2

2.1 Introduction

Water resources management is one of the most important issues in almost every country. The observed climate changes make control of available resources more and more difficult. Apart from ensuring the supply of drinking water, there is also a need to protect the population from harmful effects of the excess of water which results in floods, destructive outflows and erosion of the land surface. The problems are solved by construction of expensive hydrotechnical structures. Despite differences in approach to the estimation whether retention reservoirs are structures which comply with the principles of shaping and conservation of the natural environment, we must be aware that they always mean some loss in environmental resources. This applies in particular to interference with the natural environment and the ecosystem functioning on the basis of natural river flows, hence also the flows at flooding stages. The following chapter discusses the use of non-renewable mineral resources occurring in reservoir basins.

Depending on the type of hydrotechnical structures: dams, water reservoirs, polders, canals, etc., the scale and the scope of assessment of geological and engineering conditions are different. This chapter focuses on analysis of retention reservoirs and dry polders. They are constructions which include dams, reservoir basins—periodically or permanently flooded and embankments which are related to management of mineral resources. To locate the dam, the narrowest part of the river valley is usually found. This is mainly due to a potential saving of material (rock raw materials, concrete) necessary to construct it. Such a narrow is usually preceded by a wide part of the valley which is rich in deposits of natural aggregates. Such situation seems to be very beneficial from the point of view of management of natural deposits. However, it almost always happens that the narrow is a gorge whose genesis is associated with young uplifting movements. It usually results in difficult and complicated geological structure of the bed, a cracked rock massif caused by dislocations which eventually incurs additional expense of materials used to protect the dam base (concrete sealing injections). Hence, it turns out that the conditions which are favourable from a hydrotechnical point of view can mean difficult or even unfavourable geological conditions. Detailed analysis of the geological structure is the basis for considerations when constructing dams. The rational choice of a dam location requires investigation of (Olszamowski 1999; Dziewański 1999; Sroczyński 1999):

- geological conditions in the place where foundation of the construction is to be located including a type of the bed; for example in the case of a rocky bed, it is essential to determine strength of the rock, on a clay bed to examine waterlogging and changes in volume and strength, on mineral soil to determine the bearing capacity and deformability of the bed,
- rock filtration properties associated with determination of parameters showing water permeability of the rock massif needed to assess the need for substrate sealing,

- documenting of the local resources of minerals of appropriate quality and located within a short distance which may be used as materials necessary for construction of the dam body, hydraulic concrete and accompanying objects,
- evaluation of water escapes after filling the reservoir due to large soil permeability affecting the balance of disposable water resources, assessment of the impact of the reservoir silting on its volume,
- a forecast of the shoreline changes as a result of mass movements and abrasions of the future shores.

It is assumed that mineral resources should exceed even two or three times the anticipated demand for construction materials. Gravel and sand, whose deposits are located in the Carpathian valleys within low or middle accumulation terraces, are the most demanded ones. The clay material in the overburden, if it meets the requirements, is used to build a tight core of the dam. The technology of possible improvement and ways of using mineral deposits during the construction period are also under consideration (Dziewański 1999; Sroczyński 1999).

2.2 Methodology

This chapter presents evaluation of hydrotechnical structures in terms of mineral deposits management considering thc following issues:

- construction of a water reservoir, a polder associated with exploitation of mineral resources,
- exploitation of minerals associated with maintaining water retention and used for economic purposes,
- mineral deposits with limited access resulting from construction of the water reservoir.

To obtain the goal, documents prepared for construction of water reservoirs, planning documents and associated environmental impact assessment studies as well as available data on the occurrence and mineral resources were analysed. Spatial analysis was performed by means of ArcGis. The work was focused on the balance of resources and the amount of extracted aggregate from the area of the basin of flood detention constructions and determination of factors affecting management of the mineral deposits.

2.3 Construction of Hydrotechnical Structures Associated by Mineral Exploitation

A dry polder "Racibórz Dolny" on the Odra river, Silesia voivodeship, is a hydrotechnical structure whose construction was intentionally planned with exploitation of minerals. The structure is the basic part of the Flood Protection System in the Odra River Basin in the south-western part of Poland, which is one of the areas with the highest flood risk. It was planned within the framework of the Odra Programme, which was intended to ensure safety of 2.5 million people living in the Odra Valley from Racibórz to Wrocław (Program for the Odra 2011).

The currently carried on project of construction of the dry polder "Racibórz Dolny" is based on a spatial–technical concept based on combination of a step-by-step construction of the reservoir with successive exploitation of natural aggregates deposited in the reservoir basin, which was developed in the mid-1970s. Such a process will increase the capacity of the reservoir, which is equal to 170 million m^3 now (RWMB Gliwice 2017). At the same time, it will reduce the costs of reclamation of the post-mining areas and enable further use of the agricultural and forest areas on its shore area and protection of valuable natural values—Natura 2000 sites: the Special Bird Conservation Area "Stawy Wielikąt i Las Tworkowski" (Wielikąt Ponds and Tworków Forest) and the Special Area for the Conservation of Habitats "Las koło Tworkowa" (Forest near Tworków) (Dyka 2013; Gałaś et al. 2017).

A deposit of natural aggregates "Racibórz—the lower reservoir" (Racibórz—zbiornik dolny) covering the area of 21.3 km^2 and of the volume of 123.8 million m^3, under the overburden of the volume of 58.1 million m^3 was documented within the area of the basin in 1994. It was divided into a few separate areas in the following years (Table 2.1) (Sroczyński (ed) 2002; Łagosz 2008). The industrial resources of natural aggregate within the area of the basin are equal to 110.487 thousand tonnes. The amount is much greater than the amount of material needed to build the reservoir, including the front and the side dams (7.5 million m^3) (Łagosz 2008). The total length of the embankments will exceed 22 km and their height will vary from 7 to 10 m above the ground level while the total surface area of the polder will be 2426.72 ha. The reservoir dam will be made of the material of which about 60% will comprise aggregates—loose materials, extracted from the deposits within the basin—the static body and 40% will comprise the overburden of the aggregate deposits—the sealing body. The polder will be a first-class technical object with a capacity of 185 million cubic metres and a maximum damming height of 195.20 m a.s.l. (Hydroprojekt 2009).

The first stage of the construction is constituted by Buków polder, an artificial flood protection reservoir separated by embankments, which was completed in 2002. The polder is filled up only with the Odra flows during flood waves. Its total capacity is approximately 53 million cubic metres. It is divided by an embankment into a flow part and a controlled part. The capacity of the polder gradually increases as a result of the progressive exploitation of aggregates within its both parts (Hydroprojekt 2009).

Construction of the main dry polder "Racibórz Dolny" began in 2013. By 2015, 12 companies had been granted mining concessions and 47 mining areas were exploited in that year (Fig. 2.1) (Midas 2017; Hydroprojekt 2009; Gałaś et al. 2017). Exploitation of aggregates is estimated for 40–50 years. Due to directed relocation of the overburden and after full exploitation of the aggregate deposits, the reservoir capacity will be about 300 million m^3, which will almost double the capacity of the reservoir counted from the original level of the terrain. After that period, it is

Table 2.1 Mining management in Racibórz Dolny polder, after Balance of mineral resources 2007–2015 [thousand tonnes] (PGI-NRI 2008, 2009, 2010, 2011, 2012, 2013, 2014, 2015, 2016)

No	Deposit	Initial resources	Number of documented lots/deposits	Resources in 2015	Mining together 2007–2015
1	Brzezie n/Odrą	39,650		10,423	3169
2	Racibórz I-Zbiornik	6359	1	6283	44
3	Racibórz II-Zbiornik	22,864	11	7851	3189
4	Bieńkowice -Wschód	28,895		16,665	10,222
5	Nieboczowy III	6168 prefeasibility	4	0	218
6	Lubomia III	23,649	5	27,600	8892
7	Krzyżanowice -Tworków	39,362		30,800	543

planned to convert the dry polder "Racibórz Dolny" into a multi-purpose reservoir whose shape and capacity will result from the area of the reservoir and the scope of aggregate exploitation and utilization of the overburden. Such activities give employment to the local population in mining and transport companies, while boosting construction and road industries in the same time. Mining of minerals from the polder "Racibórz Dolny" in 2015 amounted to 5825 thousand tonnes, which accounted for 61% of total mineral production in the Silesian voivodeship (PGI-NRI 2008, 2009, 2010, 2011, 2012, 2013, 2014, 2015, 2016).

The Świnna Poręba water reservoir, which has been being built since 1986, is located on the Skawa river, in the Małopolska voivodeship. It will be a multi-purpose reservoir with a total capacity of 161 million m^3 which will cover the area of about 10 km^2 (Gałaś and Gałaś 2009).

The current stage of the construction is such that the dam in Świnna Poręba was put into operation (2015), but works on landslide stabilization and construction and modernization of roads are still carried on (ME 2017).

Within the meadow terraces of the Skawa, within the area of the reservoir, five natural aggregate deposits were documented. Total resources in the basin of the reservoir had been estimated at 29 million tonnes, and after verification of the available resources, they amounted to approximately 11 million tonnes when the construction was started (1986) (Dziewański and Sroczyński 1998; Hydrogeo 1975; Hydroprojekt 1987). So far, total extraction of natural aggregates according to the last Balance of Exploitation of Gravel Resources for the Needs of the Świnna Poręba Reservoir (RWMB Krakow 2013) has amounted to approximately 9.2 million tonnes, which constitutes more than 83% of the extracted projected mineral resources. In the remaining resources, exposed resources, i.e. 0.694 million tonnes of aggregate, were determined (Fig. 2.1; Table 2.2). The remaining fraction comprises non-exploited parts of the deposits, of which less than 2% are exposed deposits with the removed

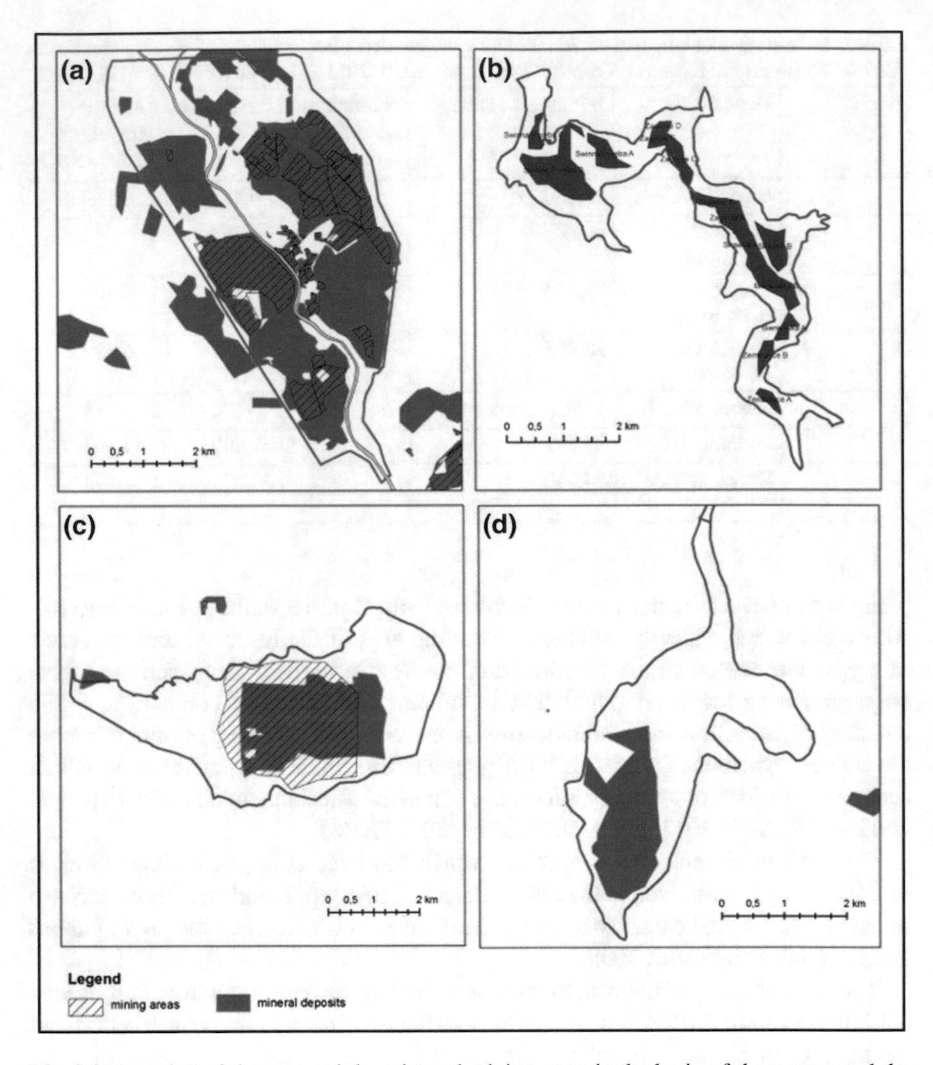

Fig. 2.1 Location of documented deposits and mining areas in the basin of the constructed dry polder Racibórz Dolny (**a**), water reservoirs: Świnna Poręba (**b**), Nysa Kłodzka (**c**) and Tresna (**d**) as of March 2017 (*Source* CGD 2017)

overburden. The rest of the resources are located under the existing buildings and infrastructure, including viaducts, fragments of a railway line. After their demolition, it will be possible to exploit the resources. Due to the prolonged time of completion of the construction of the Świnna Poręba reservoir and the need to stabilize the landslides, it is most likely that the aggregate deposits will be almost fully exploited, which from the point of view of rational management of mineral resources is an extremely desirable situation (Borowy 2017).

Table 2.2 Characteristics of mineral deposits of natural aggregates, including the projected ones (without exploitation losses) in 1986—the reference year and in 2012

Deposit name	Exploitation field	Deposit development, as for 2012	Projected resources in 1986 (thousand m^3)		Available resources in 2012 (thousand m^3)	Output usage
			1986	2012		
Świnna Poręba	A	ED	872.1	0	0	Dam
	B1	ED	2433.00	1352	424.5	Embankments, contractor's base, service roads to the dam region, sewage treatment plant and water intake, railway embankment, local roads
	B2	PED	1521.3	0	0	Dam, surface spillway, lower barrage, railway embankment
	C	ERD	1266.048	0	0	Dam
Zagórze	A	Deposit exploited before the beginning of the reservoir construction, reclaimed				
	B	ED	390	17	17	Stabilisation of landslides, local roads
	C	CED	394	48	0	Railway embankment
	D	ED	104.4	36	0	Railway embankment
	E	ED	359.3	82	82	Railway embankment

(continued)

Table 2.2 (continued)

Deposit name	Exploitation field	Deposit development, as for 2012	Projected resources in 1986 (thousand m^3)		Available resources in 2012 (thousand m^3)	Output usage
			1986	2012		
Skawce—Bieńkówka B	Skawce deposit A + Bieńkówka deposit B	CED	1266.048	325	66,5	Railway embankment
Skawce	B	ERD	1258.447	0	0	Railway embankment, local roads
Bieńkówka A	A	ERD	181.9	0	0	Local roads
Zembrzyce	A	ERD	645.8	104	104 (as for 2008)	Railway embankment, local roads, upper barrage
	B	ERD	430.158	0	0	Railway embankment
TOTAL			11,122.501	1964	694	

Also the available resources (exposed, with removed overburden) in 2012 and ways of managing the output (simplified): exploited deposit—ED ZW, exploited and reclaimed deposit—ERD, partly exploited deposit—PED, currently exploited deposit—CED ZE (Hydroprojekt 1987; RWMB Krakow 2013; Borowy 2017)

Topola and Kozielno dams on the Nysa Kłodzka are also examples of multi-purpose water reservoirs which were built together with aggregate exploitation. The construction of the reservoirs began in 1986 and it was completed in 2002. Natural aggregates are still mined from the reservoirs, increasing their capacities and areas (Łagosz 2008).

2.4 Exploitation of Minerals Associated with Maintaining Retention of the Water Reservoir and for Economic Purposes

The dam reservoir Nysa Kłodzka was built in 1971. The surface of the Nysa Lake is 2077 hectares and the total capacity of the reservoir is 123.44 million cubic metres. A deposit "Głębinów-Zbiornik" (Głębinów Reservoir), covering the area of 542.56 ha with resources of 106,053 thousand tonnes, was documented in 1971 in a complex of sand–gravel deposits at the bottom of the reservoir (Szcpietowska 1971; Szapliński 1988; Midas 2017; Awdankiewicz et al. 2004). In 2011, as requested by the investor, a decision on environmental conditions was issued by the Mayor of Nysa. The decision concerned environmental conditions of the project, located in the area of Natura 2000, which involved formation of the bottom of the Nysa Reservoir by extraction of natural aggregates from the documented deposit "Głębinów-Zbiornik" in the reservoir basin and depositing waste materials in the post-exploitation excavation (Decision 2011). Basing on that decision, a decision establishing a mining area was issued in 2012 with an expected expiry date of the concession 2037 and the area of 264 ha. In 2015, the geological resources of the deposits amounted to 86,284,000 and the industrial value was 40,523 thousand tonnes, the annual output was 764,000 tonnes (Midas 2017; Balance 2016). There are high-quality natural aggregates in the deposit which are used by building industry (Awdankiewicz et al. 2004). The mine extracting minerals offers for sale: sand, gravel, sandy gravel and decorative pebbles.

A local land use plan (Local plan 2011) was adopted for the projected mining area "Głębinów Zbiornik II" (Głębinów Reservoir II) to ensure integration of activities undertaken for the needs of the concession, general security and protection of the environment. The local plan indicated necessity to protect the archaeological sites which were entered in the voivodeship register of monuments, establishing the area where mining works are not allowed. Exploitation carried on under the water table, due to the type of the activity, is considered to be a type of a planned project which can always significantly affect the environment in the light of the existing legal regulations. Apart from that, the site is located within the area of special protection of birds Natura 2000 "Nysa Reservoir" PLB 160002.PLB 160002 in the Otmuchów—Nysa protected landscape area and in the vicinity of the Main Underground Water Reservoir (GZWP 338—Sub-reservoir Paczków Niemodlin) and in the area of the intermediate protection zone of water intake for the city of Wrocław (Local plan 2011). It is determined in the local land use plan that preliminary processing of the raw materials extracted from the deposit should consider possibility of depositing unnecessary mineral material in the Nysa Reservoir and that the management of the overburden rock (clay, sands) and mineral waste resulting from the process of initial refining of the mineral should include formation of shallows and islands, as a refuge for waterfowl, and spawning grounds for fish. Following principles of the environmental protection and landscape formation, including sustainable development, and taking into account local and supra-local public objectives are the basis for granting a concession for exploitation of natural aggregate deposits.

Extraction of natural aggregates from the reservoir basin, apart from the fact that it enables formation of the reservoir basin, can also cause landslides and river bed degradation. This was the case which occurred during exploitation of the "Zembrzyce" natural aggregate deposit on the Skawa River, when the increased river flow caused backward erosion on the slope of the excavation. As a result, movement of rubble on the bottom of the river bed occurred and clastic material from under the pillars of the bridge in Zembrzyce was removed.

2.5 Mineral Deposits with Limited Access Due to Construction of a Water Reservoir

In 1966, as a result of completion of the Tresna reservoir (Żywiec Lake) on the Soła River in Tresna, the Silesian Voivodeship, over 16 million tonnes of gravel documented in the Żywiec Tresna deposit were flooded. The deposit covering the area 195 ha located in the southern part of the reservoir is built of gravel deposits under a 0.8 m thick overburden. Żywiec Lake tributaries show high transport capacity and the bottom of the reservoir is filled quickly with silt. This results in a rapid increase of silting and increase of the overburden thickness. Due to deterioration of the operating conditions, the deposit is not suitable for exploitation. The exploitation is only carried on in the western part of the deposit, in the vicinity of the estuary of the Żylica River, under the reclamation project of deepening the Tresna reservoir. The reclamation involves removal of the accumulated bottom sediments and shaping the reservoir basin to recreate the planned capacity of the reservoir considering its usable and flood protection aspects and to limit, unfavourable from the qualitative and recreational point of view, shallows, exposures and isolated still water bodies in backwater of the reservoir. The excavated dredged material is segregated, crushed and sorted in order to obtain a full-value aggregate. The reclamation process allows to obtain a positive ecological and economic effect by limiting environmental degradation in watercourses and maintaining the stability of the river bed (Midas 2017; Prognoza 2010; Żywieckie Kopalnie Kruszyw 2017).

A conflict between management of mineral deposits and the existing economic function can be observed also in the case of the "Klimkówka" dam reservoir on the Ropa River, completed in 1994. The "Klimkówka" deposit of Magura sandstone, located in the shore area of the reservoir, was recognized as a high conflict, unmanageable deposit and the "Łosie" sandstone deposit was removed from the Balance Sheet of Mineral Resources (Lis et al. 2004) in 2003 due to location in the backwater of the reservoir.

2.6 Conclusions

The mineral resources management should be carried out in the most rational way, which means the emphasis on the most complete extraction of minerals, which is essential in the case of planned flood protection reservoirs and dry polders, which will be flooded (temporarily or permanently), as their basins often contain minerals that can be used as raw rock building materials. In the analysed cases, there have been different ways to achieve comprehensive use of the resources. Construction of a hydrotechnical object with associated mineral exploitation seems to be the most rational solution considering sustainable development. The local occurrence of mineral deposits, their resources and their quality are important when deciding on the location of the future reservoir, dam type, shape of the reservoir basin and accompanying investments (roads, bridges, resettlement). The planned rational extraction of deposits from the basin of the constructed reservoir allows to (Sroczyński (ed) 2002; Łagosz 2008):

- shape consciously the reservoir basin,
- provide construction material and enable rational use of the raw materials, which would be irretrievably lost as a result of flooding the reservoir basin,
- carry on properly oriented reclamation of post-exploitation areas,
- protect other mineral deposits as non-renewable resources,
- limit administrative work related to acquiring new areas for exploitation,
- and reduce also the cost of prospecting and exploration of new deposits.

Control of the level of extraction before flooding should be an important aspect of proper management of mineral resources of the deposits. Major losses occur at the stage of dividing the deposits into smaller lots, which is due to operation of many business entities in a small area. Such situation results in the loss due to setting the barrier pillars for transport ways and dumping grounds. The efficiency of mineral extraction depends on the applied extraction technology and the equipment being used. The number of companies and the number of concessions issued, improperly carried out balance of the resources and the volume of the extracted minerals, as well as the subcontractor's lawless operation have negative influence on the comprehensive extraction of the available mineral resources. The investor should be responsible for all such losses as it is in their interest to control management of mineral resources within the area. The primary objective of managing environmental resources is to ensure the sustainability of raw materials supply for present and future generations while preserving the environment (Gałaś and Gałaś 2012; Zeleňáková and Zvijáková 2017).

Acknowledgement The study was supported by AGH 11.11.140.626—Economic geology analyses and environmental management.

References

Awdankiewicz H et al (2004) Objaśnienia do Mapy Geośrodowiskowej Polski 1:50,000. Arkusz Otmuchów. Państwowy Instytut Geologiczny, Warszawa, p 39

Balance (2016) The balance of mineral resources deposits in Poland as of 31.12.2015. Polish Geological Institute—National Research Institute

Borowy E (2017) Ocena wydobycia kruszywa naturalnego z czaszy zbiornika retencyjnego Świnna Poręba (Assessment of natural aggregates extraction from the Świnna Poręba reservoir's basin). Thesis AGH, Kraków, p 84

CGD (Central Geological Database) (2017) Polish Geological Institute—National Research Institute. http://baza.pgi.gov.pl. Accessed Apr 2017

Decision (2011) Decision no. ROŚ.ŚR.7624.DS/47/10 was issued by the Mayor of Nysa. Urząd Miejski w Nysie

Dyka M (2013) Kopanie w zbiorniku: Wydobycie kopalin w obrębie obiektów gospodarki wodnej (Digging in the reservoir: Mineral extraction within water management facilities), Surowce i maszyny budowlane, vol 3. Wydawnictwo BMP, Racibórz, Poland, pp 56–60

Dziewański J (1999) Badania geologiczne masywów sklanych podłoża obiektów hydrotechnicznych. Dziewański J (ed) (1999) Sozologiczne problemy w budownictwie wodnym (Sozological problems in hydrotechnical constructions), Studia, Rozprawy, Monografie, vol. 62, Wydawnictwo Instytutu Gospodarki Surowcami Mineralnymi i Energią PAN, Kraków, pp 53–71, 157

Dziewański J, Sroczyński W (1998) Wykorzystanie żwirów z czaszy zbiornika Świnna Poręba (Use of gravel from the Świnna Poręba reservoir's basin). In: Paulo A (ed) Sozologia na obszarze antropopresji na przykładzie zbiornika Świnna Poręba, pp 153–164. Kraków

Gałaś S, Gałaś A (2009) Assessment of ecological stability of spatial and functional structure around Świnna Poręba water reservoir. Pol J Environ Stud 18(3A):83–87

Gałaś S, Gałaś A (2012) Protection of mineral resources as a part of spatial planning in Poland and in Slovakia. Pol J Environ Stud 21(5A):73–77

Gałaś S, Gałaś A, Zeleňáková M (2017) Environmental resources management in the light of construction of a reservoir and a polder exemplified by construction of the reservoirs in Świnna Poręba and Racibórz Dolny, Poland. In: Conference proceedings. Book. International Multidisciplinary Scientific GeoConference SGEM 27 June–6 July, 2017 Albena, Bulgaria (in press)

Hydrogeo (1975) Dokumentacja geologiczno - inżynierska do ZTE zbiornika wodnego na rzece Skawie w Świnnej Porębie. Część 6—złoża (Geological-engineering documentation for a water reservoir on the Skawa river in Świnna Poręba), PGBW Hydrogeo Kraków

Hydroprojekt (1987) Technologie eksploatacji złóż. Zbiornik wodny Świnna Poręba na rzece Skawie (Mining technology. Świnna Poręba water reservoir on the River Skawa) Centralne Biuro Studiów i Projektów Budownictwa Wodnego Hydroprojekt w Warszawie

Hydroprojekt (2009) Budowa zbiornika przeciwpowodziowego Racibórz Dolny na rzece Odrze, województwo śląskie - polder, Raport o oddziaływaniu na środowisko (Construction of the flood reservoir Racibórz Dolny on the Oder River, Silesian Voivodeship—Polder), Hydroprojekt Sp. z o.o., Warszawa

Łagosz R (2008) Powiązanie budowy obiektów hydrotechnicznych z górnicza eksploatacją kruszyw naturalnych. Surowce i Maszyny Budowlane, Wydawnictwo BMP, Racibórz, Poland 5:77–84

Lis J et al (2004) Objaśnienia do Mapy Geośrodowiskowej Polski 1:50 000, Arkusz Gorlice. Państwowy Instytut Geologiczny, Warszawa, p 39

Lokal plan (2011) Uchwała NR XII/196/11 Rady Miejskiej w Nysie z dnia 27 października 2011 r. w sprawie uchwalenia miejscowego planu zagospodarowania przestrzennego dla projektowanego terenu górniczego „Głębinów Zbiornik II" znajdującego się w granicach Zbiornika Nyskiego w części należącej do gminy Nysa

ME (2017) Ministry of the environment. www.mos.gov.pl. Accessed Apr 2017

Midas (System of management and protection of mineral resources in Poland) (2017) Polish Geological Institute—National Research Institute. http://geoportal.pgi.gov.pl/portal/page/portal/midas. Accessed Apr 2017

Olszamowski Z (1999) Rola geologii w projektach hydrotechnicznych. In: Dziewański J (ed) (1999) Sozologiczne problemy w budownictwie wodnym (Sozological problems in hydrotechnical constructions), Studia, Rozprawy, Monografie, vol 62. Wydawnictwo Instytutu Gospodarki Surowcami Mineralnymi i Energią PAN, Kraków, pp 33–39, 157

PGI-NRI (Polish Geological Institute-National Research Institute) (2008, 2009, 2010, 2011, 2012, 2013, 2014, 2015, 2016) The balance of mineral resources deposits in Poland as of 31.12.2007, 31.12.2008, 31.12.2009, 31.12.2010, 31.12.2011, 31.12.2012, 31.12.2013, 31.12.2014, 31.12.2015

Prognoza (2010) Prognoza oddziaływania na środowisko dla projektu zmiany studium uwarunkowań i kierunków zagospodarowania przestrzennego Miasta Żywca, Żywiec, INPLUS Spółka z o.o. Olsztyn

Program for the Odra (2011) Program for the Odra 2006—update 2011. The Government Plenipotentiary for the Program for the Odra, p 183

RWMB Gliwice (The Regional Water Management Board in Gliwice (2017) http://www.rzgw.gliwice.pl/modules.php?name=News&file=print&sid=51. Accessed Apr 2017

RWMB Krakow (The Regional Water Management Board in Krakow) (2013) Bilans zasobów eksploatacyjnych żwirów dla potrzeb Zbiornika Świnna Poręba na koniec 2012 (The balance of gravel exploitation resources for the needs of the Świnna Poręba Reservoir made in 2012)

Sroczyński W (1999) Rola karpackich gruntów pokrywowych (czwartorzędowych) w budownictwie wodnym—aspect geoekologicky. In: Dziewański J (ed) (1999) Sozologiczne problemy w budownictwie wodnym (Sozological problems in hydrotechnical constructions), Studia, Rozprawy, Monografie, vol 62. Wydawnictwo Instytutu Gospodarki Surowcami Mineralnymi i Energią PAN, Kraków, pp 133–143, 157

Sroczyński W (ed) (2002) Uwarunkowania geologiczne realizacji zbiornika przeciwpowodziowego Racibórz Dolny na Odrze (Geological conditions of the realization of the Racibórz Dolny flood reservoir on the Oder River). IGSMiE PAN, Kraków, Poland, p 88

Szapliński A (1988) Dodatek nr 1 do dokumentacji geologicznej w kat. C1 + B złoża kruszywa naturalnego "Głębinów—Zbiornik", gmina Nysa, woj. Opole

Szepietowska H (1971) Dokumentacja geologiczna w kategorii C1 + B złoża kruszywa naturalnego Głębinów-Zbiornik

Zeleňáková M, Zvijáková L (2017) Risk analysis within environmental impact assessment of proposed construction activity. Environ Impact Assess Rev 62:76–89

Żywieckie Kopalnie Kruszyw (2017) Żywieckie Kopalnie Kruszyw Sp. z o.o. http://www.zkk.pl/onas. Accessed Apr 2017

Chapter 3
The Impact of Rainwater Harvesting System Location on Their Financial Efficiency: A Case Study in Poland

Agnieszka Stec and Daniel Słyś

Abstract Natural water resources of Poland are among the lowest in Europe. In addition, the intensive development of urbanized areas and the associated increase in water demand necessitate the need to look for alternative sources. However, limiting the amount of resources available for use does not go hand in hand with the development of ecological awareness of society, which has the greatest attention still attached to the financial criterion. Considering this, the studies have been conducted to determine the cost-effectiveness of the rainwater harvesting system (RWHS) in a single-family house located in selected Polish cities where rainfall varies in height. Financial analysis for four different variants of the water supply system in the building in question has been done using the Life Cycle Cost (LCC) Methodology. The results show that RWHS financial performance varies widely, but it has also been found that the variant in which rainwater will be used to flush toilets, wash, and water the garden is characterized by the lowest LCC costs irrespective of tank capacity, number of users, and the location of RWHS system. The study also examines the impact of the capacity of the rainwater storage tank on the tap water savings. Depending on the installation variant these savings ranged from 11– 40% for Zakopane, 10–25% for Warsaw and Katowice, and 10–28% for Koszalin.

Keywords Stormwater management · Rainwater harvesting · Sustainable urban drainage · Life cycle cost

A. Stec (✉) · D. Słyś
Department of Infrastructure and Water Management, The Faculty of Civil and Environmental Engineering and Architecture, Rzeszow University of Technology, al. Powstańców Warszawy 6, 35-959 Rzeszów, Poland
e-mail: stec_aga@prz.edu.pl

M. Zelenakova (ed.), *Water Management and the Environment: Case Studies*, Water Science and Technology Library 86,
https://doi.org/10.1007/978-3-319-79014-5_3

3.1 Introduction

Water is one of the most important environmental resources that determine human existence. However, over the years, freshwater resources have been overexploited due to anthropogenic activities. This has led in many regions of the world to a state in which their quantity and quality are not adequate to ensure proper social and economic development. Existing water shortages, which are caused not only by the poor management of its resources, but also by growing demand, changing climate, and intensive urbanization, are now becoming one of the world's major problems (Li et al. 2010). Climate change, which is the result of natural factors and human activity (Stern and Kaufmann 2014), has influenced precisely the quantity and intensity of precipitation resulting in an increased occurrence of extreme weather events (Kaźmierczak and Kotowski 2014). Significant amounts of water, which, as a result of intense rainfall, fall into a given area in a short time, only slightly supply that part of the water resources that can then be used. Such precipitation, combined with the sealing of the terrain resulting from urbanization, creates intensive surface runoff, which contributes to an increase in flood risk (Todeschini 2016; Lu et al. 2014; Du et al. 2012). Intensity of surface runoff also causes significant hydrological changes in the catchments and hydraulic ones in sewage systems (Pochwat et al. 2017; Kim et al. 2015; Słyś and Stec 2013).

The actions are taken, especially in urban areas, for an introduction and an implementation of sustainable water and wastewater management (Hoang and Fenner 2016; Willuweit and O'Sullivan 2013). This is a strategy whose primary purpose is to maintain water resources in a state of order that is economically and socially possible for present and future generations (Water Frame Directive). The most commonly used "end-of-pipe" rainfall management model is not compatible with this strategy. There is a tendency toward a more integrated and sustainable approach that takes into account the changes in the rainwater flow regime, the protection of natural water resources, and the need to adapt technical infrastructure to modern urban water management standards (Palhegyi 2010; Mitchell 2006; CIRIA 2000). It is implemented, among others, by using facilities and equipment that are part of the sustainable urban drainage system (SUDS) and low impact development (LID) (Campisano et al. 2017; Fletcher et al. 2013). These solutions are based mainly on the processes of retention and infiltration of precipitation into the ground. The use of such objects are retention reservoirs (Starzec et al. 2015; Stec and Słyś 2014; Słyś and Dziopak 2011) green roofs (Burszta-Adamiak and Stec 2017; Poorova et al. 2016; Czemiel-Berndtsson 2010), bio-retention systems, infiltration trenches and basins (Liu et al. 2014; Hirschman et al. 2008; Hatt et al. 2009).

Searching for alternative sources of water, in terms of the rapidly increasing population in the world and the intensification of the urbanization process, is becoming a key issue to ensure the right quantity and quality of water to meet hygienic and human health needs (Ait-Kadi 2016). Nearly 54% of the world's population now lives in urban areas, but it is projected to increase to 66% in 2050 (UN 2014), thus increasing water demand by 55% (OECD 2012). The most urbanized regions in the world are the USA (82%), South America (80%), and Europe (73%). Taking into account that cities absorb almost 70% of the world's resources, the measures are needed to reduce their excessive use. Modern urban water management should be based on sustainable consumption, based not only on available freshwater resources but also on alternative sources of water such as rainwater and gray water (An et al. 2015; Hyde 2013). In recent years, the concept of sustainable homes with reduced demand for water and energy has become increasingly popular (Stec et al. 2017; Kaposztasova et al. 2016; Zelenakova et al. 2014; Stec and Kordana 2015).

Of the possible alternative sources of water, rainwater for hygienic reasons is the most socially acceptable source (Marleni et al. 2015; Hurlimann and Dolnicar 2010). Rainwater harvesting systems have been used all over the world for many years, both for potable and non-potable use (Lopes et al. 2017; Gwenzi et al. 2015; Fewkes 2006). However, in the vast majority of cases, rainwater replaces non-potable water, particularly for toilet flushing (Jones and Hunt 2010; Słyś and Stec 2014; Devkota et al. 2015a), greening and arable fields (Devkota et al. 2015b; Unami et al. 2015; Ghimire et al. 2014), cleaning and laundering work (Morales-Pinzón et al. 2014; Angrill et al. 2012). The effectiveness of these systems depends on many factors, including rainfall, roof size, demand for non-potable water, building type, and tank capacity, which is the main component of RWHS (Vieira et al. 2014; Santos and Taveira-Pinto 2013; Imteaz et al. 2012; Słyś et al. 2012; Ghisi 2010). Commercial rainwater utilization systems are often used in combination with the gray water recycling system, which, due to the uneven distribution of rainfall over the year, has a positive effect on the improvement of the water supply system from alternative water sources (Fonseca et al. 2017; García-Montoya et al. 2015; Morales-Pinzón et al. 2015; Proença and Ghisi 2013).

Rainwater harvesting is not a new technology, but because of the environmental and financial benefits of its use, it is in constant interest among many researchers in the world. In recent years, the cost-effectiveness of using rainwater harvesting systems has been analyzed by, among others (Ghisi and Ferreira 2007; Liang and Van Dijk 2011; Rahman et al. 2010; Ghisi et al. 2014; Wang and Zimmerman 2015). The tap water savings that can be obtained by replacing it with rainwater are very different depending on the climatic conditions and the technical and hydraulic parameters of the RWHS and the building (Ghisi et al. 2007; Abdulla and Al-Shareef 2009; Haque et al. 2016; Markovic et al. 2014; Palla et al. 2012; Khastagir and Jayasuriya 2010). For example, Kuller et al. (2015) set tap water saving at 58% for a large Amsterdam airport. Ghisi (2006), in turn, presented research results showing tap water reductions ranging from 48 to 100% for residential buildings in different regions of Brazil. A similar study was conducted for petrol stations where rainwater was used to wash cars. Water savings in this case ranged from 9.2 to 57.2, 32.7% on average (Ghisi et al. 2009a, b). Ward et al. (2012) have set the tap water savings for toilets in a large office building located in the UK which is 79%.

Taking into account that rainwater harvesting systems are rarely used in Poland, and this is mainly due to the public's belief in the ineffectiveness of installing these solutions, in the paper, the research has been conducted to determine the financial efficiency of RWHSs for a single-family building located in various cities in Poland.

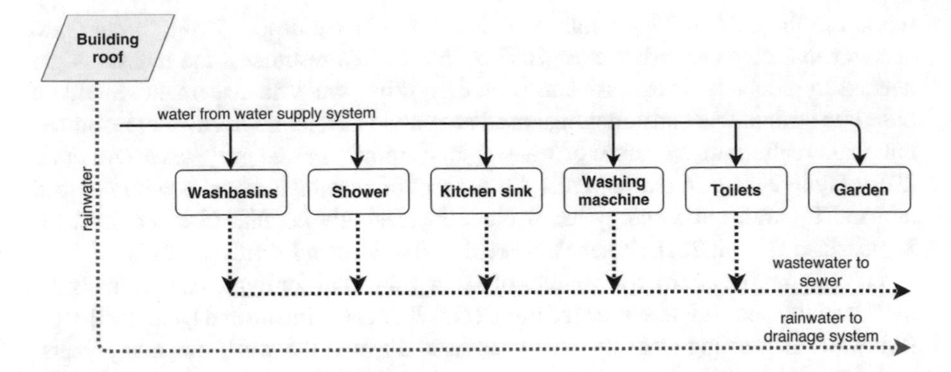

Fig. 3.1 Diagram of system operation in Variant 0 (*Source* Authors)

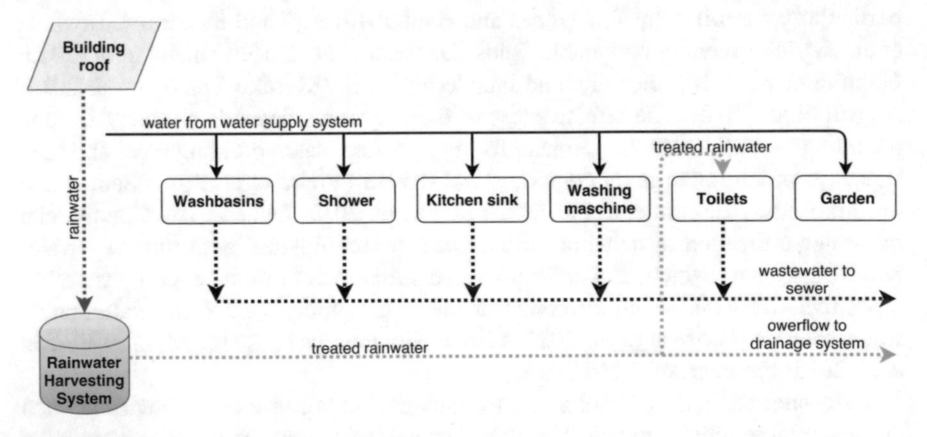

Fig. 3.2 Diagram of system operation in Variant 1 (*Source* Authors)

3.2 RWHS Simulation Model

The simulation model developed by Słyś (Słyś 2009) was used to determine the use of rainwater in the analyzed building. The model algorithm is based on daily water balance. It was assumed that the rainwater from the roof would flow through the pipe system to the retention tank located in the vicinity of the building. Then, by means of a pumping system rainwater from the tank will be transported to the internal installation for use as non-potable water.

In order to determine the influence of changes in operating parameters of the internal water supply and sewerage system on the financial effectiveness of the project, the cost-effectiveness studies of the analyzed variants of the installation were made for different values. The study included a variable number of inhabitants and different capacities of the retention tank. Installation systems that have been analyzed are shown in Figs. 3.1, 3.2, 3.3, 3.4, and 3.5.

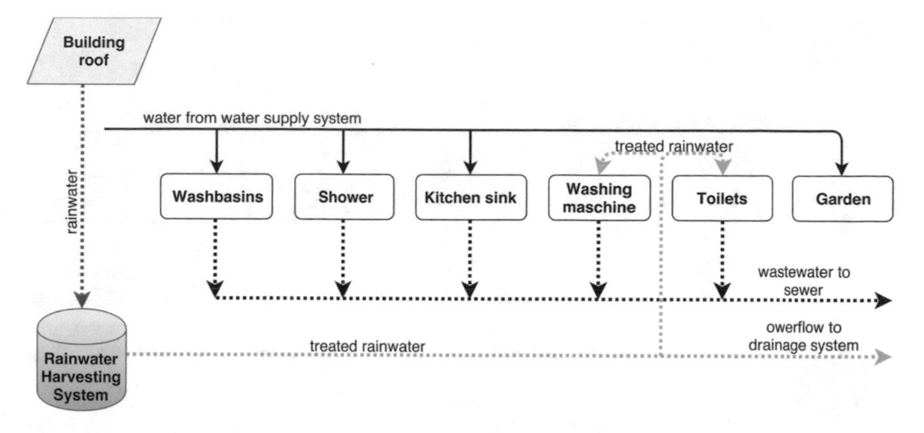

Fig. 3.3 Diagram of system operation in Variant 2 (*Source* Authors)

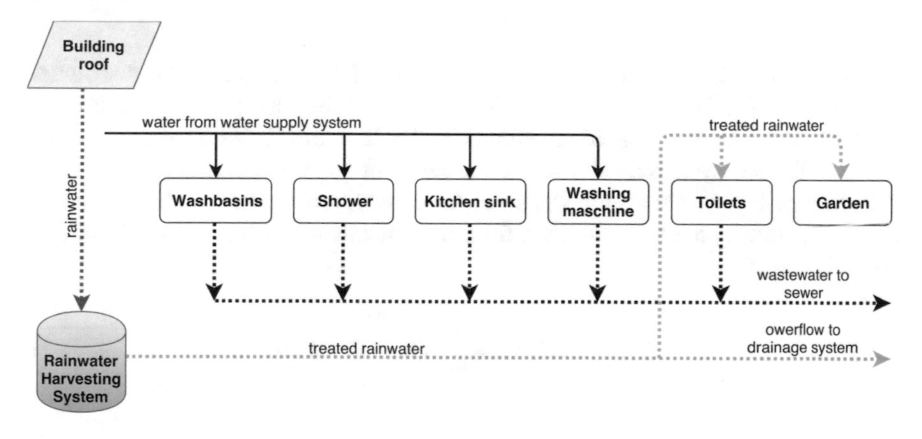

Fig. 3.4 Diagram of system operation in Variant 3 (*Source* Authors)

3.3 Financial Analysis

Making investment decisions solely on the basis of the initial investment outlays can lead to the choice of a wrong solution that will generate high operating costs in the future. Therefore, in the research the method of financial effectiveness assessment has been applied, which allows to determine costs in the whole life cycle. The use of Life Cycle Cost (LCC) analysis in evaluating different investment options enables to compare capital-intensive comparisons and thus allows the selection of the optimal solution, the implementation of which requires the lowest cost over the life cycle of the investment. The Life Cycle Cost Methodology takes into account the initial investment expenditure INV incurred in year 0, the KE operating costs resulting from the use of the solution over a longer period of time as well as the residual value of the RV which is the remaining value at the end of the study period (Fuller

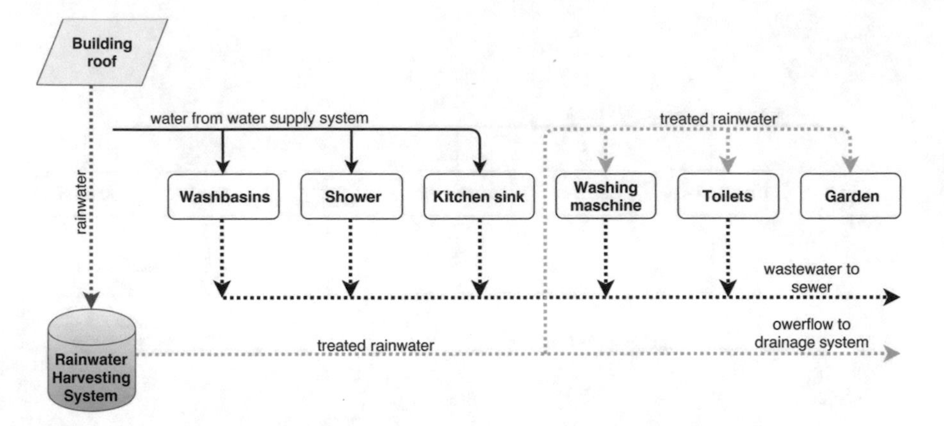

Fig. 3.5 Diagram of system operation in Variant 4 (*Source* Authors)

and Petersen 1996). Cash flows occurring in the following years are discounted. Due to the difficulty of determining the RV, especially during long periods of the analysis, according to the guidelines contained in the work (DOE 2014), the residual value of the RV system need not be determined in quantifiable terms. Considering this, financial costs were omitted, as was the case with other authors (Rahman et al. 2012). Therefore, the LCC analysis for *k* investments variants performed using the formula (1).

$$\mathrm{LCC}_k = \mathrm{INV}_k + \sum_{t=1}^{T} (1+r)^{-t} \cdot \mathrm{KE}_{kt} \quad (1)$$

where:

LCC_k Total cost of *k*-variant of installation, €;
INV_k Investments of *k*-variant of installation, €;
KE_{kt} Operating costs in the year *t* of *k*-variant of installation, €;
T Duration of the LCC analysis, years;
r Constant discount rate;
t Another year of the system use

In the studies conducted, the initial investment outlays of INV_k were estimated on the basis of cost estimates for each of the variants, which included the purchase and assembly costs of the individual components. In turn, the annual operating costs of KE_{kt} which included the cost of purchasing water from the water supply network, the costs of discharging sanitary sewage, and rainwater to the sewage system were calculated for each of the investment options analyzed using the current unit prices set by the network managers in each city. The variants that include the use of the RWHS system in the cost of the $\mathrm{KE}_{\mathrm{kt}}$ also determine the cost of purchasing the electricity used to drive the rainwater pump from the tank to the installation. The data for the

Fig. 3.6 Location of case study cities in Poland (*Source* Authors)

tests are summarized in Table 3.1. The discount rate r was set at 5%, which is in line with the assumptions adopted by other research authors who have analyzed RWHS financial performance (Roebuck et al. 2011; Ghisi and Oliveira 2007).

3.4 Study Case

The climate of Poland is defined as the transitional climate of temperate warm zone. It is characterized by high variability of weather and a significant variation of the seasons in successive years. Precipitation shows a high dependence on surface configuration. The average rainfall in the country is about 600 mm, but the rainfall ranges from less than 500 mm in the central part of Poland to almost 800 mm on the coast and over 1000 mm in the Tatras. The highest sums of precipitation fall in the summer months and in this period are 2–3 times higher than in winter. Taking into account the variation in precipitation levels in Poland, four cities located in different parts of the country were selected for the research. Their location is shown in Fig. 3.6.

On the basis of the daily rainfall totals from the period 2003–2012, which was used for simulation studies, the average annual precipitation for each of the selected cities was determined. Table 3.2 summarizes the values of these data.

Research aimed at determining the financial efficiency of the rainwater harvesting system (RWHS) depending on the location of the system was carried out for a single-family house located in four Polish cities. It is a one-storey building with a shower,

Table 3.1 Data used in the calculation of LCC costs (*Source* Authors)

Parameter	Parameter value
Investments	
The cost of purchasing and installing the RWHS with the tank 2 m^3 $INV_{RWHS\text{-}2}$	€1.631
The cost of purchasing and installing the RWHS with the tank 3 m^3 $INV_{RWHS\text{-}3}$	€1.938
The cost of purchasing and installing the RWHS with the tank 4 m^3 $INV_{RWHS\text{-}4}$	€2.151
The cost of purchasing and installing the RWHS with the tank 5 m^3 $INV_{RWHS\text{-}5}$	€2.600
The cost of purchasing and installing the sanitary systems INV_0	€1.891
Operating costs	
The annual increase in electricity prices i_e	4%
The annual increase in the prices of purchase of water from the water-pipe network i_w	6%
The annual increase in the prices of rainwater discharge to the sewage network i_r	4%
The annual increase in the prices of sanitary sewage discharge to the sewage system i_s	6%
The cost of purchasing electricity in the year 0 c_e	0.139 €/kWh
The cost of purchasing water from the water-pipe network in Katowice in the year 0 c_w	1.375 €/m^3
The cost of sanitary sewage discharge to the sewage network in Katowice in the year 0 c_s	1.960 €/m^3
The cost of purchasing water from the water-pipe network in Koszalin in the year 0 c_w	0.844 €/m^3
The cost of sanitary sewage discharge to the sewage network in Koszalin in the year 0 c_s	1.173 €/m^3
The cost of purchasing water from the water-pipe network in Warszawa in the year 0 c_w	1.073 €/m^3
The cost of sanitary sewage discharge to the sewage network in Warszawa in the year 0 c_s	1.638 €/m^3
The cost of purchasing water from the water-pipe network in Zakopane in the year 0 c_w	0.565 €/m^3
The cost of sanitary sewage discharge to the sewage network in Zakopane in the year 0 c_s	1.577 €/m^3
Analysis period T	20 years
The discount rate r	5%

Table 3.2 Amount of rainfall in the years 2003–2012 in selected cities (*Source* Authors)

City	Year										
	2003	2004	2005	2006	2007	2008	2009	2010	2011	2012	Average
	Annual rainfall H, mm										
Katowice	510	512	694	727	464	550	574	552	459	412	545
Zakopane	924	1098	1156	875	1166	1111	1250	1600	980	806	1097
Warszawa	545	520	514	482	593	547	653	789	601	537	578
Koszalin	614	834	743	611	972	738	757	801	696	824	759

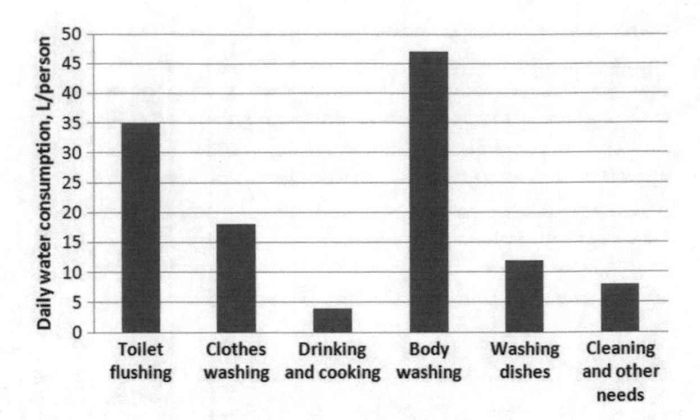

Fig. 3.7 Daily structure of water consumption in the analysed building (*Source* Authors)

2 washbasins, 2 toilet bowls, a washing machine, and a sink. It was assumed that the characteristics of water consumption in individual cases were the same for all users of the installation. The daily structure of water consumption in this building is shown in Fig. 3.7. It was also assumed that in the period from May to September three times a week the garden of 500 m^2 would be watered in the amount of 2.5 dm^3/m^2.

In order to determine the optimum capacity of the retention tank to achieve the greatest savings in tap water, tanks of 2, 3, 4, and 5 m^3 were considered for the research. Simulation studies were conducted using 10-day historical rainfall data (2003–2012) collected from meteorological stations located in four selected Polish cities (Fig. 3.6). Different RWHS location allowed to determine the impact of precipitation on the reduction of water consumption from the water supply and the cost-effectiveness of the system in Poland.

The following data was used in the study:

- Building roof surface $F = 150\ m^2$;
- Garden surface $F_g = 500\ m^2$;
- Number of inhabitants $M = 3, 4, 5$ persons;
- Average unit water requirement for toilet flushing $q_t = 35$ L/person/day;
- Average unit water requirement for washing $q_w = 18$ L/person/day;
- Average unit water requirement for garden watering $q_g = 2.5$ L/m^2/day (May to September, 3 times a week);
- Daily water requirement for toilet flushing $V_t = q_t \cdot M$;
- Daily water requirement for washing $V_w = q_w \cdot M$;
- Daily water requirement for garden watering $V_g = q_g \cdot F_g$;
- Runoff index of a drained surface $\psi = 0.9$;
- Number of days of water retention in a tank during a period of drought $t = 7$ days.

Table 3.3 LCC calculations for a building located in Katowice (*Source* Authors)

Tank capacity V, m^3	Variant	Life Cycle Cost, €		
		3 Occupants	4 Occupants	5 Occupants
2	0	22,198	25,539	28,879
	1	21,826	24,671	27,639
	2	21,117	23,995	27,019
	3	19,915	23,055	26,334
	4	19,695	22,924	26,190
3	0	22,198	25,539	28,879
	1	21,971	24,700	27,573
	2	21,091	23,869	26,858
	3	19,699	22,813	26,070
	4	19,439	22,651	25,926
4	0	22,198	25,539	28,879
	1	22,074	24,741	27,548
	2	21,091	23,797	26,757
	3	19,574	22,682	25,938
	4	19,302	22,521	25,802
5	0	22,198	25,539	28,879
	1	22,444	25,073	27,833
	2	21,398	24,027	26,976
	3	19,784	22,890	26,152
	4	19,513	22,743	26,034

3.5 Results and Discussion

The obtained LCC values for different rainwater harvesting system locations are shown in Tables 3.3, 3.4, 3.5, and 3.6. When analyzing these results, it can be seen that the variant in which rainwater will be used to flush toilets, wash, and water the garden (Variant 4) is characterized by the lowest LCC costs irrespective of the volume of the tank, the number of inhabitants, and the location of the RWHS system. This is due to the largest reduction in water consumption from the water supply and the resulting savings. It was also found that the traditional variant (Variant 0) in any of the analyzed cases was not the most financially advantageous solution, even though its use was associated with the lowest initial investment INV_0.

Comparing the total LCC values for the location of the RWHS system in different Polish cities, the highest costs were obtained for the investment located in Katowice (Table 3.3). These results from the fact that the unitary water supply and sewage disposal costs in the city are the highest among all analyzed cities.

Table 3.4 LCC calculations for a building located in Koszalin (*Source* Authors)

Tank capacity *V*, m^3	Variant	Life Cycle Cost, €		
		3 Occupants	4 Occupants	5 Occupants
2	0	14,972	16,991	19,011
	1	15,334	16,998	18,707
	2	14,797	16,420	18,129
	3	13,637	15,451	17,392
	4	13,407	15,270	17,168
3	0	14,972	16,991	19,011
	1	15,535	17,139	18,813
	2	14,928	16,474	18,141
	3	13,602	15,398	17,327
	4	13,348	15,188	17,072
4	0	14,972	16,991	19,011
	1	15,687	17,238	18,882
	2	15,006	16,515	18,138
	3	13,590	15,375	17,302
	4	13,321	15,160	17,029
5	0	14,972	16,991	19,011
	1	16,083	17,616	19,212
	2	15,352	16,824	18,430
	3	13,883	15,669	17,576
	4	13,596	15,435	17,304

It was also noted that only for this system location when comparing Variant 0 and Variant 1 the second solution, regardless of tank capacity and number of installation users, was more profitable. Despite the fact that in this city the lowest rainfall (annual rainfall H = 545 mm) and the economic use of rainfall waters are limited, the amount of unitary charges c_w and c_s affect the increase of financial efficiency of Variant 1. In this case, increased investment expenditure, which depends on the volume of the tank were 46–58% higher than those of Variant 0, was compensated by the lower operating costs spent on the purchase of tap water and the discharge of sewage into the sewage system during the 20 years of operation of the RWHS. The exception is a case in which the installation is used by three inhabitants and the tank capacity is 5 m^3. In this situation, due to the low demand for non-potable water and the resulting negligible tap water savings and the high capital expenditure required to install RWHS with such a high capacity tank, Variant 0 is more financially advantageous.

In turn, the lowest cost of LCC was the use of RWHS in Koszalin (Table 3.4). This is mainly due to the relatively low unit costs for water supply and sewerage and an increased rainfall in this area (annual rainfall $H = 759$ mm). This has resulted in a decrease in the financial performance of Variant 1 in favor of Variant 0 for cases

Table 3.5 LCC calculations for a building located in Zakopane (*Source* Authors)

Tank capacity V, m^3	Variant	Life Cycle Cost, €		
		3 Occupants	4 Occupants	5 Occupants
2	0	16,125	18,270	20,415
	1	16,629	18,310	20,032
	2	15,928	17,502	19,192
	3	13,790	15,575	17,711
	4	13,496	15,420	17,397
3	0	16,125	18,270	20,415
	1	16,851	18,506	20,339
	2	16,106	17,601	19,193
	3	13,473	15,230	17,355
	4	13,152	15,045	16,994
4	0	16,125	18,270	20,415
	1	17,001	18,635	20,299
	2	16,227	17,691	19,228
	3	13,290	15,030	17,154
	4	12,951	14,836	16,768
5	0	16,125	18,270	20,415
	1	17,397	19,020	20,663
	2	16,597	18,032	19,550
	3	13,447	15,177	17,295
	4	13,093	14,976	16,898

where the installation is used by three or four people. Only when the installation is used by five people, Variant 1 is more cost-effective than the variant in which the installation is designed in the traditional way (Variant 0).

A similar relationship was observed for a building located in Zakopane (Table 3.5). Despite the fact that the highest precipitation reaches 1097 mm per year in the area, and their height could indicate the highest savings, the very low unit cost for purchasing water from the water supply network c_w reduces the financial efficiency of the RWHS system. The obtained LCC values for this location are close to the LCC costs set for the RWHS located in Koszalin, where the annual rainfall is significantly lower. However, comparing the LCC costs between Variant 0 and Variant 4, it was found that, depending on the capacity of the reservoir, the differences in these costs ranged from 9 to 10% for Koszalin and from 15 to 18% for Zakopane. An increase in the difference is due to the high fees for discharging rainwater into the sewage system that is required to be incurred at Variant 0 during the 20 years of building use. This is due to the high, compared to Koszalin, rainfall occurring during the year in Zakopane.

Table 3.6 LCC calculations for a building located in Warszawa (*Source* Authors)

Tank capacity V, m^3	Variant	Life Cycle Cost, €		
		3 Occupants	4 Occupants	5 Occupants
2	0	18,526	21,242	23,957
	1	18,672	20,941	23,291
	2	18,015	20,284	22,698
	3	16,928	19,439	22,103
	4	16,714	19,310	21,946
3	0	18,526	21,242	23,957
	1	18,853	21,070	23,598
	2	18,111	20,292	22,654
	3	16,806	19,312	21,959
	4	16,579	19,152	21,784
4	0	18,526	21,242	23,957
	1	18,983	21,159	23,413
	2	18,179	20,313	22,634
	3	16,729	19,235	21,874
	4	16,496	19,065	21,697
5	0	18,526	21,242	23,957
	1	19,363	21,510	23,741
	2	18,520	20,604	22,898
	3	16,973	19,478	22,104
	4	16,726	19,290	21,943

In the case of the location of the RWHS system in Warsaw, its financial effectiveness is primarily influenced by the amount of precipitation whose average annual amount in this area is 578 mm. Such low value limits the economical use of precipitation water and consequently reduces the cost-effectiveness of RWHS in the building is being analyzed. The obtained results show that differences in LCC values between Variant 0 and Variant 4 are insignificant and vary from 8 to 9%, respectively (Table 3.6). It also turned out that only if the installation was used by three people, then the variant in which rainwater was used only to toilet flush (Variant 1) was less profitable than the traditional installation solution. In all other cases, Variant 0 had the highest LCC costs.

In the research, the impact of RWHS tank capacity on tap water savings was also analyzed. Their size is affected primarily by the demand for non-potable water resulting from the number of inhabitants and the amount of daily precipitation that depends on the location of the rainwater harvesting system. The results of this study are shown in Figs. 3.8, 3.9, and 3.10.

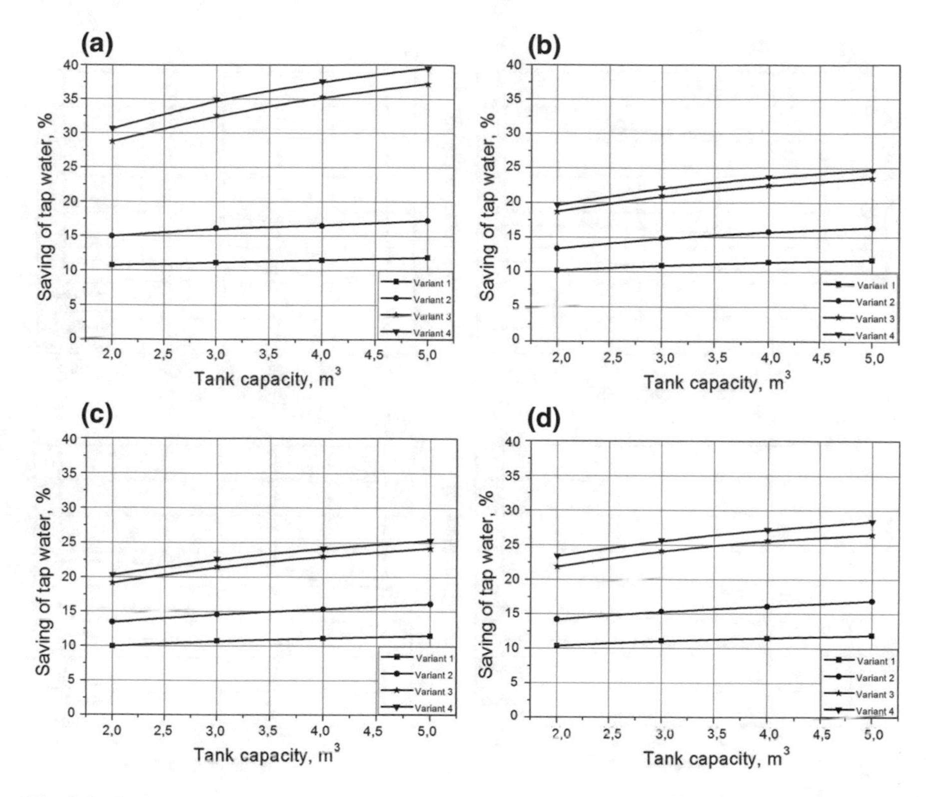

Fig. 3.8 Savings of tap water depending on the capacity of the tank used in the rainwater harvesting system for the case where the plant is used by three people **a** Zakopane, **b** Katowice, **c** Warsaw, **d** Koszalin (*Source* Authors)

It was noted that with an increase in the number of users of the system, the efficiency of using rainwater in the analyzed building for Variant 3 and Variant 4 was decreasing, irrespective of its location in Poland. This is due to the high demand for non-potable water in those variants that rainwater cannot cover, even in the case of the RWHS system located in Zakopane, where the highest precipitation occurs. In the case of Variant 1 and Variant 2, tap water savings increased slightly as the number of inhabitants increased. The highest tap water savings were obtained for the RWHS system located in Zakopane and ranged 25 to almost 40% for Variant 4, 23 to 37% for Variant 3, 15 to 19% for Variant 2, and 11 to 14% for Variant 1. Due to the comparable precipitation rates for Katowice and Warsaw, the savings on tap water were very similar in the range of 15–25% for Variant 4, 15–23% for Variant 3, 13–16% for Variant 2, and 10–13% for Variant 1. The average annual rainfall of about 760 mm in Koszalin area resulted in savings of 19–28% for Variant 4, 17–26% for Variant 3, 14–17% for Variant 2, and 10–14% for Variant 1.

The study also showed that with the increase in the capacity of the retention tank, the tap water savings increases. This tendency was especially noticeable for the

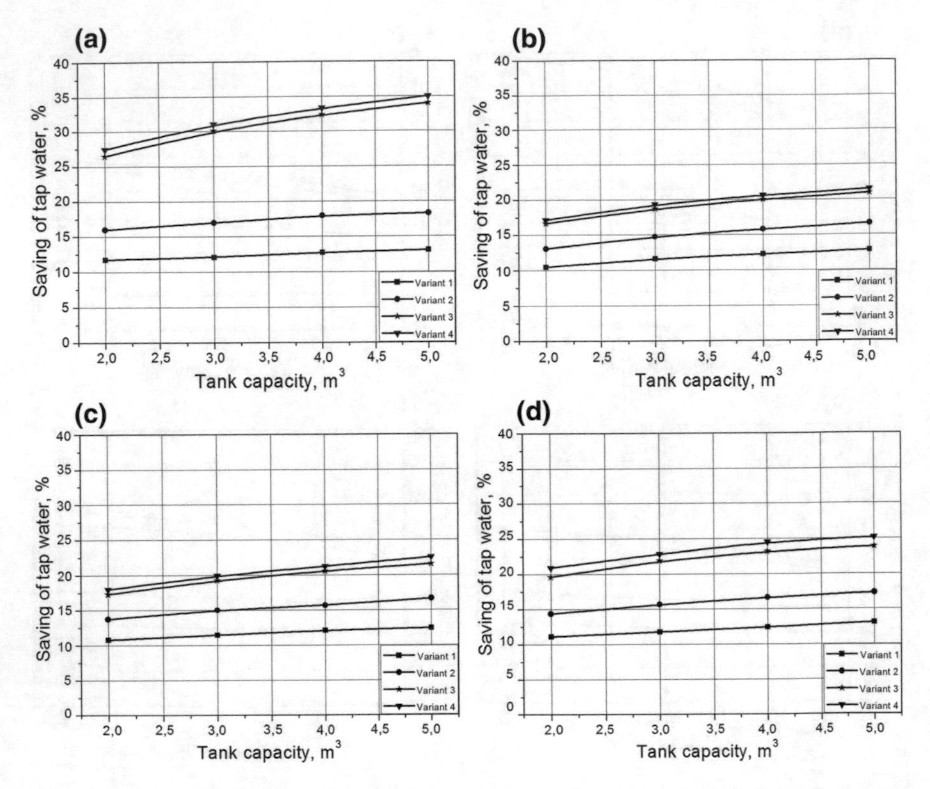

Fig. 3.9 Savings of tap water depending on the capacity of the tank used in the rainwater harvesting system for the case where the plant is used by four people **a** Zakopane, **b** Katowice, **c** Warsaw, **d** Koszalin (*Source* Authors)

location of the RWHS in high rainfall areas and for the variants where the demand for rainwater was significant (Variant 3 and Variant 4). For example, for Zakopane, Variant 4 and the case where the installation was used by three people (Fig. 3.8a), the increase in the tank capacity from 2 to 5 m^3 determined the increase in water savings by almost 10%. For the same case of calculation but localization of the RWHS system in Katowice or Warsaw this increase was about 5% (Fig. 3.8b and c). Taking into account the same cities and the situation when the installation is used by four people (Fig. 3.9) and five people, an increase of the capacity of the tank resulted in an increase in tap water savings of about 7% for Zakopane (Fig. 3.10a) and 4.5% for Katowice and Warsaw (Fig. 3.10b and c). These results have shown that the impact of increased RWHS users on tank capacity and associated tap water savings is noticeable in areas with high rainfall, while in other cities the impact was almost imperceptible.

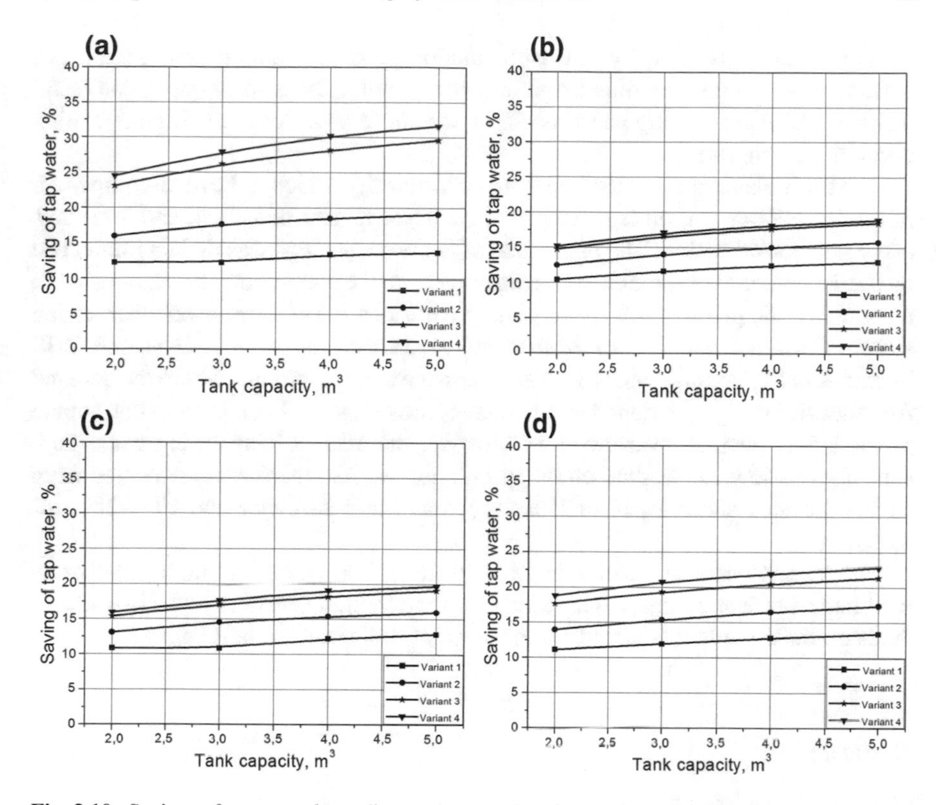

Fig. 3.10 Savings of tap water depending on the capacity of the tank used in the rainwater harvesting system for the case where the plant is used by five people **a** Zakopane, **b** Katowice, **c** Warsaw, **d** Koszalin (*Source* Authors)

3.6 Conclusion

In the chapter, the research was conducted to investigate the cost-effectiveness of the rainwater harvesting system, depending on local climatic conditions. Four cities were located in different parts of Poland. The analysis was performed using the Life Cycle Cost Methodology. The results of these studies showed that the RWHS performance under varying climatic conditions was very varied, but it was also found that the variant in which rainwater was used to flush toilets, wash, and water the garden (Variant 4) was characterized by the lowest LCC cost regardless of the volume of the tank, number of inhabitants, and location of RWHS system. This was due to the largest reduction in water consumption from the water supply and the resulting savings. It was also found that the traditional variant of the installation (Variant 0) in none of the analyzed cases was the most financially advantageous solution, even though its use was associated with the lowest initial expenditure. It confirms the validity of using the Life Cycle Cost methodology to evaluate different investment options, as selecting a solution based only on the initial investment outlay can result in wrong decision making and choosing a variant that will generate high operating costs in the long run.

The study also examines the impact of tank capacity, which is the main component of the RWHS system, on tap water savings. The capacity of 2, 3, 4, and 5 m^3 was taken into consideration. The magnitude of tap water savings was mainly influenced by the demand for non-potable water resulting from the number of inhabitants and the amount of daily precipitation that depends on the location of the rainwater harvesting system. It was noted that as tap water capacity increased, tap water savings for RWHS locations in high rainfall areas such as Zakopane and for variants where the demand for rainwater is high (Variant 3 and Variant 4) also went up. In cities with low annual precipitation, such as Warsaw and Katowice, the effect of increasing water tank capacity was low. Depending on the installation variant, these savings ranged from 11 to 40% for Zakopane, 10 to 25% for Warsaw and Katowice and 10 to 28% for Koszalin.

The research carried out and their results are not only of scientific but also practical importance and may provide guidance for potential investors in the investment decision-making process already at the stage of designing the buildings.

References

Abdulla FA, Al-Shareef A (2009) Roof rainwater harvesting systems for household water supply in Jordan. Desalination 243:195–207

Ait-Kadi M (2016) Water for development and development for water: realizing the sustainable development goals (SDDs) Vision. Aquat Procedia 6:106–110

An KJ, Lam YF, Hao S, Morakinyo TE, Furumai H (2015) Multi-purpose rainwater harvesting for water resource recovery and the cooling effect, Water Res 86:116–121

Angrill S, Farreny R, Gasol CM, Gabarrell X, Viñolas B, Josa A, Rieradevall J (2012) Environmental analysis of rainwater harvesting infrastructures in diffuse and compact urban models of Mediterranean climate. Int J Life Cycle Assess 17:25–42

Aquafit4use (2010) http://www.aquafit4use.eu/mainmenu/home.html

Burszta-Adamiak E, Stec A (2017) Impact of the rainfall height on retention and delay from green roofs. J Civ Eng Environ Archit 64(1):81–95

Campisano A, Butler D, Ward S, Burns M, Friedler E, DeBusk K, Fisher-Jeffesf L, Ghisi E, Rahman A, Furumai H, Han M (2017) Urban rainwater harvesting systems: Research, implementation and future perspectives. Water Res 115:195–209

CIRIA (2000) Sustainable urban drainage systems—design manual for Scotland and Northern Ireland, CIRIA report no. C521. Dundee, Scotland

Czemiel-Berndtsson J (2010) Green roof performance towards management of runoff water quantity and quality: a review. Ecol Eng 36:351–360

Devkota J, Schlachter H, Apul D (2015a) Life cycle based evaluation of harvested rainwater use in toilets and for irrigation. J Clean Prod 95:311–321

Devkota JP, Burian SJ, Tavakol-Davani H, Apul DS (2015b) Introducing demand to supply ratio as a new metric for understanding life cycle greenhouse gas (GHG) emissions from rainwater harvesting systems. J Clean Prod (in press)

DOE (2014) Life Cycle Cost Handbook. Guidance for Life Cycle Cost Estimation and Analysis. Office of Acquisition and Project Management, U.S. Department of Energy, Washington. Available: http://www.energy.gov/sites/prod/files/2014/10/f18/LCC%20Handbook%20Final%20Version%209-30-14.pdf

Du J, Qian L, Rui H, Zuo T, Zheng D, Xu Y, Xu CY (2012) Assessing the effects of urbanization on annual runoff and flood events using an integrated hydrological modeling system for Qinhuai River basin, China. J Hydrol 464–465:127–139

Fewkes A (2006) The technology, design and utility of rainwater catchment systems. In: Butler D, Memon FA (eds) Water demand management. IWA Publishing, London

Fletcher TD, Andrieu H, Hamel P (2013) Understanding, management and modelling of urban hydrology and its consequences for receiving waters: a state of the art. Adv Water Resour 51:261–279

Fonseca CR, Hidalgo V, Díaz-Delgado C, Vilchis-Francés AY, Gallego I (2017) Design of optimal tank size for rainwater harvesting systems through use of a web application and geo-referenced rainfall patterns. J Clean Prod 145:323–335

Fuller S, Petersen S (1996) Life Cycle Costing Manual for the Federal Energy Management Program/National Institute of Standards and Technology. NIST Handbook 135, the U.S. Department of Energy. Available: http://www.fire.nist.gov/bfrlpubs/build96/PDF/b96121.pdf

García-Montoya M, Bocanegra-Martínez A, Nápoles-Rivera F, Serna-González M, Ponce-Ortega JM, El-Halwagi MM (2015) Simultaneous design of water reusing and rainwater harvesting systems in a residential complex. Comput Chem Eng 76:104–116

Ghimire SR, Johnston JM, Ingwersen WW, Troy R (2014) Life cycle assessment of domestic and agricultural rainwater harvesting systems. Hawkins Environ Sci Technol 48:4069–4077

Ghisi E (2006) Potential for potable water savings by using rainwater in the residential sector of Brazil. Build Environ 41:1544–1550

Ghisi E (2010) Parameters influencing the sizing of rainwater tanks for use in houses. Water Resour Manage 24(10):2381–2403

Ghisi E, Ferreira DF (2007) Potential for potable water savings by using rainwater and greywater in a multi-storey residential building in southern Brazil. Build Environ 42(2007):2512–2522

Ghisi E, Oliveira S (2007) Potential for potable water savings by combining the use of rainwater and greywater in houses in southern Brazil. Build Environ 42:1731–1742

Ghisi E, Lapolli Bressan D, Martini M (2007) Rainwater tank capacity and potential for potable water savings by using rainwater in the residential sector of Southeastern Brazil. Build Environ 42(4):1654–1666

Ghisi E, Tavares D, Rocha VL (2009a) Rainwater harvesting in petrol stations in Brasília: Potential for potable water savings and investment feasibility analysis. Resour Conserv Recycl 54:79–85

Ghisi E et al (2009b) Rainwater harvesting in petrol stations in Brasília: Potential for potable water savings and investment feasibility analysis. Resour Conserv Recycl. https://doi.org/10.1016/j.resconrec.2009.06.010

Ghisi E, Rupp RF, Triska Y (2014) Comparing indicators to rank strategies to save potable water in buildings. Resour Conserv Recy 87:137–144

Gwenzi W, Dunjana N, Pisa C, Tauro T, Nyamadzawo G (2015) Water quality and public health risks associated with roof rainwater harvesting systems for potable supply: review and perspectives. Sustain Water Qual Ecol 6:107–118

Haque MM, Rahman A, Samali B (2016) Evaluation of climate change impacts on rainwater harvesting. J Clean Prod 137:60–69

Hatt BE, Fletcher TD, Deletic A (2009) Hydrologic and pollutant removal performance of biofiltration systems at the field scale. J Hydrol 365:310–321

Hirschman D, Collins K, Schueler TR (2008) Technical memorandum: the runoff reduction methods. Center for Watershed Protection, Ellicott

Hoang L, Fenner RA (2016) System interactions of stormwater management using sustainable urban drainage systems and green infrastructure. Urban Water J 13(7):739–758

Hurlimann A, Dolnicar S (2010) Acceptance of water alternatives in Australia. Water Sci Technol 61(8):2138–2142
Hyde K (2013) An evaluation of the theoretical potential and practical opportunity for using recycled greywater for domestic purposes in Ghana. J Clean Prod 60:195–200
Imteaz MA, Rahman A, Ahsan A (2012) Reliability analysis of rainwater tanks: a comparison between South-East and Central Melbourne. Resour Conserv Recycl 66:1–7
Jones MP, Hunt WF (2010) Performance of rainwater harvesting systems in the southeastern United States. Resour Conserv Recy 54:623–629
Kaposztasova D, Vranayova Z, Rysulova M, Markovic G (2016) Water management options-portfolios for safe water utilization in buildings. J Civ Eng Environ Archit 64(1):81–95
Kaźmierczak B, Kotowski A (2014) The influence of precipitation intensity growth on the urban drainage systems designing. Theor Appl Climatol 118:285–296
Khastagir A, Jayasuriya N (2010) Optimal sizing of rain water tanks for domestic water conservation. J Hydrol 381(3–4):181–188
Kim Y, Kim T, Park H, Han M (2015) Design method for determining rainwater tank retention volumes to control runoff from building rooftops, KSCE. J Civ Eng 19:1585–1590
Kuller M, Dolman NJ, Vreeburg JHG, Spiller M (2015) Scenario analysis of rainwater harvesting and use on a large scale—assessment of runoff, storage and economic performance for the case study Amsterdam Airport Schiphol. Urban Water J 14:237–246
Li Z, Boyle F, Reynolds A (2010) Rainwater harvesting and greywater treatment systems for domestic application in Ireland. Desalination 260:1–8
Liang X, Van Dijk MP (2011) Economic and financial analysis on rainwater harvesting for agricultural irrigation in the rural areas of Beijing. Resour Conserv Recycl 55:1100–1108
Liu J, Sample D, Bell C, Guan Y (2014) Review and Research Needs of Bioretention Used for the Treatment of Urban Stormwater. Water 6:1069–1099
Lopes VAR, Marques GF, Dornelles F, Medellin-Azuara J (2017) Performance of rainwater harvesting systems under scenarios of non-potable water demand and roof area typologies using a stochastic approach. J Clean Prod 148:304–313
Lu HW, He L, Du P, Zhang YM (2014) An Inexact Sequential Response Planning Approach for Optimizing Combinations of Multiple Floodplain Management Policies. Pol J Environ Stud 23:1245–1253
Markovič G, Káposztásová D, Vranayová (2014) The analysis of the possible use of harvested rainwater and its potential for water supply in real conditions. WSEAS Trans Environ Dev 10:242–249
Marleni N, Gray S, Sharma A, Burnc S, Muttil N (2015) Impact of water management practice scenarios on wastewater flow and contaminant concentration. J Environ Manage 151:461–471
Mitchell VG (2006) Applying integrated urban water management concepts: a review of Australian experience. Environ Manage 37(5):589–605
Morales-Pinzón T, Lurueña R, Gabarrell X, Gasol CM, Rieradevall J (2014) Financial and environmental modelling of water hardness—implications for utilizing harvested rainwater in washing machines. Sci Total Environ 470–471:1257–1271
Morales-Pinzón T, Rieradevall J, Gasold CM, Gabarrell X (2015) Modelling for economic cost and environmental analysis of rainwater harvesting systems. J Clean Prod 87:613–626
OECD (2012) Environmental outlook to 2050: the consequences of inaction. Retrieved from: http://www.oecd.org/environment/indicators-modelling-outlooks/oecdenvironmentaloutlookto2050theconsequencesofinaction.htm
Palhegyi GE (2010) Designing storm-water controls to promote sustainable ecosystems: science and application. J Environ Eng 15:504–511
Palla A, Gnecco I, Lanza LG, La Barbera P (2012) Performance analysis of domestic rainwater harvesting systems under various European climate zones. Resour Conserv Recycl 62:71–80
Pochwat K, Słyś D, Kordana S (2017) The temporal variability of a rainfall synthetic hyetograph for the dimensioning of stormwater retention tanks in small urban catchments. J Hydrol, Available online 18 April 2017 (in press)

Poorova Z, Vranay F, Al Hosni MS, Vranayova Z (2016) Importance of Different Vegetation Used on Green Roofs in Terms of Lowering Temperature and Water Retention. Procedia Eng 162:39–44
Proença LC, Ghisi E (2013) Assessment of potable water savings in office buildings considering embodied energy. Water Resour Manage 27(2):581–599
Rahman A, Dbais J, Imteaz M (2010) Sustainability of rainwater harvesting systems in multistorey residential buildings. Am J Eng Appl Sci 3:889–898
Rahman A, Keane J, Imteaz MA (2012) Rainwater harvesting in Greater Sydney: Water savings, reliability and economic benefits. Resour Conserv Recycl 61:16–21
Roebuck RM, Oltean-Dumbrava C, Tait S (2011) Whole life cost performance of domestic rainwater harvesting systems in the United Kingdom. Water Environ J 25(3):355–365
Santos C, Taveira-Pinto F (2013) Analysis of different criteria to size rainwater storage tanks using detailed methods. Resour Conserv Recycl 71:1–6
Słyś D (2009) Potential of rainwater utilization in residential housing in Poland. Water Environ J 23:318–325
Słyś D, Dziopak J (2011) Development of mathematical model for sewage pumping-station in the modernized combined sewage system for the town of Przemysl. Pol. J. Environ. Stud. 20:743–753
Słyś D, Stec A (2013) Effect of development of the town of Przemysl on operation of its sewerage system. Ecol Chem Eng S 20:381–396
Słyś D, Stec A (2014) The analysis of variants of water supply systems in multi-family residential building. Ecol Chem Eng S 21:623–635
Słyś D, Stec A, Zeleňáková M (2012) A LCC analysis of rainwater management variants. Ecol Chem Eng 19:359–372
Starzec M, Dziopak J, Alexeev MI (2015) Effect of the sewer basin increasing to necessary useful capacity of multichamber impounding reservoir. Water Ecol 1:41–50
Stec A, Kordana S (2015) Analysis of profitability of rainwater harvesting, gray water recycling and drain water heat recovery systems. Resour Conserv Recycl 105:84–94
Stec A, Słyś D (2014) Optimization of the hydraulic system of the storage reservoir hydraulically unloading the sewage network. Ecol Chem Eng S 21(2):215–228
Stec A, Kordana S, Słyś D (2017) Analysing the financial efficiency of use of water and energy saving systems in single-family homes. J Clean Prod 151:193–205
Stern DI, Kaufmann RK (2014) Anthropogenic and natural causes of climate change. Clim Change 122:257–269
Tam VWY, Tam L, Zeng SX (2010) Cost effectiveness and tradeoff on the use of rainwater tank: an empirical study in Australian residential decision-making. Resour Conserv Recycl 54:178–186
The Water Framework Directive 2000/60/EC
Todeschini S (2016) Hydrologic and environmental impacts of imperviousness in an industrial catchment of Northern Italy. J Hydrol Eng 21. https://doi.org/10.1061/(asce)he.1943-5584.0001348#sthash.hhx9hciv.dpuf
Unami K, Mohawesh O, Sharifi E, Takeuchi J, Fujihara M (2015) Stochastic modelling and control of rainwater harvesting systems for irrigation during dry spells. J Clean Prod 88:185–195
United Nations (2014) World urbanization prospects: the 2014 revision. Department of Economic and Social Affairs, Population Division
Vieira AS, Beal CD, Ghisi E, Stewart RA (2014) Energy intensity of rainwater harvesting systems: a review. Renew Sustain Energy Rev 34:225–242
Wang R, Zimmerman JB (2015) Economic and environmental assessment of office building rainwater harvesting systems in various U.S. cities. Environ Sci Technol 49(3):1768–1778
Ward S, Memon FA, Butler D (2012) Performance of a large building rainwater harvesting system. Water Res 46:5127–5134
Willuweit L, O'Sullivan JJ (2013) A decision support tool for sustainable planning of urban water systems: presenting the Dynamic Urban Water Simulation Model. Water Res 47(20):7206–7220
Zaizen M, Urakawa T, Matsumoto Y, Takai H (2000) The collection of rainwater from dome stadiums in Japan. Urban Water 1:355–359
Zeleňáková M, Markovič G, Kaposztásová D, Vranayová Z (2014) Rainwater management in compliance with sustainable design of buildings. Procedia Eng 89:1515–1521

Chapter 4
Participatory Management for Rainwater Harvesting in Patan, Nepal

Zuzana Boukalová, Jan Těšitel, Binod Das Gurung and Daniel Kahuda

Abstract EUREKA Project STORAGE—Sustainable TOols for gRoundwater manAGEment optimisation and water scarcity mitigation, whose research serves like the base for our paper, deals with the water supply problems affecting local communities in the area of Thapa hiti and Tagal hiti, Patan, Nepal. The innovation of our research and piloting is in an implementation of managed aquifer recharge not only for a few local users within a square or a school. Instead, we have devised a rainwater infiltration system encompassing the whole local shallow aquifer, without being tied to a particular end-user. For this, strong participative management (i.e. involving local communities and interest groups) is needed, and the close cooperation with local residents is of high importance to initiate integrated water management at least on a local scale of several neighbouring communities.

Keywords Managed aquifer recharge · Participatory management · Integrated water resource management · Nepal

Z. Boukalová (✉) · D. Kahuda
VODNÍ ZDROJE, a.s., Jindřicha Plachty 16, 150 00 Praha 5, Czech Republic
e-mail: zuzana.boukalova@vodnizdroje.cz

D. Kahuda
e-mail: kahuda@vodnizdroje.cz

J. Těšitel
METCENAS o.p.s., Tleskačova 16, 323 00 Plzeň, Czech Republic
e-mail: jan.tesitel@metcenas.cz

B. D. Gurung
CISD, Kathmandu, Nepal
e-mail: binod.gurung@cisd.org.np; binodg@gmail.com

M. Zelenakova (ed.), *Water Management and the Environment: Case Studies*, Water Science and Technology Library 86,
https://doi.org/10.1007/978-3-319-79014-5_4

4.1 Introduction

The groundwater status in Patan, Nepal, is affected by several direct and indirect drivers (the most important being various unsustainable land-use activities, engineering works, increase of immigration of the people to Kathmandu Valley and the climate change). These drivers cause changes in groundwater recharge and dynamics, leaching of pollutants and groundwater quality. Mainly because of increasing immigrant rate, improper use and mismanagement of water resources started. Kathmandu (incl. Patan area), capital city of one of the richest countries of the world in water resources, suffers from a severe drinking water supply crisis, particularly in the dry seasons of every year. Quality of water is one of the serious problems, but it is not concerns of people because of severity of quantity.

The original, historical water supply system for Patan consisted of the following main components:

- Distant source of water (the Lele River) which was connected through the Royal Canal with a system of ponds (pokharies) and flowed into Lagankhel, the water source area for the whole Patan;
- Pokharies—i.e. artificial "lakes" that were used for water storage and as well as the groundwater recharge: their partly permeable base allowed infiltration of the water to shallow aquifers that saturated local sources of the water—stone spouts (hities) and some wells (GISIDC 2011).

Hiti was connected to man-made drains, which were bringing water from an original source through a system of pokharies, or they were directly supplied with water from a local shallow aquifer. The entire system was terminated in an outlet taking water from the stone spouts to drains that emptied mostly into the Bagmati River. Problems with water supply began in Patan in the 1950s, when the royal dynasty of Nepal, inspired by the western civilisation, started to build water pipelines throughout Kathmandu and its surroundings, including Patan. At the same time, pokharies were being dried and its connection with hities interrupted, due to intensive construction works in the area (UMES 2012).

In the beginning of this century, the water from water mains started to be insufficient to satisfy the needs of the inhabitants of Kathmandu and the people initiated to solve this situation by digging new wells for water use, usually to the same aquifer, without sufficient hydrogeological information. This activity, supported by extensive groundwater withdrawal, caused serious groundwater level declination in the whole Patan (Center Bureau of Statistics, Government of Nepal 2011). Changes in the groundwater system had impacts on the functions that groundwater provides to socio-economical uses and ecosystems (Boukalova et al. 2014).

Fig. 4.1 Nahiti water source (*Source* Author)

4.2 Possibilities of Rainwater Recharge Sites in Patan

Patan—one of the tree historical towns that today is a part of the Kathmandu valley agglomeration, separated from Kathmandu city only by the Bagmati River—is the ideal pilot area for managed aquifer recharge and training of the stakeholders in the sustainable water management. Patan is facing serious water problems from the beginning of the year 2008 (Boukalova and Hrkal 2010); the first attempt to solve these troubles was the initial water management activity of the Thapa hiti community—the construction of the accumulation and infiltration wells that started in Nahiti (Thapa hiti source area) in the year 2010 (see Fig. 4.1).

This water management activity and well digging in Nahiti is the first common undertaking of the successful participatory management of several Patan's communities to ensure sufficient water supply for local people. Usually, the communities act separately, attempting to overcome the water supply shortage by digging new wells to a maximum depth of 10 m below the ground level. The public commitment of the Yashodhara Buddha Secondary school headmaster made possible an agreement and common undertaking that demonstrate the ability of the communities to join and flexibly react to the real necessities as far as they are clearly and comprehensibly defined. The participatory management depends on a clear articulation of the problem and its solution, based on the historic experience as well as technical possibilities of the local people (CISD 2012).

However, this "enlightened" procedure is far from being commonly accepted in Patan, particularly due to the fact that local people do not understand why to infiltrate

Fig. 4.2 Water management training for the Thapa hiti community in the year 2014 (*Source* Author)

water into the geological environment, as it is needed for immediate consumption. According to their reasoning, the infiltrated water gets lost somewhere in the underground and eventually will be used by someone else, if at all. For them, the best solution is to drill new shallow wells or—in the few instances they accept the rain harvesting method—to convey rainwater from the gutters through an infiltration (cleansing) unit, which in most cases is a barrel with sand, into their own separate wells (see Fig. 4.10).

The area lacks consistent environmental education programme for local people that would demonstrate the natural relations of the water cycle, explain in simple terms the harmfulness of water contamination as well as the need to provide water for a sufficient number of infiltration facilities, and present the local hydrogeological conditions and interrelations in easily comprehensible images and comic strips. Water management training of the communities is the best instrument (Fig. 4.2). The water shortage problems concerning individual bahals or schools cannot be treated separately—instead, they must be seen as parts of the whole, or at least as concerning several neighbouring communities. Such environmental education should also include an exhortation to start at least a simple integrated water management programme in a larger area under cooperation of several communities to prevent uncontrolled sinking of shallow wells into the aquifer.

The shallow groundwater system in the Patan area is divided into several smaller aquifers separated by clayey layers. However, there is no detailed hydrogeological map available. As a result, new wells that are drilled in an uncoordinated manner in Patan reach the same aquifer and thus contribute to the decline in groundwater levels,

Fig. 4.3 Prayagh pokhari—pit n. 1 (*Source* Author)

as the exploitation of the aquifer has a multiplying effect without proper recharge (infiltration). In the past, rainwater harvesting in the area was made possible by the permeable surface of the city and ensured by infiltration facilities built in every larger bahal to drain the rainfall excess during the monsoon period. At the present time, however, most of the Patan's surface is covered with impermeable materials (concrete, built-up areas) and the surface storage facilities are not properly maintained, being in most cases clogged with various run-off materials, and occasionally rubbish is found in their vicinity that may cause bacteriological contamination of the water.

The example of the successful participative management of rainwater harvesting, that could be repeated in the other parts (bahals) of Patan, was an establishment of «Prayagh pokhari prototype»—the rainwater harvesting pilot plant organised under the STORAGE project in the area of destructed Prayagh pokhari artificial lake (where the school is now situated). This plant consists of two infiltration pits (that are simulating the original function of artificial water body and enrich groundwater flowing directly to Tagal hiti and Thapa hiti). The detail of the pit is presented on the Fig. 4.6. The first infiltration pit is situated in the former pond, that is now cemented and impermeable. The other pit (n. 2) is outside of pokhari, close to the wall of the school, in the grass section of the school ground. See Figs. 4.3, 4.4 and 4.5.

The infiltration pits were sunk to a depth of approximately 4 m bgl. into permeable sandy layers. Plastic pipes perforated in the sections, where infiltration of rainwater filtered through layers of sand, gravels and round stones would take place, were set in the pit bottoms. In the part above the filter layers, the pipes are impervious and sealed in their upper parts to exclude contaminants. In constructing the pits, the final users'

Fig. 4.4 Prayagh pokhari—pit n. 2 (*Source* Author)

Fig. 4.5 Prayagh pokhari—pit n. 2 (*Source* Author)

requirements as well as the local experts' technological proposals for the sinking and equipping of the pits were respected, so that the procedure could be easily repeated and technically feasible in all of Patan.

The resulting design for the infiltration pit equipment was then based on the determined geological setting and the expected filtration coefficient of the site. The

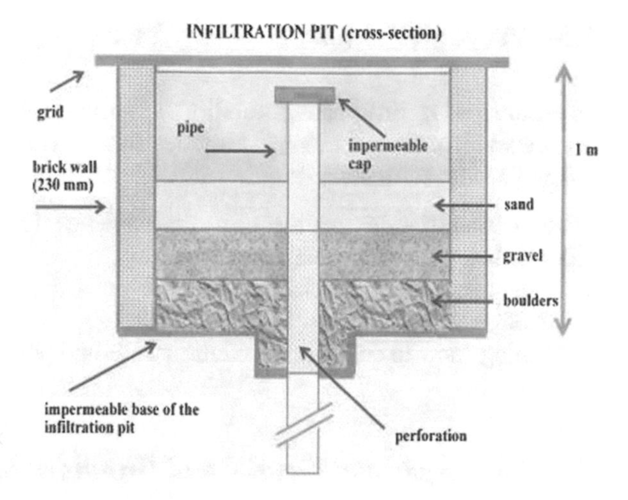

Fig. 4.6 Detail of the pit (*Source* Author)

pits were sunk into sandy gravels of very high permeability, found below fill layers and bituminous "black" clays. They were further secured from mechanical contamination (covers with dense grates, replaceable geotextiles). The entire facility was handed over to the owner of the premises, the Lalitpur District Sports Development Committee who, by signing a joint memorandum, expressed their commitment to ensuring its proper use. The handing-over ceremony was organised as a little celebration involving schoolchildren as well as communities directly neighbouring the Prayagh pokhari, so that maximum visibility was attained and contact to local people was established.

In this stage of "launching" the infiltration pits, an analysis of the key actors in the area was made to ensure social sustainability of the project. There are three main types of these actors. The first group consists of people who are directly connected to the site and affected by the project (the communities of Tangal and Thapa hiti, the school on whose premises the «Prayagh pokhari prototype» was implemented, and the owner of the premises). The second group involves individuals or organisations that develop activities similar to ours (HIMCCA, UEMS, CISD) and with which it is advisable to agree on the definition of roles in future collaboration and above all on the ways of ensuring the sustainability of rainwater harvesting in Patan. The third group consists of actors providing context for the project implementation—either local public administration officials (the municipality), donor organisations, etc., or schools and training organisations. The analysis focused on the definition of the roles of every actor both in the maintenance of the existing facility and their potential for the project replication planned for other sites in Patan.

In application of the «Prayagh pokhari prototype» also in other parts of Patan, it would be possible to revive the original water supply system, raise awareness of the local residents of the importance of cooperation and initiate integrated water management at least on a local scale of several neighbouring communities (Fig. 4.2).

4.3 Prayagh Pokhari In Situ Works

To ensure the viability and replicability of the «Prayagh pokhari prototype», various test measurements and training activities were carried out at the site of Prayagh pokhari and its surroundings:

- Monitoring of water quality and groundwater level fluctuation;
- Modelling of groundwater flow at the site;
- Model parameter input characterisation;
- Tracing test;
- Training "how to deal with infiltration wells and pits clogging".

4.3.1 Groundwater Quality and Groundwater Level Measurements

The groundwater quality on the site of Prayagh pokhari and its surroundings was measured in the period of April 2013–September 2014. Sampling was alternately made on the following sites: Thapa hiti, Minath bahal (well on the square), well in the Prayagh Pokhari, Thaina well, well in the Yashodhara Bauddha Secondary school, Machhindra bahal (old well on the square), and Nahiti (the new well). All of the mentioned wells are used for individual supply both of non-potable and potable water for the local people. The laboratory analysis of the water samples showed that all samples were contaminated with coliform and *E. coli* bacteria, and without proper treatment, they were found unsuitable for drinking purposes. In addition, increased contents of nitrates, nitrites, and ammonia were detected (especially in Thapa hiti and Minath bahal), in the Thaina well, an increased content of iron was found, whereas *E. coli* did not occur (or only in minimum amounts), and the coliform bacteria content was significantly lower as compared with the hiti. The well in Nahiti is generally the least contaminated—this infiltration well is situated in an enclosed grass-covered area where secondary contamination is minimised. The laboratory analyses show that only the bacteria concentrations exceed the limits. In comparison with the other wells, however, this excess is found to be very small—the water can be treated easily. In periods of drought, the well in Nahiti feeds water (through drains) into Thapa hiti and Tagal hiti. However, the sampling in the hiti shows that the water gets contaminated with nitrates, nitrites and also with coliform and *E. coli* bacteria during conveyance. In addition, slightly reduced pH was detected in the wells in periods of drought. The laboratory analysis results were handed over to the respective communities, while the need for the protection of wells from contamination was pointed out. In this respect, the advantages of rainwater infiltration were presented as well, as rainwater is treated to become water suitable for drinking purposes, potentially without organic pollutants, by passing through geotextiles, filter layers in the pit and the rock itself before entering the aquifer, where it may dilute the already contaminated water of the shallow groundwater system.

The on-site groundwater level measurements clearly showed the groundwater fluctuation ranges during the dry and rainy seasons. The table below illustrates the groundwater level fluctuations in the period of April–September 2014 (see Table 4.1).

4.3.2 *Modelling of Groundwater Flow in the Area of Prayagh Pokhari*

Modelling of the distribution of an inert tracer within the Quaternary aquifer was carried out in order to predict the right dosage to be used and to estimate the time necessary for the detection of the tracer in the observation well. Based on the experience from the Prayagh pokhari site, a single infiltration application of 20 l of the solution of a maximum expected concentration of 100 g/l in the infiltration pit was designed. For this model, an unsteady saturated groundwater flow of a constant gradient was assumed. For the range of saturated hydraulic conductivity values, three scenarios were taken into consideration:

- MIN—minimum permeability and groundwater flow velocity, $K = 3.11\text{E}{-}04$ m/s;
- AVR—average permeability and groundwater flow velocity, $K = 7.93\text{E}{-}04$ m/s;
- MAX—maximum permeability and groundwater flow velocity, $K = 2.18\text{E}{-}03$ m/s.

The model area schematically covers the distance between infiltration pit 1 (Prayagh pokhari) and the observation well (Thaina)—approximately 100 m. The transportation model is assumed as an advection–dispersion model—given the inert character of the tracer, neither the effect of sorption nor chemical or biological decomposition are assumed. The mathematical model is based on the USGS-MODFLOW NWT standard. The numerical calculation for the governing groundwater flow equation was performed by the finite volume method, which is by an application of the continuity equation on a finite volume represented by a cell on the mesh used. For the subsequent calculation of the advection–dispersion contaminant migration at a conservative flow rate, the MT3D99 module was used. The aim of the modelling was to test the designed dosage of the tracer, so that it could reach the observation area in the expected, visually discernible concentration of 5 mg/l within 30 days. The conductivity values K [m/s] were interpreted on the basis of the infiltration experiments performed in the saturated zone of the rock medium. According to Darcy's law, the expected velocity ranges of the tracer migration for the expected groundwater level gradient $i_{\text{hd}} = 0.01$ at the distance of 100 m between the Prayagh pokhari and Thaina were subsequently determined. Effective porosity (the same as for the modelling) was assumed as $n_{\text{ef}} = 0.25$. The resulting values are the migration time estimates (Table 4.2) for different saturated hydraulic conductivity values K [m/s].

Table 4.1 Groundwater level fluctuations (*Source* Author)

Site	April				May				June			
	Temperature in °C	Conductivity in µS/cm	pH	Water level in m	Temperature in °C	Conductivity in µS/cm	pH	Water level in m	Temperature in °C	Conductivity in µS/cm	pH	Water level in m
Thapa hiti	18	646	4.6	–	15	738	6.6	–	20	655	7	–
Tangal hiti	–	–	–	–	–	–	–	–	–	–	–	–
Prayagpokhari well	20	939	4.1	10	16	1077	6.5	8.7	22	993	6.9	9.6
Thaina well	20	636	5.1	11.6	16	686	7.2	10.4	23	645	6.5	10.6
Minath bahal	19	636	5.5	9.7	16	757	6.4	2	22	694	7.1	9.4
Site	July				August				September			
	Temperature in °C	Conductivity in µS/cm	pH	Water level in m	Temperature in °C	Conductivity in µS/cm	pH	Water level in m	Temperature in °C	Conductivity in µS/cm	pH	Water level in m
Thapa hiti	22	646	7.2	–	21	795	6.3	–	–	–	–	–
Tangal hiti	–	–	–	–				–	21	858	6.2	–
Prayagpokhari well	22	1023	6.6	8.7	20	1108	6.6	8.5	–	–	–	
Thaina well	25	642	7.7	10	20	1018	7	8.2	–	–	–	–
Machhindra bahal	22	736	6.6	8.4	–	–	–	–	22	778	6.4	7.2
Nahiti	–	–	–	–	–	–	–	–	22	577	6.5	–
Buddha School well	22	646	7.2	–	–	–	–	–	21	759	6.5	6.5

Table 4.2 Expected range of migration times (*Source* Author)

K=	8.56E−04	2.18E−03	3.11E−04	7.93E−04	m/s conductivity
vD=	8.56E−06	2.18E−05	3.11E−06	7.93E−06	m/s Darcy's velocity
V_{real}=	3.43E−05	8.74E−05	1.24E−05	3.17E−05	m/s flow velocity
t_{tr}=	33.79	13.24	93.14	36.51	days migration time

4.3.3 Discretisation of the Model Area and Model Parameter Inputs

The groundwater flow modelling was performed for the schematic section of the tracer migration pathway between Prayagh pokhari and Thaina, having an area of 150 × 30 m corresponding to the distance between the facilities and the assumed effect of transversal dispersivity on the migration. The discretisation was done using a resolution of 1 m; i.e., the model area was composed of 4500 square elements.

The groundwater hydraulic gradient $i_{hd} = 0.01$ was set by constant levels as first-type boundary conditions on the edges of the area of interest. Effective porosity was assumed as $n_{ef} = 0.25$, the thickness of the aquifer $t_a = 10$ m and the proportion of transversal dispersivity $d_t = 0.1$. The initial condition for the migration model was a single input of 20 l of the tracer having a concentration of 100 g/l, entering the aquifer for a period of 1 h for the sake of the calculation. No other infiltration event was included in the model simulation.

Modelling was performed for three scenarios based on the range of saturated hydraulic conductivity values. In the MIN scenario, the observation area would not be reached within the expected 30 days (Fig. 4.7). On the other hand, in the MAX scenario the diluting effect prevails and the minimum visually detectable concentration of 5 mg/l may not be ensured (Fig. 4.9). Thus, it is the AVR scenario that presents the optimal migration rate for a successful outcome of the tracing test (reaching the observation area within 14 days). The resulting graph for the AVR scenario is given in Fig. 4.8.

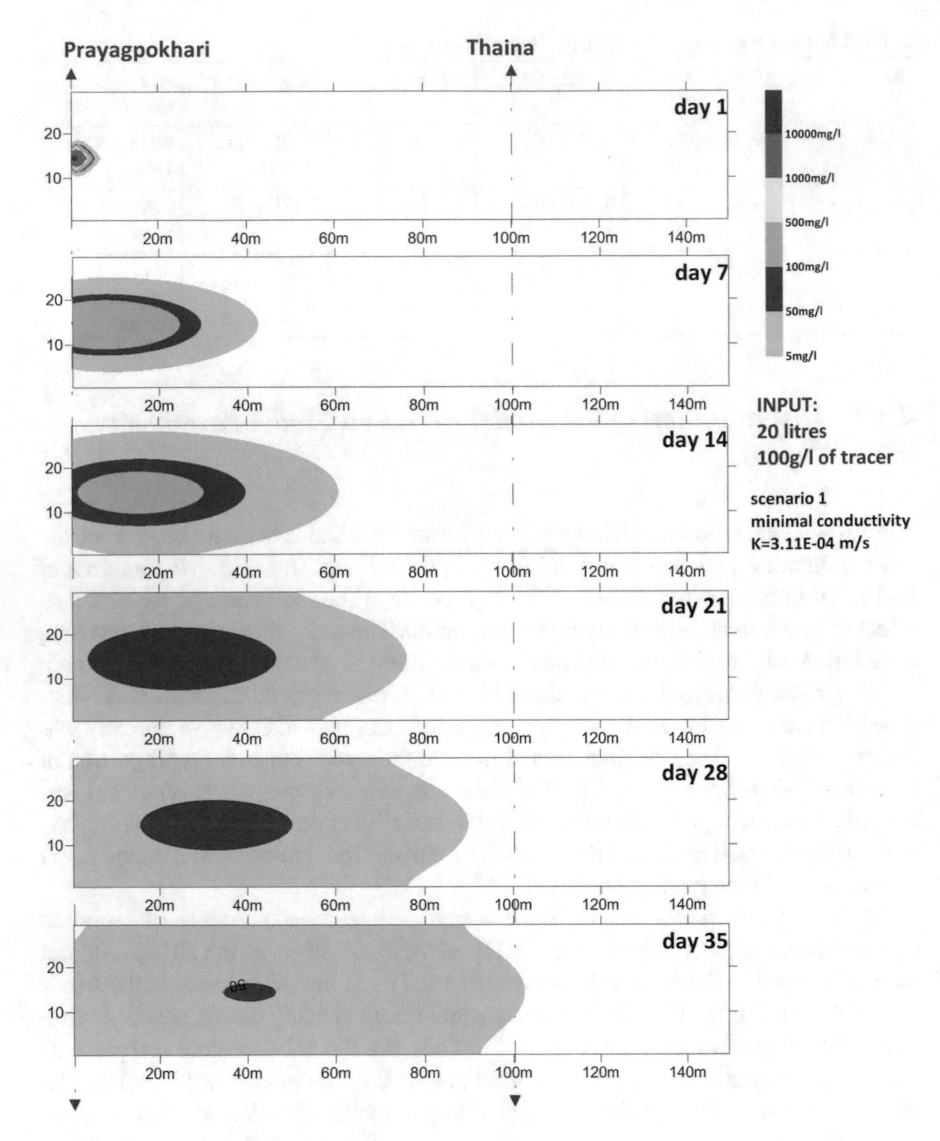

Fig. 4.7 Simulation of tracer migration for the MIN scenario with marked detectable concentration (*Source* Author)

4.3.4 Tracing Test

Tracing tests along with environmental tracers are virtually the only technique to enable the observation of the migration of water particles and therefore of the substances diluted in water or transported by it. In contrast to environmental tracers, tracing test, thanks to much higher concentration differences, enables pinpointing

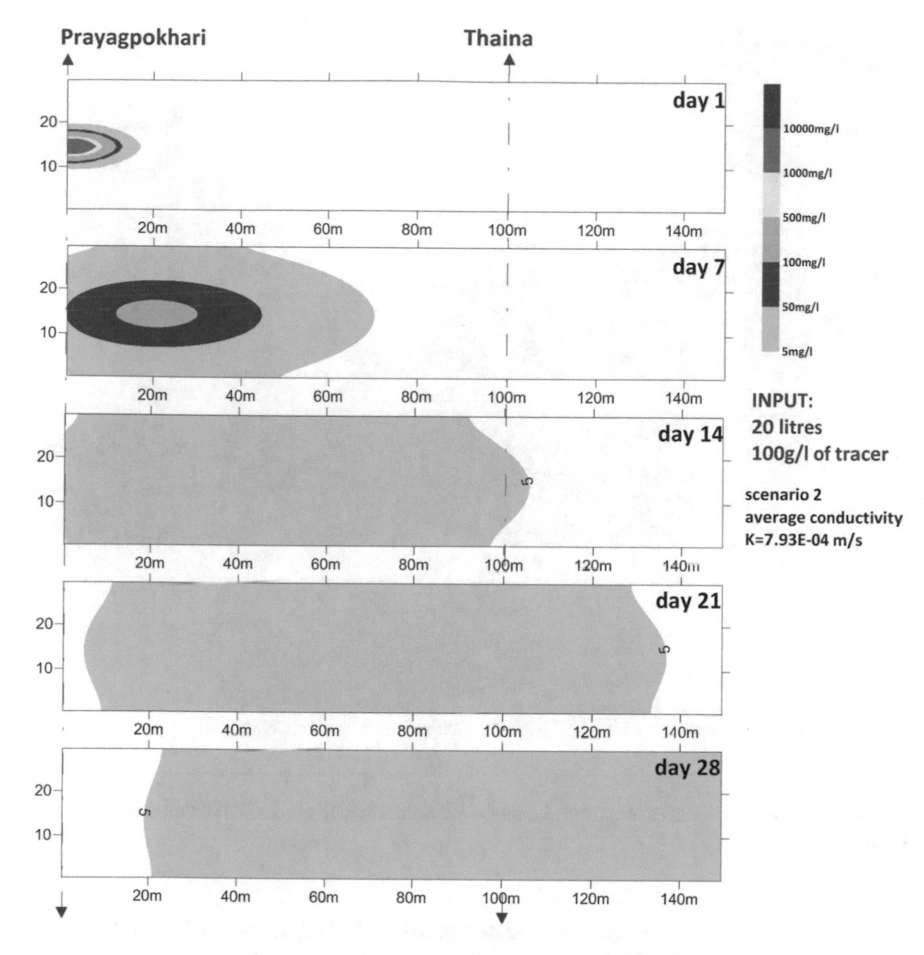

Fig. 4.8 Simulation of tracer migration for the AVR scenario with marked detectable concentration (*Source* Author)

and quantifying even minute quantities of water inflows within the total mixture. The tracing test on the site of Prayagh pokhari was carried out based on the results of a particle-size analysis of the rock of the aquifer found in the area of the infiltration pits, on a preliminary infiltration test and a preliminary groundwater migration model. In the case of the Prayagh pokhari, it had two research purposes:

- to support the participative management and to enhance communication with the communities connected to the infiltration site and the potential final users of the infiltrated water by examining their behaviour to determine the most suitable way of communication;
- to assess the preferential groundwater flow pathway and determine the flow velocity and the time of delay between two facilities, to assess the transportation param-

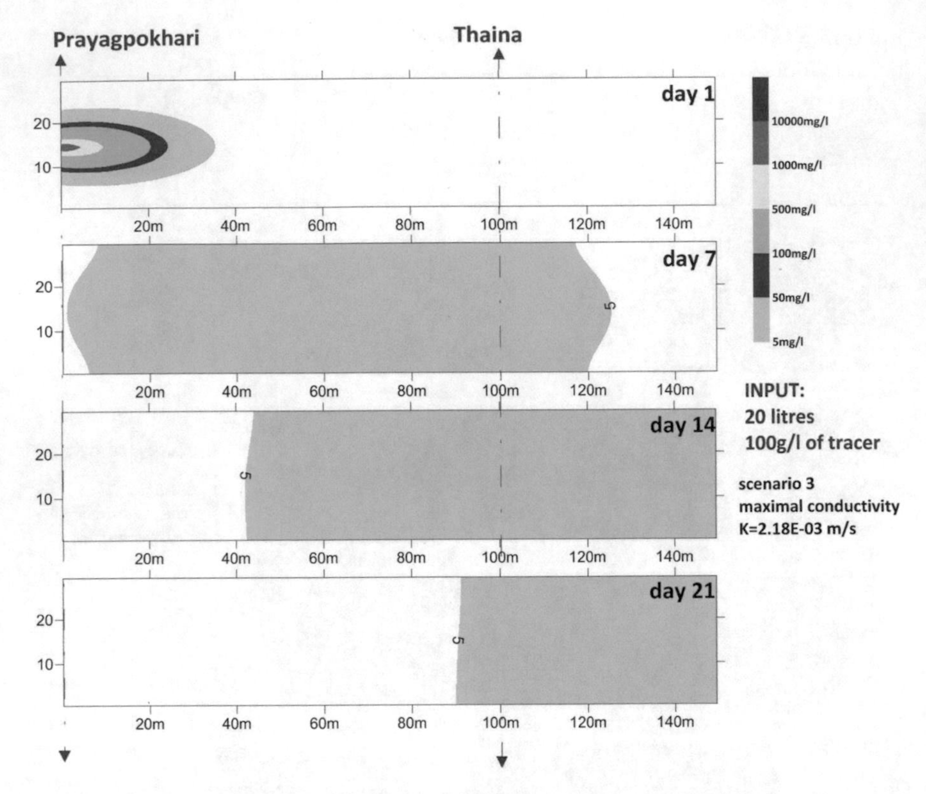

Fig. 4.9 Simulation of tracer migration for the MAX scenario with marked detectable concentration (*Source* Author)

eters of the medium such as flow velocity, time of delay as well as to perform the calibration and verification of the numerical model.

The infiltration test—which for the local people meant "pouring red dye into the infiltration well" which, as we assumed, was recharging the aquifer of Tangal and Tapa hiti—was a sensitive operation from a social point of view. Therefore, it had to be properly discussed with the potentially affected communities. This was done by workshops in each of the communities under the guidance of a Nepalese coordinator who explained to the representatives of the communities in an acceptable way all about the test to be done: what it is good for, what dye will be used for the tracing test and what risks and benefits it involves. During the discussion, illustrative images were shown and promotional materials in Nepalese were presented. At the end, an open debate took place followed by an invitation to the infiltration test in Prayagh pokhari.

The infiltration test itself was preceded by "tuning" of the situation that took about an hour. The test became a social affair. In particular, the stage of dissolving the dye in water by stirring was very important for establishing communication and involving the onlooking locals into what was happening. Through this, the idea of rainwater infiltration into the aquifer and the model of sustainable groundwater exploitation became more familiar to them. The "stirring" stage turned out to be socially significant and at the same time opened the possibility of closer communication between the researchers and the community—local people felt free to ask questions about the quality of water and the infiltration system. In particular, the women from Thapa hiti lightened the atmosphere—each one finally took the piece of lath we used for stirring the tracer and stirred the solution in the bucket for a moment, while others were taking photographs of her. As soon as the first brave woman (apparently one of the highest ranking in the community) dared to stir the coloured solution, the others could not resist following her. An important part of the programme was demonstrative drinking of the coloured water. The dye used for the infiltration test was being dissolved in water that had been drawn from the local well on the school premises. It was fine for the test itself, but from the results of the chemical analysis, we knew that this water complied neither with the European nor Nepalese drinking water standards due to elevated bacteria concentrations, and so it was not suitable for such demonstrative tasting. Therefore, we had brought a bottle of drinking water, in which we dissolved a little dye. This, for greater credibility, was done by our Nepalese coordinator, who was also the first to drink followed by the European researchers. Then, the bottle was passed around. All of the women took a sip, then some of the men. At the time, the tracer was ready for infiltration, the former Minister of the Environment Ganesha Shah arrived at the site of Prayagh pokhari and a boisterous greeting took place among him, the representative of the Honorary Consul of the Czech Republic and the representatives of several local businesses and organisations. Thanks to this, the test gained social prestige and became an opportunity to be publicly seen. It was instructive to watch how the various "important" participants tried to take advantage of the situation. However, by doing this they became "accomplices" in the test and indeed facilitated its social acceptance. While the participants were socially engaged in talking about personal affairs (and no longer technical specifications), the red solution was poured into the infiltration pit and the Nepalese coordinator explained to the representatives of the communities how they might assist in the groundwater monitoring.

Thanks to a thorough preparation (long-lasting communication with the representatives of the local communities started at the very beginning of the project along with informational workshops), the infiltration test was successfully accomplished. Given its happening-like character, we may even consider it a promotional event. Thanks to this, the pilot plant in Prayagh pokhari became well-known, which means more readily transferable to other similar areas in Patan.

According to the original plan, the tracing test should be done at both pits at the same time. However, after the reconnaissance of the situation, it was decided to perform the infiltration of dyed water at pit No. 2 (outside pokhari) only. This was done in a concentration of 20 kg per 100 l recommended by the groundwater flow model, using a food colour Ponceau 4R produced by: *M/s. SUN FOOD TECH. Indi; W-15, J/21, Western Avenue, Sainik Farms, New Delhi—110062, INDIA*.

The aim of the tracing test was to infiltrate as much dye as quickly as possible into the aquifer. The shallow local aquifer is probably not large (we estimate its thickness to be up to several metres of highly permeable sandy gravels separated from the deeper aquifer by a massive layer of clayey rock reaching a thickness of up to 100 m, possibly with local intercalations of more permeable materials) and highly permeable, which means quicker transport of the dye but unfortunately its quicker dilution as well, so that there is greater probability that the dye will be diluted beyond detection before it reaches the monitoring well. The arrival of the dye in the surrounding wells and storage facilities was expected between 14 and 30 days after the infiltration of the tracer. It was also assumed that at the end of this period the dye might not be recognisable in the water due to the fact that it will have been diluted beyond visual detection in the groundwater. After the tracing test was carried out, water at four monitoring locations situated at 30–160 m from the infiltration pit was monitored with the assistance of local people. The monitoring campaign was conducted by a Ph.D. student in Environmental Science at Kathmandu University who personally examined the monitoring wells every morning and afternoon from the 7th–30th day after the infiltration test was performed, with the assistance of local people who examined the groundwater colour in other wells in the broader area. The times of the student's monitoring visits followed the power supply shutdown schedule for the site—without electricity, the pump in the monitoring wells could not be used. On the 25th day after the start of the tracing test, pinkish groundwater was detected in the Thaina well. At the other locations, no tracer was detected. The results of the tracing test showed that the groundwater probably flows as expected—in the direction to Thapa hiti and the Thaina well. However, potential impact on Tangal hiti was not demonstrated by the tracing test.

4.3.5 Clogging

Clogging (a process of gradual sealing of the aquifer in the proximity of an artificial hydrogeological object during water filtration) in the time of managed aquifer recharge in the third world is a major problem that influences both the qualitative and quantitative parameters of infiltration, and it determines the selection of the technical type of infiltration wells and pits, their efficiency, as well technical and technological requirements for the whole implementation. And, last but not least, it has a great impact on the economy of the whole project. The clogging is a result of attachment and accumulation of mechanically suspended solids or substances precipitated from water as a consequence of physical, chemical or biological processes resulting from the interaction of water and rock medium or of infiltrated water and groundwater.

Considering the fact that the infiltration pits in Prayagh pokhari are intended for rainwater infiltration, it can be assumed that potential clogging of these pits will be caused rather by mechanical sealing with sludge, rain-washed colloids and suspensions than by chemical and microbiological processes (Boukalova et al. 2014). Nevertheless, it is necessary to carry out continuous monitoring of physical, physico-chemical, chemical and microbiological parameters of the infiltrated water (precipitation) and the hydraulic parameters of the infiltration pits. Based on the information

Table 4.3 List of analytical methods to be applied within the monitoring campaign of the infiltration pit (*Source* Author)

Parameters	Analytical method
Physical parameters of infiltrated water	Water colour Turbidity Content of undissolved substances
Physico-chemical parameters of infiltrated water	Temperature pH Conductivity Redox potential Dissolved oxygen
Chemical parameters of infiltrated water	CHSK(Mn) Fe^{2+} Fe_{tot} Mn NO_3^- NH_4^+ PO_4^{3+}
Microbiological parameters of infiltrated water	Number of organisms in 1 ml *E. Coli* Possibly saprobity
Hydraulic parameters of the infiltration plant	Infiltration test

available, it was recommended to the final end-users of the infiltration plant to carry out ideally at least one monitoring campaign in three months (and after all events in the monsoon period), that will comprise the analyses stated in Table 4.3.

To minimise rain-wash of waste and mud into infiltration pits during the monsoon period, the Prayagh pokhari pits have been double-protected with grids of different meshing sizes and geotextiles.

Clogging issues concerning infiltration devices constitute one of the main problems affecting artificial infiltration. They determine the length of the infiltration cycles and, in case of infiltration using bores, may even lead to their complete hydraulic failure. For these reasons, much attention has been paid to methods of restoring the hydraulic function of artificial infiltration devices. In case of infiltration using pits, the solution is relatively simple and consists in removing the upper filtration layer of the pit and replacing it with new material.

Fig. 4.10 Rainwater harvesting to individual well (*Source* Author)

Fig. 4.11 Thapa hiti (*Source* Author)

4.4 Conclusions

The example of the participative management for rainwater harvesting in Patan is the building of the «Prayagh pokhari prototype» in the area. This tool consists in the implementation of managed aquifer recharge not only for a few local users within a square or a school (Fig. 4.11). Instead, we have devised a rainwater infiltration system encompassing the whole local shallow aquifer, without being tied to a particular end-user. For this, strong participative management (i.e. involving local communities and interest groups) is needed, and the essential role is assigned to the local coordinator, who is taking care to be the "interface" which is in the close contact with the local communities and end-users and who is the best guarantee that the water for communities could be assured, via participative "thinking" and close cooperation with the experts responsible for a technology implementation.

Infiltration of rainwater, as it is presented at the pilot site Prayagh pokhari (both technically and socially), is one of the possibilities of sustainable participative rainwater harvesting, that is substituting the defunct or destroyed parts of the historical water supply system and strengthening the capacity of stone spouts (hities) (Fig. 4.10).

Acknowledgements This chapter was written thanks to the project STORAGE (Sustainable TOols for gRoundwater manAGEment optimisation and water scarcity mitigation), financed by the Ministry of Education, Youth and Sports, Czech Republic, and the company VODNÍ ZDROJE, a.s.

References

Boukalova Z, Hrkal Z (2010) Solution of problems of groundwater quality and quantity in the region of Kathmandu (Nepal) on the basis of experience of Czech hydrogeologists in providing drinking water for Praha. City Hall Prague, CR

Boukalova Z, Těšitel J, Hrkal Z, Kahuda D (2014) Artificial infiltration as integrated water resources management tool, water pollution XII; ISBN: 978-1-84564-776-6, eISBN: 978-1-84564-777-3, pp 201–210. Water Pollution conference 2014, the Algarve, Portugal

Center Bureau of Statistics, Government of Nepal (2011) Major Highlights, Nepal

CISD—Center for Inclusive Social Development (2012) Water for Machindra Bahal community (Patan, Kathmandu, Nepal)

GISIDC—Geographic Information System & Integrated Development Center, Ltd (2011) Integrated water resources management in Nepal—pilot project Water for Kathmandu, Nepal

UEMS (2012) Action research on "Implication of harvested rainwater in recharging shallow ground water (GW) aquifer and its quality", Nepal

Chapter 5
Geomorphologic Hazard in Romania. Typology and Areal Distribution

Florina Grecu

Abstract The general physiographic arrangements of the relief are characterized by a quasi-concentric disposal of altitudes, and the characteristics of the geomorphologic hazards are shaped accordingly. An essential feature of the geomorphologic hazards is their cohabitation with the pedologic processes, deriving long-lasting effects on elements at risk through land degradation. In Romania, the hills and the plateau regions are the most prone to geomorphologic hazards, particularly to landslides and erosion. The specificity of the geomorphologic hazards is given by their genesis and long-term evolution; thus, their impact upon the elements at risk manifests less direct with risk effects (human and material damages, respectively). In Romania, one can distinguish geomorphic hazards induced by slope processes, by river channel dynamics and by other various processes. The gravitational hazards include avalanches and rock falls landslides, collapses, suffusion and compaction phenomena. In their turn, the pluvial–torrential systems are highly frequent for the lands deprived of wood vegetation in sedimentary hilly, plateau and mountain areas. The hazards induced by channel processes occur mainly in the hills, lowlands and mountain depression areas. They are associated with high silting, channel wandering, braiding, side erosion and meandering in plains and subsidence areas. The hazards induced by various processes refer to the marine geomorphic system of the Black Sea coast and to the aeolian systems developed on sandy areas and at high altitudes. Geomorphologic hazards affect the Romanian territory with various degrees of intensity, thus contributing to land and soil degradation, with long-lasting effects on population.

Keywords Geomorphologic hazard · Typology · Slope processes · Landslides Pluvio–torrential · Land degradation · Romania

F. Grecu (✉)
Department of Geomorphology, Pedology and Geomatics, Faculty of Geography, University of Bucharest, 1 Bd. Nicolae Bălcescu, 010041 Bucharest, Romania
e-mail: grecu@geo.unibuc.ro

M. Zelenakova (ed.), *Water Management and the Environment: Case Studies*, Water Science and Technology Library 86,
https://doi.org/10.1007/978-3-319-79014-5_5

5.1 Introduction

Romania over an area of 238,391 km^2 placed between 43°37′ and 48°15′ N, and 20°15′ and 20°41′ E is characterized by a noteworthy complexity of its natural landscape and with large variety of triggering factors. The landscape is developed on a complex geology, dominated by the alpine Carpathian orogen, neighboured by several pre-alpine platforms, and completed with the North Dobrogea orogen. Sedimentary, igneous and metamorphic rocks stretch out over considerable areas. As a consequence, a large variety of structural and petrographic forms of relief develop, and the geomorphologic dynamics can be remarkable (Badea et al. 1983; Bălteanu et al. 2012). The specificity of the geomorphologic hazards is given by their genesis and long-term evolution; thus, their impact upon the elements at risk manifests less direct with risk effects (human and material damages, respectively). The geomorphologic hazards result from the action of geomorphologic processes (erosion, transport, accumulation) upon the ground, generating landforms with potential negative effect on the environment or society, which represents, in other words, the morphological modifications of the ground surface with a direct impact on the environment and an indirect one on the society. Examples are the surface erosion, suffusion and subsidence in loess, coastal hazards, all with a prolonged action period and with cumulated effects on the environment and the society.

5.2 General Features and Conditioning Factors

For the Romanian territory, the occurrence and dynamics of geomorphological hazards are influenced by geology (structure, petrography and seismicity), climatic conditions, hydrological features and anthropogenic factors. The general physiographic arrangement of the relief is characterized by a quasi-concentric disposal of altitudes, declining from the centre (Moldoveanu Peak, 2544 m) to the periphery (5–6 m in the Romanian Plain and 0 m on the Black Sea coast), and the characteristics of the geomorphologic hazards are shaped accordingly (Fig. 5.1). An essential feature of the geomorphologic hazards is their cohabitation with the pedologic processes, deriving long-lasting effects on elements at risk through land degradation (Grecu 2009). In Romania, the hills and the plateau regions are the most prone to geomorphologic hazards, particularly to landslides and erosion (Surdeanu 1999; Florea 2003). While these areas are the most densely populated (more than 15 settlements per square kilometre) (Grecu 2009), the land cover and use impact severely on the geomorphologic processes.

The geological evolution of the Romanian territory led to the formation of two large morphostructural units: the Carpathian orogen (a) and the platform (b). The orogenic units are characterized by a great variety of rocks, structures and geotectonic features (Fig. 5.2). At the same time, the Carpathian orogen, made up predominantly of crystalline schists, is characterized by the incidence of large molasse areas,

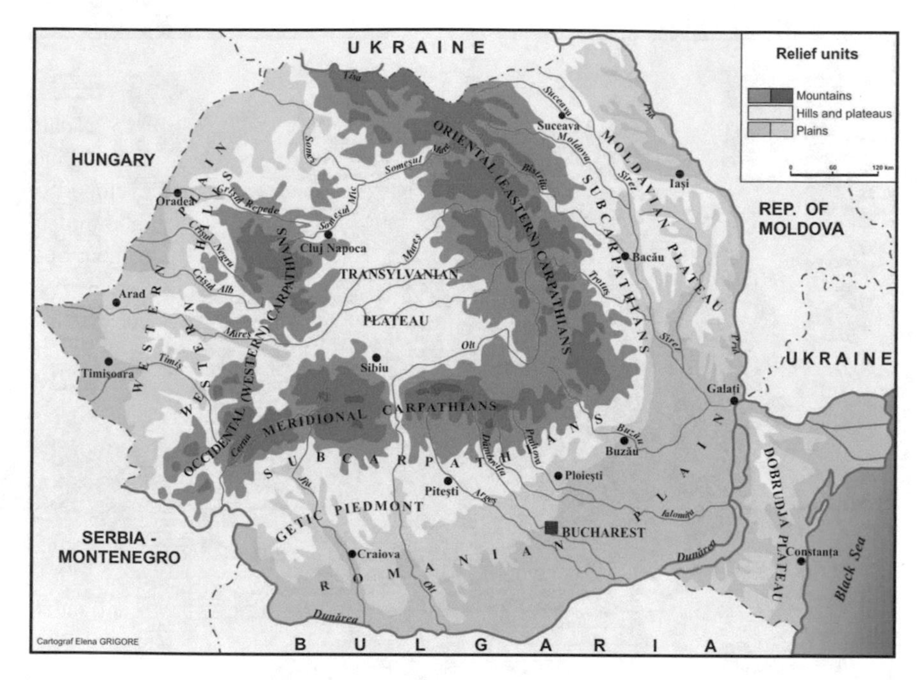

Fig. 5.1 Map of main relief units in Romania. *Source* Grigore

with intricate tectonics, and by a great expanse of the flysch. The Carpathian flysch sub-unit composed of detritic rocks (sandstones, marls, clays, conglomerates, marly limestones and limestones) is disposed in lithological assemblages (Sinaia strata, Comarnic strata, Fusaru strata etc.), and its largest extension is in the eastern part of the Eastern Carpathians. Also, it occupies almost the entire area of the Bend Carpathians.

The platform units underlie the plateaus and plains lying outside the Carpathian range, and a sedimentary cover overlays their Precambrian bedrock. Thus, most of the Moldavian Plateau overlaps the south-eastern part of the East European Platform. The surface sedimentary deposits (Sarmatian and Pliocene) show a monoclinic structure with an inclination along the NNW–SSE direction, and they are made up of sandstones, limestones, marls, clays, sands and other weak rocks. The Romanian Plain corresponds to the northern part of the Moesian (or Wallachian) Platform, which sinks in front of the Carpathians. The surface deposits belong to the Quaternary: loess, loess-like deposits and alluvial. The Dobrogea Plateau is covered by quaternary loess and loess-like deposits (Badea et al. 1983).

The vertical disposal of climate strongly influences the spatial distribution of the geomorphologic hazards. The location of the Carpathians in the middle of the country and their altitude impose substantial alterations of the temperate continental climate of Romania. The moving air masses are blocked on either side of the Carpathians, while the altitudes determine vertical differentiation of the climate. The torrential

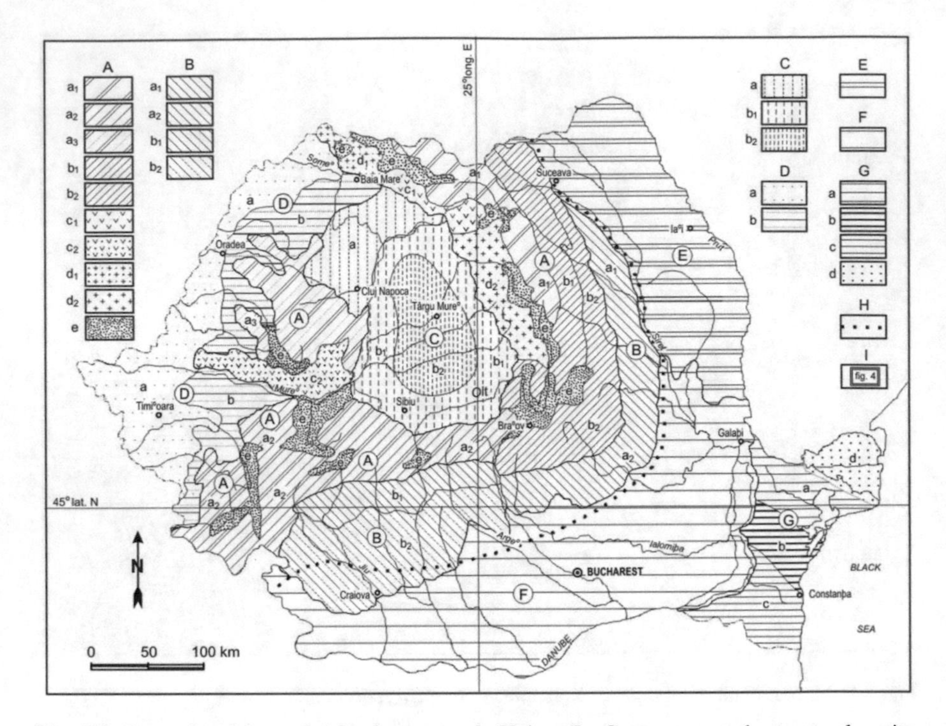

Fig. 5.2 Romania. Map of Morphostructural Units. I. Orogene morphostructural units: **A** Carpathian Mountains a: crystalline-Mesozoic sub-units (a_1: Eastern Massif; a_2: Southern Massif; a_3: Western Massif); b: flysch sub-units (b_1: internal flysch; b_2: external flysch); c: volcanic–sedimentary sub-units (c_1: Ţibleş-Bârgău Mts.; c_2: Metalliferous Mts); d: neovolcanic sub-units (d_1: Oaş-Gutâi-Văratec Mts; d_2: Călimani-Gurghiu–Harghita Mts.); e: intra-montane depressions (tectonic or barrage). **B** Pericarpathian Hills: a: Subcarpathians of Moldavia and of the Bend area (a_1: Moldavian Subcarpathians; a_2: Bend Subcarpathians); b: sub-units of the Getic Subcarpathians and Piedmont Hills (b_1: Getic Subcarpathians; b_2: Getic Piedmont); **C** Intracarpathian depression of Transylvania; a: sub-units of Someşan Plateau; b_1: sub-units of Transylvanian Tableland (sector of diapir folds); b_2: central and dome sector; **D** Plain (a) and Hills (b) of Banat and Crişana. II. Platform morphostructural units: **E** Moldavian Plateau; **F** Romanian Plain; **G** Dobrogea: a: Northern Dobrogea sub-unit; b; Central Dobrogea sub-unit; c: Southern Dobrogea sub-unit; d: Danube Delta sub-unit. **H** Dotted line—platform-to-orogene boundary. *Source* Badea et al. (1983), in Grecu (2002a)

character of precipitation is the main factor in triggering of the hazards. In this context, an indicator for the susceptibility to geomorphologic processes depending on water (i.e. erosion) is Angot index. Analyses the spatiotemporal variability of Angot index in Romania (from 30 meteorological stations located in all the natural units of Romania, for the time interval of 1961–2000) to reveal the most intense susceptibility to erosion is specific to June in the mountains, hills and plateaus from the eastern side of the country in relation to the pluviometric regime and, thus, to baric centres' dynamics over Europe and to the barrier formed by the Carpathian (Grecu et al. 2014).

The anthropogenic evolution of Romania explains why the forest covers 28–29% of the Romanian Territory. In the early nineteenth century, the forest covered about 55–60% of the country. Its distribution according to the altitudinal levels is the following: 52% in the mountains, 37% in the hills and the remaining 11% in the plains (Ielenicz 2007).

Most of the recent scientific papers are using GIS techniques for analysing each factor (coefficient) of landslide activation and for pluvial erosion evolution (Armaș 2006; Grecu et al. 2013; Jurchescu and Grecu 2015). For example, the landslide susceptibility assessment was done using the method described in the Romanian legislation H.G. 447/2003 (mapping methodology and content of landslide and flood risk maps) (României 2003). In this case, the estimation of value and spatial distribution of each coefficient were made individually for the lithologic, geomorphologic, structural, hydrologic and climatic, hydrogeologic, seismic, sylvic and anthropic factors (Manea and Surdeanu 2012). Alternatively, the use of logistic regression for the integration of landslide predictors coefficients reveals important characteristics of landslides manifestation (Măgărint et al. 2013). The use of this method proves that landslide susceptibility is essentially connected mostly with more specific factors such as slope angle, land use, slope height and lithology.

5.2.1 Areal Distribution and Typology

In Romania, one can distinguish geomorphic hazards induced by slope processes, river channel dynamics and by other various processes (Grecu 2002a). The gravitational hazards include avalanches and rock falls landslides, collapses, suffusion and compaction phenomena. In their turn, the pluvial–torrential systems are highly frequent for the lands deprived of wood vegetation in sedimentary hilly, plateau and mountain areas. The hazards induced by channel processes occur mainly in the hills, lowlands and mountain depression areas. They are associated to high silting, channel wandering, braiding, side erosion and meandering in plains and subsidence areas. The hazards induced by various processes refer to the marine geomorphic system of the Black Sea coast and to the aeolian systems developed on sandy areas (about 540,000 ha) and at high altitudes (Grecu 2002b) (Fig. 5.3).

5.2.1.1 Gravitational Hazards

Avalanche hazard has been tackled mainly in the Southern Carpathians (Transylvanian Alps), being reported more studies especially in the last two decades. Avalanches have the highest frequency in the Făgăraș Massif, occurring on most valleys and slopes (Voiculescu 2002). The greatest number of casualties has been recorded on the Bâlea Valley (60 people, of whom 35 are dead). Many tourist chalets (Bâlea Lac, Turnuri, Valea Sâmbăta) are frequently affected by avalanches, as is also the case of the high part of the national high-altitude road called Transfăgărașan. In the Piatra

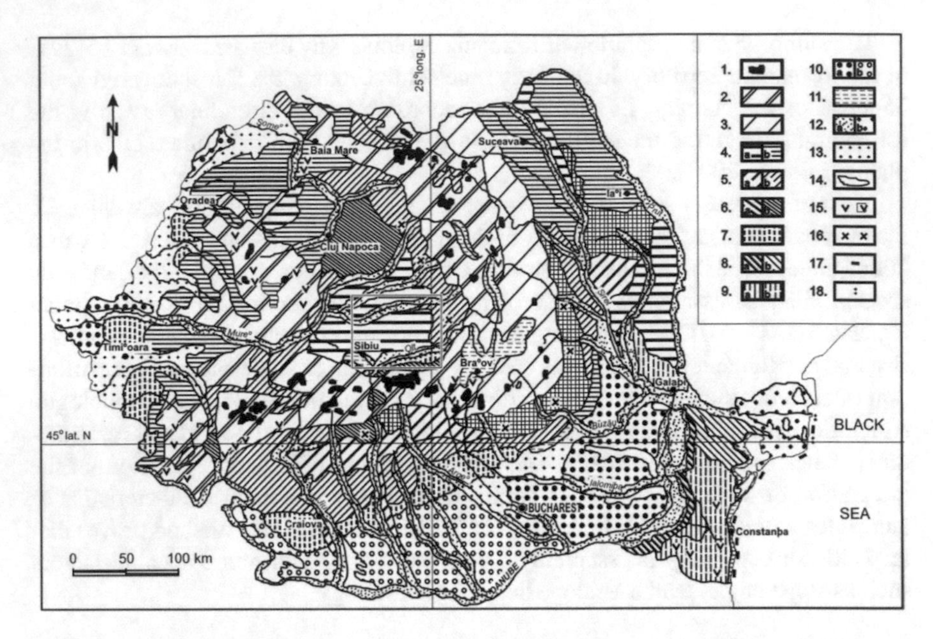

Fig. 5.3 Romania. Map of present-day processes/hazards. 1: Cryonival, aeolian and rilling processes; 2: fluvial-gullying, rockfalls and topples; 3: fluvial-gullying, landslides and mudflows; 4: sheet erosion and gullying associated with mass movements (*a*, intense; *b*, moderate); 5: gullying and sheet erosion associated with mass movements; 6: mass movements associated with gullying and sheet erosion (*a*, intense; *b*, moderate); 7: mild and moderate sheet erosion only in the valley side. 8: sheet erosion and gullying, associated with piping: intense (*a*); moderate (*b*); 9: sheet erosion associated with piping: intense (*a*); moderate(*b*); 10: down-sagging and piping (*a*) associated with mild erosion in the valley sides (*b*); 11: colluvial, proluvial and alluvial accumulation; 12: fluvial accumulation (*a*), fluvial, mineral and biogene (*b*); 13: fluvial accumulation during exceptional floodwaves, local water stagnation; 14: wind-induced deflation and accumulation; 15: dissolution of carbonate rocks; 16: dissolution of salt; 17: abrasion; 18: coastal accumulation. *Source* Badea et al. (1983), in Grecu (2002a)

Craiului Massif, avalanches take place on most valleys originating near the ridge and crossing the forests (Munteanu et al. 2012). In the Bucegi Mts., avalanches occur in the valleys originating in the alpine and subalpine level. Most victims were caught in avalanches sliding down the valleys carved in the cliff slope the Prahova River (especially on the Morarului and Coştila valleys).

Rock falls and collapses are frequent in the mountain areas, especially on hard rock slopes (crystalline schists, eruptive rocks, sandstones and conglomerates), massively influenced by works or maintenance interventions (Săndulache 2010). On the Olt transversal valley, experimental research and Rockfall Hazard Rating System (RHRS) showed that collapses and rock falls are relatively common, especially for those with volumes up to 500 m^3, which account for about 67% of the cases (41 over the period 2003–2009) (Ilinca 2009).

Landslides are the most diverse geomorphologic hazards in terms of resulting landforms and spatial distribution (Fig. 5.4). Their genesis and type are influenced mainly by geology, morphology, precipitation characteristics, land cover and anthropogenic activities. Landslides affect the hilly and plateau units, but also the sedimentary chain of the Eastern Carpathians. The landslide risk is very high in the Moldavian Plateau, Transylvania Plateau, Subcarpathians and Banat-Crişana Hills.

In the Central Moldavian Plateau, the landslides affect thick diluviums and move very quickly (1.5 m per month and sometimes 6 m/h) (Niacșu et al. 2008). The maps of landslide distribution, drawn at 1:5000 scale for each catchment, illustrate that half of the area is covered by landslides, in some shape or age. Most landslides are stable (inactive), and the active ones have a very slight occurrence, averaging today only 2.4% of the total landslide area. Ioniță (2000) reported that after the much wetter precipitation period 1968–1973, active landslides averaged 21.4% of the total landslide area.

Under the circumstances, in the Tutova watershed, the areas with degraded lands increased from 2731 ha in 1969–1972 to 9119 ha in 1983–1992 (Niacșu et al. 2008). In the Banat and Crișana Hills, landslides affect all the slopes in the Barcău Valley (Josan and Sabău 2004). In the Tășnad and Codru Hills (Satu Mare County), active landslides are less frequent, but the reactivated older ones' account for about 50% of the settlements' area (Driga 2007).

The landslides occur frequently in the Transylvanian Depression influenced by large areas occupied by Sarmatian sedimentary rocks (Morariu et al. 1964; Irimuș 2006). The glimee are indeed old, rotational landslides, generally stabilized, but their genesis is complex. The specific of the areas with glimee is given by the alignments of glimee and the depressionary trough between these alignments. The secondary/associated hazard is represented by erosion, slides on the glimee/hills slopes and by the humidity excess from the microdepressions from between the landslide alignments, which lead to terrain degradation, affecting agricultural use. Massive landslides, locally called *glimee*, which are maximum frequency is in the Transylvanian Plain, while the biggest dimensions are in the Hârtibaciu Hills (Fig. 5.4). The specific of the areas with glimee is given by the alignments of glimee and the depressionary trough between these alignments. The secondary/associated hazard is represented by erosion, slides on the glimee/hills slopes and by the humidity excess from the microdepressions from between the landslide alignments, which lead to terrain degradation, affecting agricultural use. The complex analysis of the glimee relief is made by Gârbacea (2013). The main characteristics of the *glimee-type* landslides in the Transylvanian Depression have been approached in various papers (Gârbacea 1964; Grecu 1983; Irimuș 2006 and recent regional study in doctoral thesis). These processes deeply affect the regolith and geologic substratum. The mean area of *glimee*-affected spots is 50–150 ha, but they can exceed 600 ha, i.e. Saeş—1550 ha, Movile—900 ha, Saschiz—615 ha). Step-like landslides (pseudo-terraces) have been described in the Hârtibaciu Hills (Grecu 1992), in the form of small steps or terraces. Wave-like landslides, in the form of small waves with lakes in between, are shallow (3–5 m). Tongue-shaped landslides, which occur very often, bring their contribution to present-day slope modelling too. Mudflows are processes occurring in the presence

Fig. 5.4 Gravitational hazard. **A** Deep-seated landslide affecting road and buildings in Breaza (Prahova River) (2005); **B** "Glimee"-type landslides at Movile—B1 and Apold—B2 (Hârtibaciu Plateau, Transylvanian Depression) (2009). **C** Sheet erosion and gullying associated with mass movements in the Transylvanian Depression (2009). **D** Mud volcanoes in Subcarpathians Curvature (2015) *Source* Author

of impervious rocks (marls, clays) associated with the Pannonian sands which have a large extension in the diapir regions (Turda, Praid, Ocan Sibiului, Ocna Mureşului) (Irimuș 2006). Shallow landslides occupy extended areas especially in the central part of the Transylvanian Tableland and affect the grass cover after long rainfalls.

For the Curvature Subcarpathians, several studies show an intensification of the hazards, especially landslides, with regard to frequency and affected area, likely due to precipitation and tectonics (Bălteanu 1983; Sandu and Bălteanu 2005). For example, 1997 was a very wet year in the Prahova Subcarpathians, with heavy precipitation in April (130.9 mm compared to 59.7 mm average amount), July (142.3 mm compared to 101.2 average amount), but mostly in August, when the total amount of 274.4 mm was 189.8 mm higher above the monthly average (Grecu et al. 2008, 2010a). The triggered landslides and floods that hit the Prahova County in 1997 affected severely 23 settlements (Cioacă and Dinu 1998). The year 2005 brought about a climatic record especially because of the annual amount of precipitation (1238.0 mm). In December 2005, in the Prahova County, 90% of the 103 communes reported various hazard events, as follows: 63 landslides, 19 floods and 14 other types of hazards.

5.2.1.2 Pluvial–Torrential Geomorphic Hazards

The national assessment of the total amount of eroded material deriving from agricultural lands, accomplished by Moțoc (1984), indicates a higher contribution of the sheet erosion (54%), while 46% is accounted for by gully erosion and landslides. The total annual erosion varies from 1 to 5 t/ha in the plain units, to 20–30 t/ha in the Central Moldavian Plateau and the Subcarpathians, and to 30–45 t/ha in the Curvature Carpathians and Subcarpathians, partly confirmed in experimental research (Ioniță 2000; Rădoane and Rădoane 2005).

In the experimental areas of the Putna watershed (Vrancea Subcarpathians) Constandache and Nistor (2006) reported a reduction of the sheet flow and soil erosion due to the direct influence of the forest vegetation, from more than 50 t/ha/year to less than 1 t/ha/yr (in 15–20 years). It was also found that the rainfalls totaling less than 30 mm (50–55% of precipitation amount) generate less than 15% of the eroded material. On the other hand, the rain events exceeding 30 mm determine more than 85% of the eroded material. In July 2005, the forest plantations covering degraded lands in the Putna watershed proved their anti-erosion efficiency by reducing the sheet flow to 2.5% of the precipitation volume and by decreasing the soil erosion to less 0.7 t/ha/year (Constandache and Nistor 2006).

Under favourable petrographic and climatic conditions, the hills and plateaus are highly prone to torrential and gully erosion which can extend rapidly. In the Moldavian Plateau, the gully erosion researches led to some spatial and typological generalizations and allowed the better understanding of these complex-shaping processes (Rădoane et al. 1999; Ioniță 2006; Rusu 2008). As regards continuous gullies, the gullying decline over the period 1961–1990 resulted from the rainfall distribution, and the increased influence of soil conservation practices. The critical period for gullying covers 4 months (mid-March–mid-July) in an area with mean annual precipitation ~500 mm. Another main finding of the 16 years of monitoring (1981–1996) was that 57% of the total gullying occurred during the cold season, especially in March due to the influence of freeze–thaw cycles, with the remainder during the warm season. Of the total gully growth, 66% results from only four years (1981, 1988, 1991 and 1996), when more precipitation fell (Ioniță 2006). The critical period for soil erosion in the Barlad Plateau covers two months, from May 15–20 to July 15–20 (Ioniță 2000).

5.2.1.3 Fluvial Geomorphic Hazards

In the Romanian Plain, the fluvial hazards induced by the erosion and accumulation occur on every river; their characteristics, frequency and intensity of such fearful events vary depending on the flow regime (Zaharia et al. 2011; Grecu et al. 2017). Upstream erosion affects both the rivers entirely flowing in the unit (Călmăţui, Neajlov, Mostiştea, etc.) and the tributaries of the streams coming from the neighbouring areas, so that high values of confluence ratio ($R_c > 5$) of the low-order streams, according to the Horton–Strahler system occur (Grecu et al. 2011, 2012). Lateral erosion leads to a meandering index of the lower rivers around two, apparently with no direct relationship with neotectonics (Grecu et al. 2010b). Lateral erosion coefficient for the highest discharges has much larger variations in comparison with the mean discharges, i.e. 0.367 compared to 0.037 for the Strei River in the Haţeg Depression (Goțiu and Surdeanu 2008; Manea and Surdeanu 2012). The relations between erosion and accumulation are well expressed in the watersheds developing both in the Carpathian and the Subcarpathian areas, such as the Trotuş watershed (4349 km^2) in a flysch area. Here, slope contribution to the total sediment yield of 1,030,000 t/ha/year is 65%, while that of river beds amounts to only 35%. The rather coarse materials lead to the formation of river islands, alluvial cones and bottom accumulations (Dumitriu 2007).

5.2.1.4 Piping and Compaction Hazards

In the Romanian Plain, Quaternary loess and loess-like deposits have encouraged piping and compaction processes and implicitly the appearance of ellipsoidal or circular negative minor landforms, with diameters of 1–2 km and depths up to 5 m known in the Romanian literature as *crov* (Morariu 1945). In the Central Bărăgan Plain, they occupy about 170 km^2, 5% of the plain area. So far, 387 microdepressions were identified, with an average area of 0.43 km^2, and a resulting density of 0.11 microdepressions/km^2 (Gherghina et al. 2008; Grecu et al. 2015). Water stagnation and land collapsibility lead to soil degradation (Parichi and Stănilă 2009). Piping and compaction processes and landforms are also found in the physiographic units developed on deposits similar to the loess, i.e. sands and gravels. The inner pores impact negatively the constructions in several cities (Brăila, Constanța, Cernavodă, etc.).

5.2.1.5 Coastal Geomorphic Hazards

The features of the Romanian Black Sea coast (245 km) have imposed the delimitation of the following sections: a delta unit (between Musura arm and the Periteașca mouth), a lagoon unit (between Periteașca and Cape Singol) and a cliff unit (between Singol Cape and Vama Veche (Posea 2002). Coastal dynamics is strongly influenced by the sediments carried by the Danube (about 58 million tons per year), as well as by seawater dynamics, water depths near the shore and continental shelf topography (Posea 2002). Long-term migration of the shoreline was influenced by marine transgressions and regressions, but also by sea-level oscillations. At present, erosion and accumulation along the Black Sea coast depend both on morphology, waves and currents dynamics, sea-level oscillations and on the improving works accomplished over the time. Along the Midia coast, the mean amount rate of materials accumulated within three gulfs between 1979 and 2005 point a rate of 3.22 mm/year, but some studies suggest that these sediments are generated by anthropogenic activities. For the Vama Veche beach, during the same period, the erosion rate under natural regime was 3 m/year (Constantinescu 2012).

5.2.1.6 Aeolian Geomorphic Hazards

Aeolian hazards are typical for the sandy areas, where they produce deflation, transport and accumulation. For the mountains, these processes are less significant and the wind creates minor landforms which do not pose significant risks. In Romania, the sandy areas occur mostly in plains (Oltenia, Bărăgan, Lower Siret, Valea lui Mihai, etc.), but they are also found in the Danube Delta (Caraorman, Letea and Sărăturile natural levees). The accumulation landforms are represented by dunes, which came into existence especially during the Würm period (Posea 2002). The deforestations of the eighteenth century made the sand invade the rural settlements. Subsequent reforestations with acacia species partly stabilized the sands, while irrigations encouraged the growing of some crops. Aeolian hazards are an extremely dynamic system, bringing visible effects in a short span of time; therefore, the sandy areas need a proper management.

5.3 Consequences and Concluding Remarks

Geomorphologic hazards affect the Romanian territory with various degrees of intensity, thus contributing to land and soil degradation, with long-lasting effects on population (Fig. 5.5).

However, the direct effect leading to casualties is low. From the studies undertaken at regional or national level, one can see the following distribution (Florea et al. 1999; Florea 2003):

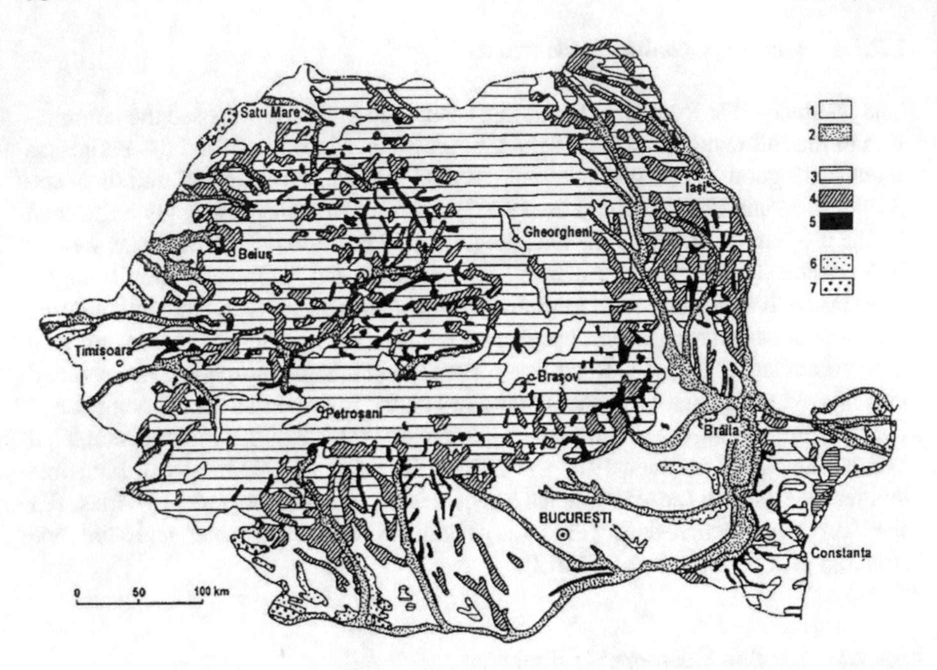

Fig. 5.5 General map of soil erosion in Romania. *Erosion-free lands*: 1 without flood risk; 2 flood and alluvial risk; *areas affected by water erosion*: 3 low erosion but accelerated erosion risk, 4 moderate strong erosion with accelerated erosion risk, 5 very strong, excessive erosion; *terrains affected by wind erosion*: 6 moderately strong erosion at risk with accelerated erosion; 7 very strong, excessive erosion. (Florea et al. 1999 with modifications)

- horizontal and relatively stable lands, with low vulnerability to complex geomorphologic hazards, but with high vulnerability to specific ones: the plain divisions affected by piping, compaction and aeolian hazards (about 40% of the country's area);
- sloping lands, with high and very high vulnerability to complex geomorphologic hazards (sheet erosion, gully erosion, torrential processes, active landslides, collapses): the hills and plateaus, the Subcarpathians, the Carpathian flysch (about 39% of the country's area);
- sloping lands, with medium vulnerability to complex geomorphologic hazards, reasonably afforested, and highly vulnerable to specific hazards: the mountain units affected by cryonival and gravitational processes (about 21% of the country's area).

On the Romanian territory, geomorphologic hazards show a wide range of types, depending on the main shaping agents (gravitational, cryonival, pluvial, fluvial, aeolian, marine).

References

Armaș I (2006) Risc și vulnerabilitate. Metode de evaluare aplicate în geomorfologie. Editura Universității din București, București

Badea L, Băcăuanu V, Posea G (1983) Relieful. In: Badea L (ed) Geografia României, vol I. Geografia fizică. Editura Academiei Române, Bucureşti, pp 64–187

Bălteanu D (1983) Experimentul de teren în geomorfologie. Aplicații la Subcarpații Buzăului. Editura Academiei, București

Bălteanu D, Jurchescu M, Surdeanu V et al (2012) Recent landform evolution in the Romanian Carpathians and Pericarpathian regions. In: Loczy D, Stankoviansky M, Kotarba A (eds) Recent landform evolution. Springer, The Carpatho-Balcan-Dinaric Region, pp 249–286

Cioacă A, Dinu M (1998) Necesitatea reabilitării unor teritorii afectate de alunecări din județul Prahova. Analele Universității Spiru Haret. Seria Geografie 1:49–56

Constandache C, Nistor S (2006) Eficiența lucrărilor de impădurire a terenurilor degradate din bazinul Putnei – Vrancea în prevenirea și combaterea inundațiilor. Revista pădurilor 131(3):41–48

Constantinescu Ș (2012) Analiza geomorfologică a țărmului cu faleză între Capul Midia și Vama Veche. Editura Universitară, București

Driga B (ed) (2007) Riscurile naturale din județul Satu Mare. Editura Arvin Press, București

Dumitriu D (2007) Sistemul aluviunilor din bazinul râului Trotuș. Editura Universității Suceava, Suceava

Florea N (2003) Degradarea, protecția și ameliorarea solurilor și terenurilor. Societatea Națională Română de Știința Solului, București

Florea N, Vespremeanu R, Parichi M, Orleanu C (1999) Soil erosion in Romanian by type of land use. In: Zăvoianu I, Walling DE, Șerban P (eds) Vegetation land use and erosion processes. Symposium proceedings, Institute of Geography, Bucharest

Gârbacea V (1964) Alunecările de teren de la Saschiz (Podișul Hartibaciului). Studia Universitatis Babeș-Bolyai, Ser Geol Geographia VIII(1):113–121

Gârbacea V (2013) Relieful de glimee. Presa Universitară Clujeană, Cluj-Napoca

Gherghina A, Grecu F, Molin P (2008) Morphometrical analysis of microdepresions in the Central Baragan Plain (Romania). Revista de geomorfologie 10:31–38

Goțiu D, Surdeanu V (2008) Hazardele naturale și riscurile asociate din Țara Hațegului. Presa Universitară Clujeană, Cluj-Napoca

Grecu F (1983) Alunecările de teren de la Movile (Podișul Hîrtibaciului). Ocrotirea naturii și a mediului înconjurător 27(2):112–117

Grecu F (1992) Bazinul Hîrtibaciului. Elemente de morfohidrografie. Editura Academiei Romane, București

Grecu F (2002a) Mapping geomorphic hazards in Romania: small, medium and large scale representations of land instability. Géomorphol Relief, Processus, Environ 2:197–206

Grecu F (2002b) Risk-prone lands in hilly regions: mapping stages. In: Allison RJ (ed) Applied geomorphology. Wiley, Chichester, pp 49–64

Grecu F (2009) Hazarde și riscuri naturale, 4th edn. Editura Universitară, București

Grecu F, Ioana-Toroimac G, Dobre R (2008) Précipitations et risques naturels durant la dernière décennie dans le département de Prahova (Roumanie). In: Actes du XXIe Colloque de l'Association Internationale de Climatologie, Montpellier, 9–13 Sept 2008

Grecu F, Comănescu L, Toroimac G, Dobre R, Săcrieru R, Mărculeț C (2010a) Slope dynamics precipitation interrelation in the Curvature Subcarpathians (Romania). Revista de geomorfologie 12:45–52

Grecu F, Ghiță C, Sacrieru R (2010b) Relation between tectonics and meandering of river channels in the Romanian Plain Preliminary observation. Revista de geomorfologie 12:97–104

Grecu F, Ghiță C, Albu M, Cîrciumaru E (2011) Geomorphometric analysis on the some riverbeds in the Romanian plain. International Journal of the Physical Sciences 6(30):7055–7064

Grecu F, Zaharia L, Ghiță C, Comănescu L, Cîrciumaru E, Albu-Dinu M (2012) Sisteme hidrogeomorfologice din Câmpia Română. Hazard – vulnerabilitate – risc. Editura Universității din București, București
Grecu F, Zaharia L, Ghiță C (2013) Hydrogeomorphological vulnerability in the Romanian Plain. Z Geomorph 57(3):3–28. https://doi.org/10.1127/0372-8854/2013/S-00141
Grecu F, Ioana-Toroimac G, Constantin (Oprea) DM (2014) Le critère pluviométrique Angot dans la détermination de la susceptibilité du terrain aux aléas géomorphologiques en Roumanie. In: Actes du XXVIIe Colloque de l'Association Internationale de Climatologie, Dijon, 2–5 July 2014
Grecu F, Eftene (Gherghina) A, Ghiță C et al (2015) The loess micro-depressions within the Romanian Plain. Morphometric and morphodynamic analysis. Revista de geomorfologie 17:5–18
Grecu F, Zaharia L, Ioana-Toroimac G et al (2017) Floods and flash-floods related to river channel dynamics. In: Rădoane M, Vespremeanu-Stroe A (eds) Landform dynamics and evolution in Romania. Springer, Cham, pp 821–844
Ielenicz M (2007) România. Geografie fizică – climă, ape, vegetatie, soluri, mediu. Editura Universitară, București
Ilinca V (2009) Rockfall hazard assessment. Case study Lotru Valley and Olt Gorge. Revista de geomorfologie 11:101–108
Ioniță I (2000) Geomorfologie aplicată. Editura Universității A. I, Cuza, Iași
Ioniță I (2006) Gully development in the Moldavian Plateau of Romania. Catena 68:133–140
Irimuș A (2006) Hazarde și riscuri asociate proceselor geomorfologice în aria cutelor diapire din Depresiunea Transilvaniei. Editura Cărții de știință, Cluj-Napoca
Josan N, Sabău NC (2004) Hazarde și riscuri naturale și antropice în bazinul Barcăului. Editura Universitătii din Oradea, Oradea
Jurchescu M, Grecu F (2015) Modelling the occurrence of gullies at two spatial scales in the Olteţ Drainage Basin (Romania). Nat Hazards 79:1–37. https://doi.org/10.1007/s11069-015-1981-6
Manea Ş, Surdeanu V (2012) Landslides hazard assessment in the upper and middle sectors of the Strei Valley. Revista de geomorfologie 14:49–56
Mărgărint MC, Grozavu A, Patriche CV (2013) Assessing the spatial variability of coefficients of landslide predictors in different regions of Romania using logistic regression. Nat Hazards Earth Syst Sci 13:3339–3355. https://doi.org/10.5194/nhess-13-3339-2013
Morariu T (1945) Câteva consideraţiuni geomorfologice asupra crovurilor din Banat. Revista geografică II(1–4):37–52
Morariu T, Diaconeasa B, Gîrbacea V (1964) Age of land-slidings in the Transylvanian tableland. Rev Roum Géol 8:149–157
Moțoc M (1984) Participarea proceselor de eroziune și a folosințelor terenului la diferențierea transportului de aluviuni în suspensie pe râurile din Romania. Bul Inf ASAS 13:221–227
Munteanu A, Nedelea A, Milian N (2012) Avalanșele – condiții, tipuri, riscuri. Editura Universitară, București
Niacșu L, Vasiliniuc I, Rusu C (2008) Deplasările de teren. In: Rusu C (ed) Impactul riscurilor hidroclimatice şi pedo-geomorfologice asupra mediului în bazinul Bârladului. Editura Performantica, Iași, pp 292–295
Parichi M, Stănilă AL (2009) Contribuții la cunoașterea solurilor din padinile situate în Câmpia Burnasului. Ser Geografie (Analele Universității Spiru Haret) 12:115–120
Posea G (2002) Geomorfologia Romaniei. Editura Fundației România de Mâine, București
Rădoane M, Rădoane N (2005) Geomorfologie aplicată. Editura Universității Suceava, Suceava
Rădoane M, Rădoane N, Ichim I et al (1999) Ravenele. Forme, procese, evoluție. Editura Presa Universitară Clujeană, Cluj-Napoca
României G (2003) HG 447/2003 – Norme metodologice privind modul de elaborare şi conţinutul hărţilor de risc la alunecări de teren. Section V–Zone de risc natural. Official Monitor 305
Rusu C (ed) (2008) Impactul riscurilor hidro-climatice şi pedo-geomorfologice asupra mediului în bazinul Bârladului. Editura Performantica, Iași
Sandu M, Bălteanu D (eds) (2005) Hazardele naturale din Carpații şi Subcarpații dintre Trotuş şi Teleajen. Studiu geografic, Editura Ars Docendi, Bucureşti

Săndulache C (2010) Hazarde și riscuri naturale în Munții Parâng. Editura Universitară, București
Surdeanu V (1999) Geografia terenurilor degradate. Editura Presa Universitară Clujeană, Cluj-Napoca
Voiculescu M (2002) Fenomene geografice de risc în Masivul Făgăraș. Editura Brumar, Timișoara
Zaharia L, Grecu F, Ioana-Toroimac G et al (2011) Sediment transport and river channel dynamics in Romania. Variability and control factors. In: Manning A.J (ed) Sediment transport in aquatic environments. InTech, Rijeka, pp 293–316

Part II
Climate Change: Floods and Droughts

Chapter 6
Backwater Floods—Case Studies with Punctual and Extremely Rare Manifestation on the Romanian Territory. A Review

Gheorghe Romanescu

Abstract Backwater as a hydrological hazard phenomenon has been scarcely studied scientifically, at both national and international levels. For this reason, this study highlights and describes five types of backwater recorded on the Romanian territory, which posed a great danger to population safety. The backwater caused in Romania is due to ice jams, large dams, misplaced dams, or too small bridges, to poorly designed spillways for large waters built on dams, on dikes in the Danube Delta, etc. The most interesting aspect of the study, at the same time, the novelty of it, is the backwater that is caused on the Buhai brook, which rose over the dam of Lake Ezer, spanned the large water spillway, and flooded the lacustrine water body. This type of backwater was named "spider flow." All types of backwater analyzed in this study were also caused by human activity, mostly by design errors (the backwater on the Buhai brook, within the Danube Delta, on the Suhu River) or by bad flow management (the backwater in the Stanca–Costesti Hydro-Energetical Junction). The backwater formed on the Bistrita River has complex causes—both natural and anthropic, and it only occurs during the winter.

Keywords Dam · Damages · Danube delta · Groundwater · Ice jams
Spider flow

6.1 Introduction

Backwater as a hydrological hazard is relatively common around the Globe. The most frequent and destructive backwater phenomena are those that occurred in the sub-Arctic areas, where glaciers melt. In this case, ice jams and ice bridges represent natural barriers against flood waves. Backwaters within man-altered areas are caused, most of the time, by the wrong placement of certain constructions or by the faulty

G. Romanescu (✉)
Department of Geography, Faculty of Geography and Geology,
Alexandru Ioan Cuza University of Iasi, Bd. Carol I, 20 A, 700505 Iasi, Romania
e-mail: romanescugheorghe@gmail.com

M. Zelenakova (ed.), *Water Management and the Environment: Case Studies*, Water Science and Technology Library 86,
https://doi.org/10.1007/978-3-319-79014-5_6

design of various hydrotechnical constructions, such as dams, dikes, road, or railway bridges. Negative effects are recorded in the populated areas because in natural areas, and they can be interpreted as a natural phenomenon. Backwater has highly diverse underlying causes, such as dams or dikes, that force excess water to deviate its course upstream, which are mostly found in inhabited sectors (Begnudelli and Sanders 2007; Chang et al. 1993; Beciu and Seteanu 2010; Gaume and Borga 2008; Gebhardt et al. 2012; Gupta and Nair 2010; He 2007; Holbach et al. 2015; Huang et al. 2007; Laurenson 1986; Mihnea 2008; Risley et al. 2006; Simm et al. 1997; Sommer et al. 2009; Sun et al. 2011; Szilagy and Laurinyecz 2014; Tockner et al. 2000; Valverde 2004; Visutimeteegorn et al. 2007; Zoppou and O'Neill 1980, and so on); ice jams that block water streams (Beltaos et al. 2000; Iturrizaga 2005; Prowse et al. 1993; Prowse and Conly 1998; Shi et al. 2011; US Army 1994; US Army CRREL 1996, and so on), peak tide, which is usually associated with strong winds from the sea, the existence of seiches, and so on (Adopo et al. 2014; Berz et al. 2001; Bondar and Podani 1979; Buschman et al. 2009; Hatheway et al. 2005; Meade et al. 1991; Nistor and Saatcioglu 2012; Radan and Radan 2011; Rahman et al. 2011; Simpson and Bland 2000, and so on); small bridges or bridges with pillars on the water stream (Barbetta et al. 2015; Charbeneau and Holley 2001; El-Alfy 2009; Hunt et al. 1999; Seckin 2007; Seckin et al. 2009; Sudhaus et al. 2008; Xia et al. 2016), and so on.

In the past few years, several floods caused by backwater were recorded, mostly in cases of rivers in the east of Romania, such as Pruth and Siret. Nationwide, studies were done only for some floods determined by the ice jam on the Bistrita River or the "plugs" formed at the foot of certain bridges. This study, therefore, analyzes and highlights the characteristics of the five types of backwater floods that occurred because of civilian buildings that are wrongfully placed in the major riverbed (Romanescu 2015), of the large dams that prevent water from advancing downstream (Romanescu et al. 2011), or of ice jams that block river mouths or that get jammed in very narrow meanders (Romanescu and Bounegru 2012), or of errors in the design of road infrastructure (Romanescu 2015), or of groundwater level elevation near the maritime littoral (Romanescu and Stoleriu 2014), and so on. Some backwater phenomena that occurred on the Romanian territory represented a unique manifestation in the world, how is for example the "spider flow" on the Buhai River (Romanescu and Stoleriu 2013). This study analyzes and classifies the importance of backwater phenomena in a general context of increasing unique flood-related phenomena.

6.2 Regional Settings

The backwater phenomena analyzed are situated in the catchment basins of Pruth (Buhai, Stanca–Costesti, Suhu), Siret (Bistrita), and the Danube (Caraorman). They have the following mathematical coordinates: Buhai 47°57′ N latitude and 26°24′ E

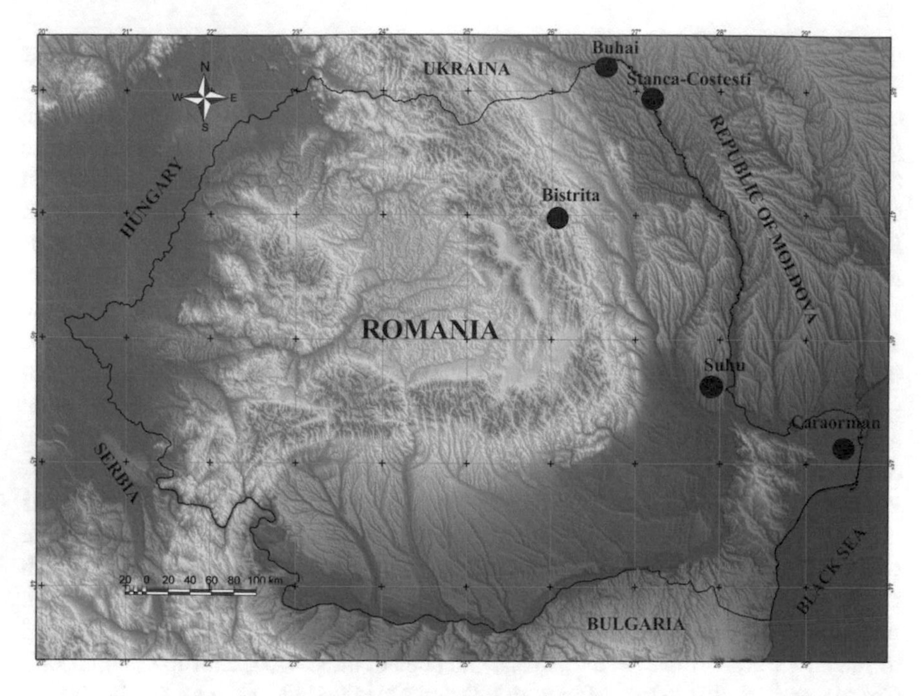

Fig. 6.1 Geographical position of the five backwater cases on the Romanian territory. *Source* Author

longitude; Stanca–Costesti 47°55′ N latitude and 27°08′ E longitude; Suhu 45°40′ N latitude and 27°43′ E longitude; Bistrita 47°34′ N latitude and 29°48′ E longitude; Caraorman 45°5′ N latitude and 29°25′ E longitude (Fig. 6.1). All locations are situated in the East of Romania.

6.3 Materials and Methods

The five floods caused by backwater were studied in different years, from the year 2005 until 2014. Observations were made during the phenomena and immediately afterward, by assessing the damage and by conducting scientific investigations. The most interesting backwater phenomena, which are also caused by human activities and errors, are also depicted.

Statistical data concerning backwater flows and levels were provided by the "Romanian Waters" National Administration, through their branches in Iasi, Bacau, and Constanta. Cartographic modeling was done using the computer software TNT-Mips (Microimage) and ArcGis (ESRI). The methodology of generating thematic maps involved the following techniques: (1) scanning the topographic support on a scale of 1:5000 (DTM—the 1953 and 1982 editions); (2) importing the maps in the desktop cartography program TNTMips (Microimage); (3) georeferencing topographic maps in the Universal Transverse Mercator rr3system; (4) thematic editing of vector layers by manual on-screen digitization, and so on. The distribution graphs and the histograms of frequencies were done by exporting vector layers from the databases and by using the histograms of thematic rasters (processed in OpenOffice) (Blistanova et al. 2016; Castillo and Gómez 2016; Hidayat et al. 2011; Kominkova et al. 2016; Le et al. 2014; Lipeme Kouyi et al. 2011; Pajici et al. 2014; Radevski and Gorin 2017; Zelenakova et al. 2015).

The field observations and measurements were performed during or immediately after the floods. They comprised of the floodable area of the river analyzed, and the daily levels of the hydrometric stations were measured directly or taken from the specialized Basin Administrations. Topographic measurements were taken upstream and downstream from the reservoirs. Field measurements were performed using the LEICA TCR 1201 Total Station, alongside the GPS LEICA 1200, which is part of the 1200 LEICA. The Topographic Directorates provided the coordinates of the topographic markings. After completing the measurements, the data were processed using the AutoCAD software. In addition, a GPS was used to establish the flooded perimeter. Hydrological analyses were also based on consulting certain materials in the Romanian literature that studied hydrological hazards, mostly related to the backwater: Birsan et al. (2014), Chendes et al. (2015), Chirila et al. (2008), Corduneanu et al. (2016), Montz and Tobin (2011), Radevski and Gorin (2017), Stancalie et al. (2012), and so on. Most satellite images that are already processed were provided by the Romanian Space Agency (ROSA), or they were taken from the PNCD12 Project (http://sigur.rosa.ro) on the Internet (free of charge).

6.4 Results and Discussion

6.4.1 The Buhai River—A Unique Type of Backwater: "Spider Flow"

The Buhai brook is a right tributary of the Jijia River which later discharges into the Pruth River, the second largest river in Moldavia. The Buhai brook springs originate from the Ukrainian territory, from an altitude of 325 m, and it joins the Jijia River at the level of Dorohoi, at an altitude of 148 m. The hydrographic basin, including the one on the Ukrainian territory, covers a surface of 134 km^2. It is a significantly forested hydrographic basin, but its longitudinal profile features a high slope (10‰) determined by its morphology and by the fact that it is not very long.

The heavy rains, which fell in the Dorohoi area between June 29 and 30, 2010, cumulated a very high value for the local climate: 205.5 mm. The maximum value in 24 h was 184.0 mm. The multi-annual average of the area is 600 mm. The precipitation that fell in the area of Lake Ezer, which is situated upstream from the backwater, on the Jijia River was valued to be 161.3 mm overall. The exceptional floods had

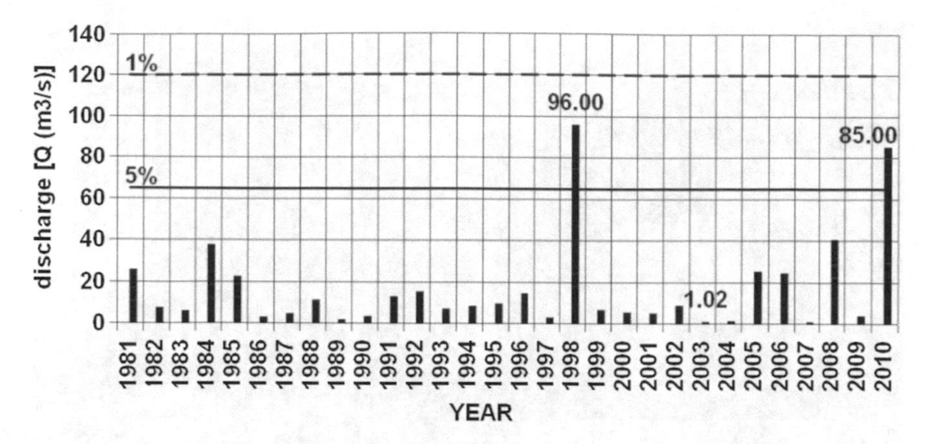

Fig. 6.2 Historic discharges recorded on the Buhai brooks within the period, 1981–2010 (the historic discharges at the Hydrometric Station of Padureni, in the years 1998 and 2010, had a 2% assurance). *Source* Author

a two-wave pattern: The first wave is between June 28 and 29, and the second one is between June 29 and 30. The discharges and levels within the upper basin of the Buhai brook and of the confluence with Jijia were extremely high. In this case, the danger limit was significantly exceeded: +220 cm on Buhai and +150 cm on Jijia (Fig. 6.2).

The major riverbed at the confluence of Buhai and Jijia was intensely altered by the building of houses and industrial units, construction of road and railway bridges, defence dams and fences, and so on. These man-made constructions have facilitated the blocking of water downstream from the Ezer reservoir and the creation of a temporary lake at the confluence (Fig. 6.3). The high-slope longitudinal profile of Buhai forced the water to flow upstream in the confluence area, toward the foot of the Ezer dam. In this case, the water blade crossed the level of the spillway for large water part of the Ezer dam, situated 153 m high (Figs. 6.4 and 6.5). For the waters that manage to climb, through the backwater, on the walls of a dam (in our case, Ezer), the term "spider flow" was used. These waters climb just like a spider, pushed by the hydrostatic force of the excess water mass from upstream (Romanescu and Stoleriu 2013). The backwater phenomenon penetrating the lake from a downstream in an upstream direction is unique; it determined the "flooding" of the lake that was constructed precisely to mitigate floods of the Jijia River (a collector of Buhai).

The backwater flood wave caused the death of six persons and the destruction of over 1500 houses (Romanescu and Stoleriu 2013). In order to avoid future floods and to eliminate the backwater phenomenon, it is mandatory to create a dredging retention (reservoir) on the Buhai brook, upstream from the city of Dorohoi, and to resize the road and railway bridges at the confluence of the brook with the Jijia River.

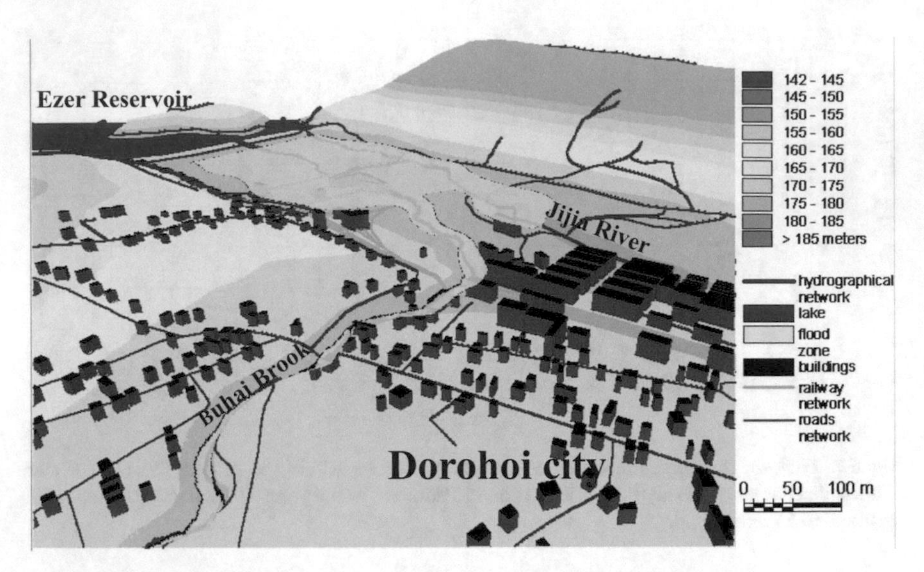

Fig. 6.3 Water accumulation in the common floodplain of Buhai and Jijia during the backwater in the summer of the year 2010. *Source* Author

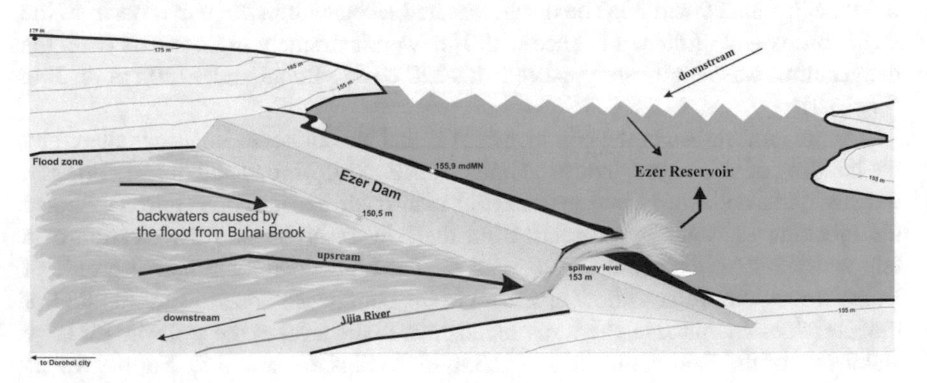

Fig. 6.4 Spider flow phenomenon in the confluence area of the rivers Buhai and Jijia: water climbing the dam and penetrating Lake Ezer through a micro-waterfall. *Source* Author

6.4.2 The Ice Jams on the Bistrita River (Eastern Carpathians) and the Winter Backwater

The Bistrita River is situated in the central-eastern sector of the Eastern Carpathians, and it represents the main right tributary of the Siret River, which is the most important on the Romanian territory. Bistrita discharges into the Siret River downstream from the city of Bacau (the altitude confluence is 134 m). The Bistrita River is 283 km long, and it gathers the waters on a surface of 7039 km^2. It has the following morphometric date and field use characteristics: average slope 5%; sinuosity coefficient 1.40;

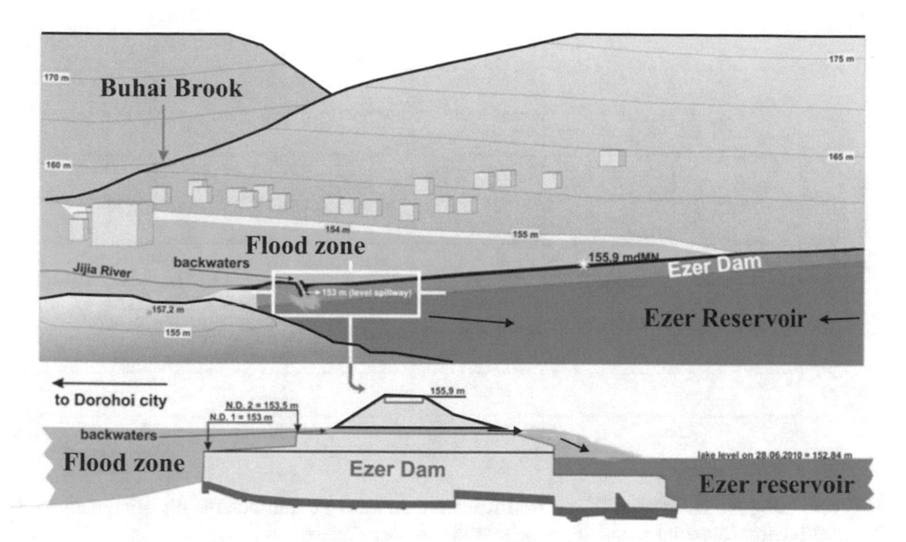

Fig. 6.5 Water penetrates Lake Ezer through backwater and the creation of a micro-waterfall upstream. The level of the temporary lake—created in front of the dam—is higher than that of Lake Ezer, situated behind the dam. *Source* Author

average altitude 919 m; surface of forest fund 424,301 ha, and so on. The Bistrita River holds the highest hydropower potential in the Eastern Carpathians, reason for which it has been significantly altered by man. At its lower half, 14 hydropower plants have been constructed, the most important of which is Stejaru. The most significant reservoir is Izvorul Muntelui, built in the year 1960, with a length of 33 km at the normal retention level (NNR). The emergence of the lake greatly favored the creation of ice jams within the middle sector (Poiana Largului—Farcasa).

The most intriguing floods on the Bistrita River struck during the winter, taking into account the climatic conditions that are typical of this mountainous river sector, and the man-made alterations made in the middle sector of the valley. The primordial factor that favors the creation of ice jams is represented by the construction of Lake Izvorul Muntelui. The lake facilitated the deposition of alluvia materials at the Bistrita waterway in the shape of a fan delta and the strong reduction of the runoff slope in the upstream sector (Fig. 6.6).

The Bistrita River valley, which is upstream from the back of Lake Izvorul Muntelui, includes numerous narrow sectors alternating with relatively narrow small depression basins, but which also favor the construction of human settlements. All localities cover depression areas, where water can no longer be actually extended to. Many bridges were built in order to facilitate circulation on the river. The lack of space and the undersizing of the opening between the pillars of roads or pedestrian bridges lead to the creation and accumulation of ice jams in the sectors with smooth slopes or behind the dams. The phenomenon is also amplified by the accumulation of logs, branches, or alluvia that is transported by the old flood waves (Piche et al. 2014). The underlying causes of ice jam backwater are natural (climatic and

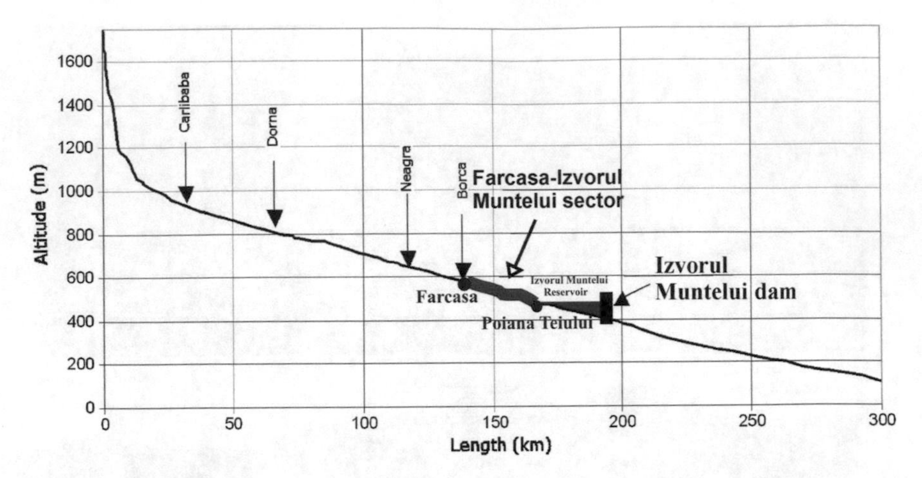

Fig. 6.6 Longitudinal profile of the Bistrita River and the ice jam emergence alignment in the Farcasa-Izvorul Muntelui sector (Poiana Teiului). *Source* Author

hydrological) and anthropogenic (bad management in the case of using minor and major riverbeds) (Cojoc et al. 2015).

Ice jams and backwater as hydrological hazard occur in the sector situated between the localities of Farcasa (upstream) and Poiana Teiului (downstream) on a length of 25–30 km. Ice jams emerge annually and the maximal thickness of the ice may even be up to 6–7 m. The first ice jam phenomena occurred between the months of December or January, mostly upstream from the locality of Popesti. During heavy winters, with temperatures <−7 °C from December to March—even an ice bridge can be noticed (for example, in the winter of the year 2002–2003) (Fig. 6.7).

The ice jam backwater produces only material damage because it forms on a slow pace and it is predictable. It takes between 2 and 3 days and then 10 days to form. The most important backwater occurred in the winter of the year 2003, when an accumulation of icicles created a 7 m thick "plug." The ice jam lasts between 90 and 95 days on average. The one formed in the winter of 2003 lasted 106 days (Stefanache 2007). The peak intensity of the backwater was recorded within the period, December 31, 2002–January 2, 2003, and it encountered a water level rise of 3 m. This backwater produced the most significant material damage: It wrecked 98 houses and it led to the loss of 3 human lives (on New Year's Eve, a guest house was destroyed too, where the three aforementioned people died). The backwater occurred in the winter of the year 2003 because of a brief warm weather, recorded upstream as well as a foehn from the peaks of the Eastern Carpathians, when temperatures recorded 3–5 °C (Romanescu and Bounegru 2012).

The ice jam formed from every frosty winter is destroyed by dynamiting it (Gaman 2015). By dredging in the upper sector of Lake Izvorul Muntelui and by eliminating the sediment materials, the slope of the Bistrita River can be extended upstream, thus avoiding the emergence of the ice jam.

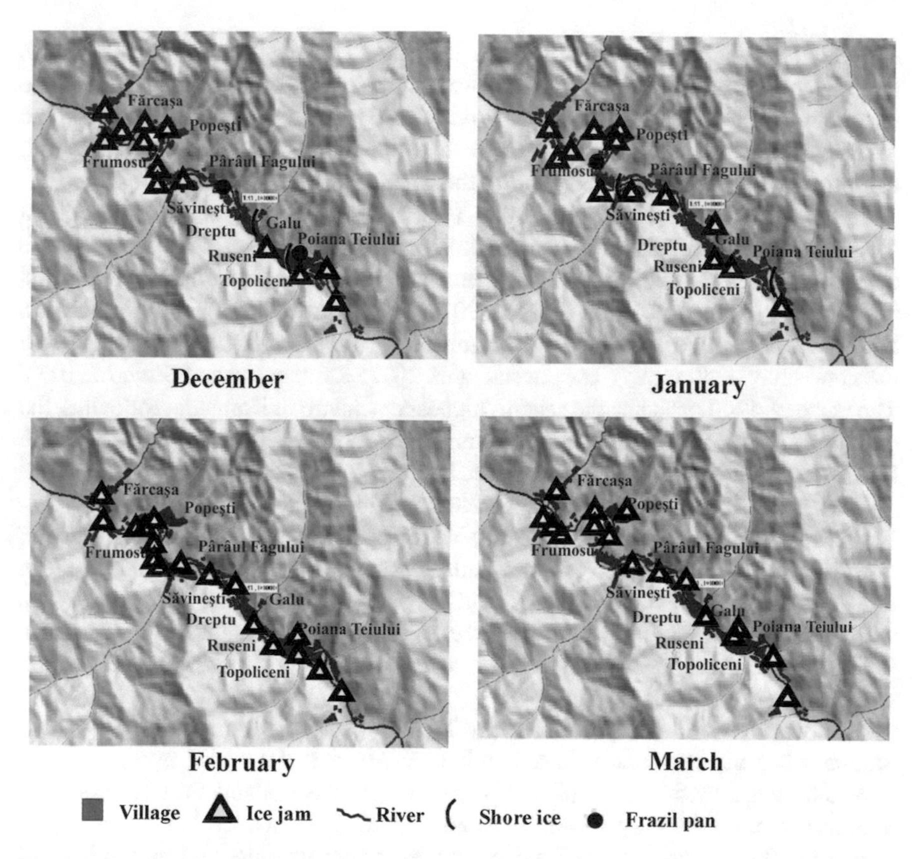

Fig. 6.7 Distribution of the ice jam on the Bistrita River in the winter of 2002–2003. *Source* Author

6.4.3 The Dam Backwater (Stanca–Costesti Reservoir) and the Flooding of an Area Covering Three States: Romania, the Republic of Moldova, and Ukraine

The Pruth River springs from the Forested Carpathians (Ukraine), and it discharges into the Danube, near the city of Galati. The first backwater phenomenon in the Romanian literature was described in the year 1716, including the way the waters of the Pruth River advanced upstream (Cantemir 1716). The Pruth catchment basin unfolds on the territory of Ukraine (upper course), Romania, and the Republic of Moldova (middle and lower courses). It covers 27,500 km^2 (10,967 km^2 on the Romanian territory, which is 4.6% of the entire surface of the country). Until Oroftiana (entry on the Romanian territory), it has a length of 211 km, a longitudinal slope of 6.4%, a sinuosity coefficient of 1.18, and a catchment basin of 8241 km^2. Its total length is 952.9 km (it is the second longest in Danube tributary). It represents the State border between Ukraine and Romania, having a distance of 31 km, and between Romania and the Republic of Moldova having a distance of 711 km. Within the catchment basin, 225 ponds were constructed as early as the reign of Stephen the Great (fifteenth–sixteenth centuries). Pruth also comprises 26 large reservoirs, the most important of which is Stanca–Costesti (volume of 1400 million m^3, making it the largest reservoir in Romania) and in the Republic of Moldova.

In the period between the year 2005–2015, historic discharges were recorded on the most important rivers on the east of Romania: Siret in the year 2005 (4650 m^3/s or 5500 m^3/s calculated discharge) (Romanescu 2005), Suceava in the year 2008 (1946 m^3/s) (Romanescu and Stoleriu 2013), Pruth in the year 2008 (7140 m^3/s, but after two years of analyses, the historic level for Romania was replaced with the new value: 4240 m^3/s) (Romanescu et al. 2011; Romanescu and Stoleriu 2013). The value of 4240 m^3/s has the second highest discharge in Romania, following the one recorded by the Siret River in the summer of the year 2005 (Romanescu 2005; Romanescu and Nistor 2011; Romanescu 2015).

The Pruth River floods and the backwater within the Stanca–Costesti Hydro-Energetical Junction (summer of 2008) were caused by the heavy rains that fell in the upper basin (Ukraine) and the middle basin (Romania). Precipitation mean indicates values up to 150–200 mm in 40 days (178.0 mm at the Botosani station or 209.1 mm at the Cotnari station). However, the highest amounts of precipitations fell on the Ukrainian territory, which was also the origin of the most important flood wave (Stoleriu et al. 2015).

The Pruth River floods started on July 24, 2008, at 6 o'clock, and at the Cernauti Station, when the flow increased suddenly to 387 m^3/s and the level rose to 314 cm, exceeding the attention level (CA) by 186 cm. At the Radauti Prut Hydrometric Station, at 8 pm, a discharge of 434 m^3/s was recorded and at a level of 290 cm (CA). Before the flood wave, the Stanca–Costesti reservoir recorded a minimum level of water due to a prolonged drought (90 m compared to the level of the Baltic Sea). The maximum values in the Radauti Prut section were 7140 m^3/s (which was replaced, two years later, by the value of 4240 m^3/s) and 1130 cm, respectively (by July 28, 2008, between 9 and 12 pm, exceeding the danger level (CP) by 530 cm (Romanescu et al. 2011). The high level of the Radauti Prut Station, situated upstream from the Stanca–Costesti Hydro-Energetical Junction, 80 km from the dam is due to the rise of water in the lake at maximum level. The Stanca–Costesti reservoir mitigated the flood wave downstream, but it also accentuated the one upstream. The reservoir maintained a high level for 20–30 days because a huge volume of water was accumulated and was distributed downstream, for a long period, because rains changed their directions to the southern sector of the catchment basin and they determined an increase in the river level toward Iasi and Galati.

The maximum level within the reservoir (F3, 98.20 m) was recorded on July 30, 2008, at 9 PM (98.21 cm, +1 F3) and it was maintained until July 31, 2008, 11 AM (98.20 cm, +0 F3). The maximum level of 98.27 cm (+7 F3) was recorded between 3 PM and 6 PM on July 30, 2008. The maximum flow that entered the reservoir was 1290 m^3/s (the maximum level for Q with 1% (percent) assurance is 98.20, and the maximum level for Q with 0.1% assurance is 99.50 m) (Figs. 6.8, 6.9 and 6.10).

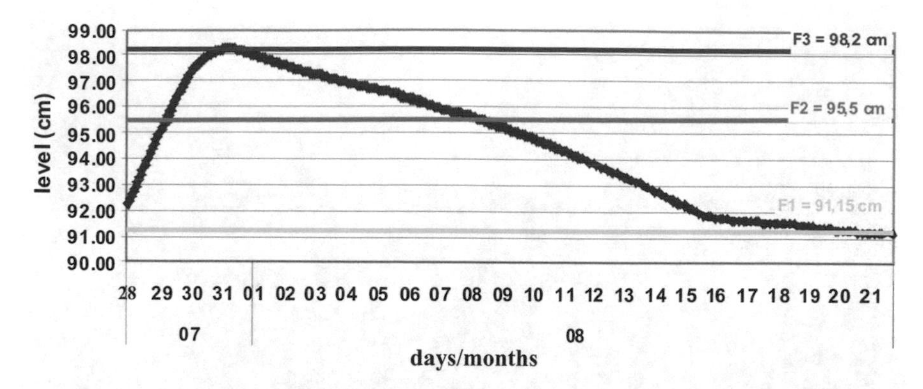

Fig. 6.8 Evolution of water level within the Stanca–Costesti Hydro-Energetical Junction during the backwater in the summer of the year 2008. *Source* Author

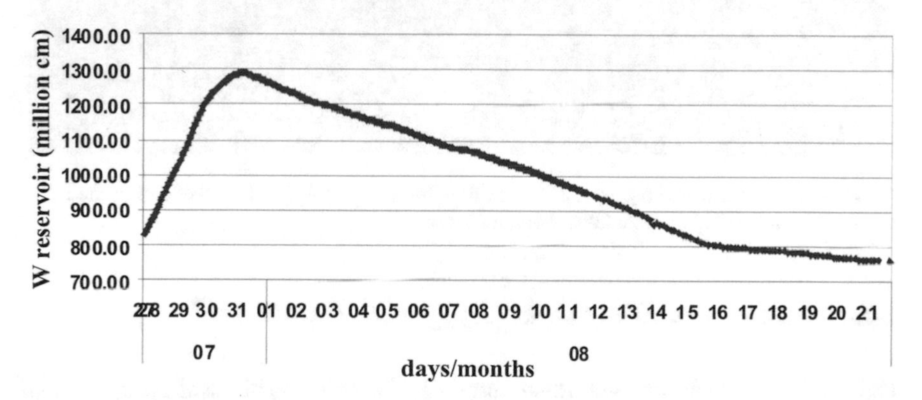

Fig. 6.9 Evolution of water volume within the Stanca–Costesti Hydro-Energetical Junction during the backwater in the summer of the year 2008. *Source* Author

The backwater produced in the Stanca–Costesti Hydro-Energetical Junction is the most spectacular one that ever occurred on a Romanian river. It moved at a distance of around 80 km and it led to a flooding of localities and agricultural fields in the lacustrine cuvette area (Fig. 6.10). It became worse by the poor management of water storage in the reservoir. The massive overflows, with a maximum value occurred within the period, July 29–August 02, 2008. For this reason, two opposite forces collided at the level of Radauti Prut: the flood wave from the Ukrainian territory and the backwater wave from the Stanca–Costesti Hydro-Energetical Junction. At the same time, the damage was worsened by the unauthorized gravel exploitation in the minor riverbed of the Pruth, by repeated deforestations within the basin, and on the alignment of longitudinal levees, et cetera (Romanescu et al. 2011).

Fig. 6.10 Stanca–Costesti Hydro-Energetical Junction at its peak flooded level during the back-water in the summer of the year 2008. *Source* Author

6.4.4 Suhu—An Earthquake-Triggering Backwater?

The catchment basin of Suhu (also known as Suhurlui) is situated on the east of Romania; its length measures 72 km; it has a mean slope of 3% (percent), (it springs from an altitude of 220 m, and it discharges into Siret at an altitude of 9 m). It covers a surface of 373 km^3 and it has an elongated shape on the N-S direction. There are eight villages within the basin, and they are as follows: Draguseni (in the upper basin), Urlesti, Suhurlui, Pechea, Cuza Voda, Slobozia Conachi, Izvoarele, and Piscu (in the lower basin). The use of agricultural fields for plant cultures determined the massive deforestation of the slopes, which are currently affected by intense landslides (Romanescu 2015). Within the catchment basin, there are also oil and natural gas exploitations.

The month of September 2013 recorded precipitations of over 100 mm/24 h in the Suhu catchment basin. The highest amounts of precipitations fell in the interval between September 11–15 and September 18–20 (Romanescu 2015). Heavy rains and low temperatures determined the stagnation of water in the river floodplain for a very long period. Even two months after the rains, small flooded areas could still be noticed; in those areas, water stagnated and houses had moist walls. Atmospheric calmness led to a very low evaporation. For this reason, landslides are activated slowly.

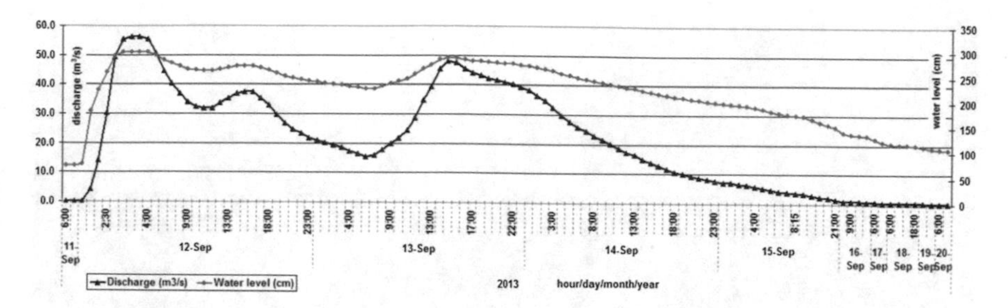

Fig. 6.11 Hydrograph of the Suhu River flood wave within the period, September 12–16, 2013 at the level of Izvoarele. *Source* Author

The flood wave hydrography featured two peaks: between September 11 and 12, the most important one; between September 13 and 14 a lower peak, which was recorded as the second historic maximum (Fig. 6.11). The maximum historic flow was 56.2 m^3/s, at a maximum historic level of 298 cm, recorded on September 11, 2013, at 3.40 and 4.50 o'clock. The second historic maximum has the value of 48.0 m^3/s, at a level of 287 cm, recorded on September 13, 2013, at 14.40 o'clock (Romanescu 2015). The basic characteristic of the flood wave is the sudden and instantaneous increase from a value of 0.12–56.2 m^3/s. In this case, the flood wave hydrograph features a right angle. The increase occurred within three hours. The high level of the flood wave within the basin, mainly in the middle and lower sector, is due to the backwater phenomenon that is caused by excess water stagnation behind the road embankment (Fig. 6.12). The pumps used to eliminate water were too small and they failed to remove all water in due time (Fig. 6.13). For this reason, five days after the flood, a decision was made to break the dams. The backwater advanced on dozens of kilometers and it affected the localities of Pechea, Slobozia Conachi, and Izvoarele. Several houses situated in the area that were not affected by previous floods were also damaged. The backwater wave coincided with water overflow from Lake Corni (Romanescu 2015).

The villages of Pechea, Slobozia Conachi, and Izvoarele that were initially affected by the flood wave and then by backwater are situated in the river floodplain, within the risk zone. The dominant land use is for agriculture, mostly for grain culture. The left slope is higher, steeper, and affected by intense landslides, threatening the village of Izvoarele. Forest was eliminated totally; perhaps a few clumps still exist, but they are not enough to stabilize the slopes (Romanescu 2015). During and after the floods, small earthquakes occurred in the Izvoarele area (400 earthquakes within two months, with a peak intensity of 4° on the Richter scale). The earthquakes core coincided with a high soil humidity. In the period, September–October 2014, 30 more earthquakes with a peak intensity of 2.4° on the Richter scale occurred. They were recorded once, when the heavy autumn rains began. This fact is evidence that weak earthquakes are caused by excess soil moisture, mostly on the right river slope, affected by landslides (Romanescu 2015).

Fig. 6.12 Maximum manifestation stage of the Suhu River backwater. *Source* Author

Fig. 6.13 Water accumulation behind the embankment and the failure of pumps (in the background) to eliminate excess water in time. *Source* Author

6.4.5 The Caraorman Village and the "Phreatic Backwater"

The area of the Rosu–Puiu lacustrine complex within the Danube Delta is situated between the arms of Sulina (to the north), St George (to the south), the fluvial–maritime levee of Caraorman (to the west) and the Black Sea coast (to the east) (Fig. 6.14). The Caraorman village (from the Turkish: cara – black; orman—forest) is situated in the central-eastern sector of the Caraorman fluvial–maritime levee (made of sandy deposits). The access from the Sulina arm is through the Crisan channel (a secondary arm is subsequently dredged). Unfortunately, the channel was oversized to facilitate the penetration of vessels that were going to carry machinery and materials for the future exploitation of quartziferous sand (that was abandoned after the year 1990). The Crisan–Caraorman channel facilitates the access of ships from the Sulina channel toward the Caraorman village and Lake Rosu, where one of the most important summer tourist resorts in the Danube Delta has developed. Therefore, transportation had to be ensured for an important number of tourists (Romanescu and Stoleriu 2014).

The Crisan–Caraorman channel was finalized in the year 1981. It has a length of 12 km, a width of 60–70 m, and maximum depths of 4–6 m. Great depths facilitated the penetration of high flows, which inevitably led to a rapid siltation of the Rosu–Puiu complex. Most channels within the Rosu–Puiu lacustrine complex were built after 1960: Sondei, Litcov-Imputita, Vatafu–Imputita, and Tataru. The Potcoava channel—that connects lakes Puiu and Rosu—was built within the period, 1952–1960, while the Busurca–Imputita channel in the period 1941–1951, connected to the Sulina arm.

The Crisan–Caraorman channel and the Black Sea coast have a direct hydrogeomorphologic relationship. The hydrotechnical works are aimed at eliminating the downsides, caused by an increased water level in the lakes. To this end, the following works were executed: a 2 m high protection dam in the coastal area, along the Tataru channel, to avoid the emergence of breaches created by spontaneous and uncontrolled runoffs (in these points, strong erosive phenomena were recorded between the mouths of Imputita and Sondei); a spillway for large waters on the coastal dam, at a crest level of +1.20 m compared to the Black Sea level and at a width of the discharging field of 200 m; a channel bordering the coastal dam, toward the inside of the delta, to discharge small and average waters toward the arms of Sulina and St George (Tataru channel). One of the most important issues when it comes to eliminating siltation is to recalibrate the access mouth of the Crisan channel. Unfortunately, this project is yet to be finalized.

Because the spillway for large waters built on the coastal dam is higher than the highest flood level caused by excess water carried through the Crisan channel, as well as water accumulated behind this barrier, and it determined the emergence of backwater. It manifests itself, even to the level of the Caraorman fluvial–maritime levee, which is flooded, mostly due to the increase in the phreatic level, favored mainly by the sandy substrate of the levee.

In order to supplement water transit on the dike, during flood waves, a spillway can be built at the 0.60 m level (r MNS) and the width of 200 m or another spillway

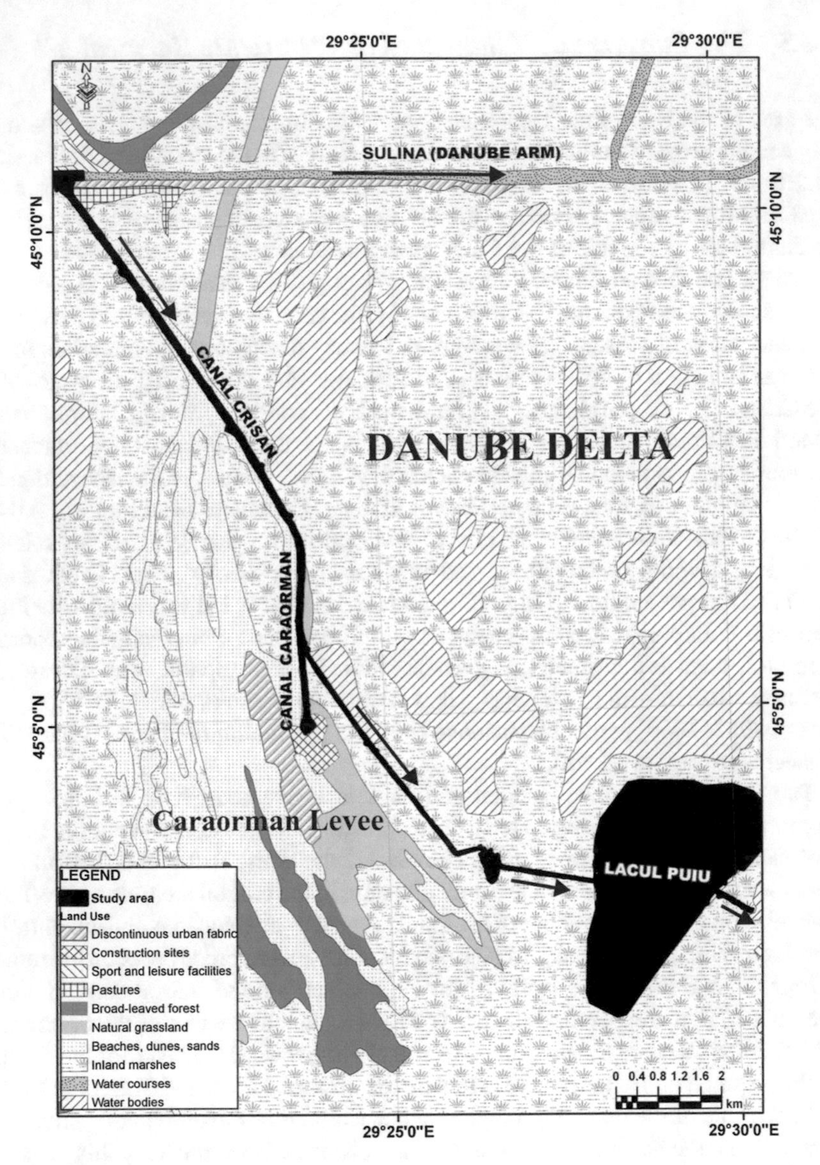

Fig. 6.14 Location of the Caraorman village, of the Crisan channel, and of Lake Rosu within the Danube Delta. *Source* Author

at the 0.80 m level (r MNS) and the width of 250 m. Additional spillways can be built at the crossroads between the Imputita channel and the Sulina–St George dike and at the level of Lake Rosu (Romanescu and Stoleriu 2014). Only an effective redesign of the two spillways can avoid the creation of backwater.

The undersizing of the spillway for large waters at the level of Sondei channel determines backwater movement on the Tataru channel and causes an increase in the

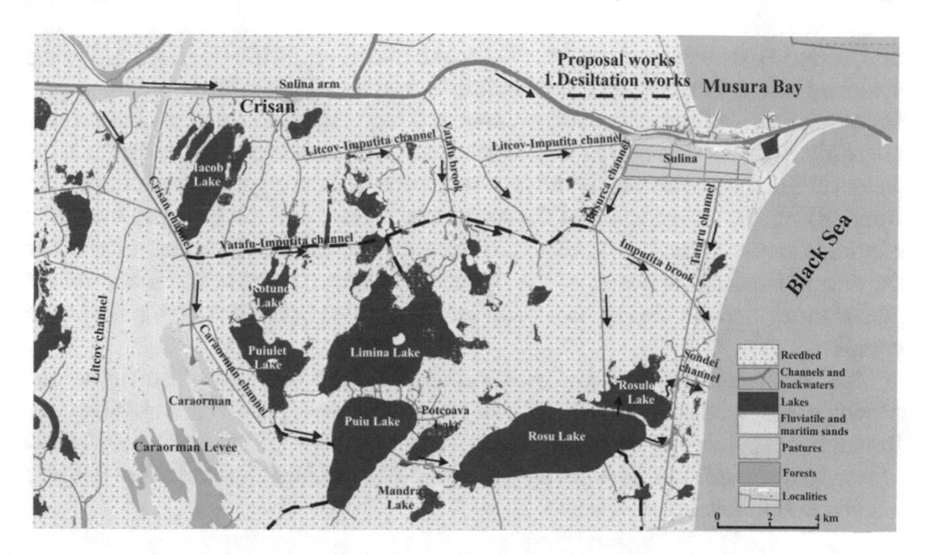

Fig. 6.15 Morpho-hydrologic situation of the Rosu–Puiu lacustrine complex and the desiltation works proposals for facilitating circulation on channels and streamlets. *Source* Author

water level within the channels and lakes or within the phreatic, even in the localities of Sulina and St George. Besides the two spillways, the Crisan–Caraorman channel mouth can be calibrated by closing the upstream mouth totally (320 m) and the downstream mouth partially (180 m), and leaving only a 20 linear-meter entry for the transit of water and vessels. On the existing spillway, there is the possibility of supplementing the discharging capacity (a length of 180 linear meter at a threshold level of +0.60 m r MNS) (Romanescu and Stoleriu 2014). Additional discharges within the Caraorman village can be mitigated through desiltation works conducted for the Caraorman channel (downstream from the confluence with the Crisan channel), the Vatafu–Imputita channel, the Vatafu–Lumina channel, the Rosu–Tataru channel, and the Erenciuc–Puiu or by dredging Lake Rosu (25,000 m^3) and Lake Puiu (15,000 m^3) (Fig. 6.15). The elimination of excess water due to phreatic backwater can also be achieved by building dredging channels in the Caraorman village area. Waters collected in the dredging channels can be evacuated in the basin of the port of the quartziferous sand exploitation.

6.5 Conclusions

Backwater as a hydrological hazard phenomenon does not manifest itself as a flood wave or as a highly intense wave. Hence, human life losses are rare, but economic damage caused can actually be significant. The backwater phenomena had complex underlying causes; however, an important role is played by mistakes made in the design of hydrotechnical works or by bad management decisions: filling the Stanca–Costesti Hydro-Energetical Junction to the maximum, while rain still fell in the upstream sector; the construction of buildings in the major riverbeds and the aggradation of Lake Izvorul Muntelui; civilian and industrial constructions built in the major riverbeds; faulty design of bridges and the existence of a spillway for large waters at a higher level than the highest level recorded during the backwater, on the Buhai River; the construction of the road embankment and the lack of a breach to eliminate excess water from the Suhu River; a spillway for large waters (installed on the dike separating the waters of the Danube Delta from those of the Black Sea) is placed lower than the highest backwater levels recorded in the Rosu–Puiu lacustrine complex.

The backwater produced on the Romanian territory can also possess a unique character: the "spider flow" on the Buhai brook. The other phenomena are unique in Romania, but commonly recorded in most States of the world. The slower movement of waters upstream is a reason for a better coordination of measures taken to mitigate or even to stop these phenomena.

Acknowledgements This work was financial supported by the Department of Geography from the "Alexandru Ioan Cuza" University of Iasi, and the infrastructure was provided through the POSCCE-O 2.2.1, SMIS-CSNR 13984-901, No. 257/28.09.2010 Project, CERNESIM and by the Partnership in Priority Domains project PN-II-PT-PCCA-2013-4-2234 No. 314/2014 of the Romanian National Research Council, Non-destructive approaches to complex archaeological sites. An integrated applied research model for cultural heritage management—arheoinvest.uaic.ro/research/prospect.

References

Adopo KL, Romanescu G, N'Guessan AI, Stoleriu C (2014) Relations between man and nature and environmental dynamics at the mouth of the Komoé river, Grand-Bassam (Ivory Coast). Carpathian J Earth Environ Sci 9(4):137–148

Barbetta S, Camici S, Moramarco T (2015) A reappraisal of bridge piers scour vulnerability: a case study in the Upper Tiber River basin (central Italy). J Flood Risk Manage. https://doi.org/10.1111/jfr3.12130

Beciu E, Seteanu I (2010) The calculus of backwater curve on the manmade channel. Universitatea Politehnica Bucuresti, Sci Bull Ser D 72(4):175–182

Begnudelli L, Sanders BF (2007) Simulation of the St. Francis dam-break flood. J Eng Mech-ASCE 1200–1212

Beltaos S, Pomerleau R, Halliday RA (2000) Ice-jam effects on red river flooding and possible mitigation methods. Ice-jam effects on red river flooding and possible mitigation methods. Report International Red River Basin Task Force International Joint Commission

Berz G, Kron W, Loster T, Rauch E, Schimetschek J, Schmieder J, Siebert A, Smolka A, Wirtz A (2001) World map of natural hazards—a global view of the distribution and intensity of significant exposures. Nat Hazards 23(2–3):443–465

Birsan MV, Zaharia L, Chendes V, Branescu E (2014) Seasonal trends in Romanian streamflow. Hydrol Process 28(15):4496–4505

Blistanova M, Zelenakova M, Blistan P, Ferencz V (2016) Assesment of flood vulnerability in Bodva River Basin, Slovakia. Acta Montanist Slovaca 21(1):19–28

Bondar C, Podani M (1979) Furtuna maritima din februarie 1979 si efectele ei asupra litoralului romanesc. Hidrotehnica 24(9):45–54

Buschman FA, Hoitink AJF, van der Vegt M, Hoekstra P (2009) Subtidal water level variation controlled by river flow and tides. Water Resour Res 45:1–12

Cantemir D (1716) Descriptio Moldaviae. Berlin

Castillo C, Gómez JA (2016) A century of gully erosion research: urgency, complexity and study approaches. Earth Sci Rev 160:300–319

Charbeneau RJ, Holley ER (2001) Backwater effects of piers in subcritical flow. Texas Department of Transportation

Chang WH, Aldam AA, Smith JA (1993) On the effects of downstream boundary conditions on diffusive flood routing. Adv Water Resour 19:259–275

Chendes V, Corbus C, Petras N (2015) Characteristics of April 2005 flood event and affected areas in the Timis-Bega Plain (Romania) analysed by hydrologic, hydraulic and GIS methods. In: 15th International Multidisciplinary Scientific GeoConference, SGEM2015, vol 1, pp 121–128

Chirila G, Corbus C, Mic R, Busuioc A (2008) Assessment of the potential impact of climate change upon surface water resources in the Buzau and Ialomita Watersheds from Romania in the Frame of Cecilia Project. Ohrid, FY Republic of Macedonia, BALWOIS, pp 1–7

Cojoc G, Romanescu G, Tirnovan A (2015) Exceptional floods on a developed river. Case study for the Bistrita River from the Eastern Carpathians (Romania). Nat Hazards 77(3):1421–1451

Corduneanu F, Vintu V, Balan I, Crenganis L, Bucure D (2016) Impact of drought on water resources in north-eastern Romania. Case study—the Prut River. Environ Eng Manage J (EEMJ) 15(16):1213–1222

El-Alfy KS (2009) Backwaterrise due to flow construction by bridge piers. In: Thirteenth International Water Technology Conference, IWTC 13 2009, Hurgada, Egypt, pp 1295–1313

Gaman C (2015) The influence of the hydraulic exchange relations in the evolution of ice jam phenomena on Bistrița river in the area between Dorna Arini (Suceava county) and Borca (Neamţ county). PESD 9(2):199–213

Gaume E, Borga M (2008) Post-flood field investigations in upland catchments after major flash foods: proposal of a methodology and illustrations. J Flood Risk Manage 1(4):175–189

Gebhardt M, Pfrommer U, Belzner F, Eisenhauer N (2012) Backwater effects of Jambor weir sill. J Hydraul Res 50(3):344–349

Gupta AK, Nair SS (2010) Flood risk and context of land-uses: Chennai city case. J Geogr Reg Planning 3(12):365–372

Hatheway D, Coulton K, DelCharro M, Jones C (2005) FEMA coastal flood hazard analysis and mapping guidelines focused study report. Flood Hazard Zones

He H (2007) Study of Sanmenxia Dam Effects on Backwater and Human Activities in Middle Yellow River, China. 18th UCOWR Conference 2007. http://opensiuc.lib.sin.edu/ucowrconfs_2007/13

Hidayat H, Vermeulen B, Sassi MG, Torfs PJJF, Hoitink JF (2011) Discharge estimation in a backwater affected meandering river. Hydrol Earth Syst Sci 15:2717–2728

Holbach A, Bi Y, Yuan Y, Wang L, Zheng B, Norra S (2015) Environmental water body characteristics in a major tributary backwater of the unique and strongly seasonal Three Gorges Reservoir, China. Environ Sci Processes Impacts 17:1641–1653

Huang S, Vorogushyn S, Lindenschmidt KE (2007) Quasi 2D hydrodynamic modelling of the flooded hinterland due to dyke breaching on the Elbe River. Adv Geosci 11:21–29

Hunt J, Brunner G, Larock B (1999) Flow transitions in bridge backwater analysis. J Hydraul Eng-ASCE 25(9):981–983

Iturrizaga L (2005) Historical glacier-dammed lakes and outburst floods in the Karambar valley (Hindukush-Karakorum). GeoJournal 62–63(1–4):1–47

Kominkova D, Nabeikova J, Vitvar T (2016) Effects of combined sewer overflows and storm water drains on metal bioavailability in small urban streams (Prague metropolitan area, Czech Republic). J Soils Sediments 16(5):1569–1583

Laurenson E (1986) Friction slope averaging in backwater calculations. J Hydraul Eng-ASCE 112(12):1151–1163

Le HM, Petrovic D, Verbanck MA (2014) The semi-sewer river: hydraulic backwater effects and combined sewer overflow reverse flows in Central Brossels reduce deoxygenation impact further downstream. Water Sci Technol 69(4):903–908

Lipeme Kouyi G, Bert P, Didier JM, Chocat B, Billat C (2011) The use of CFD modelling to optimise measurement of overflow rates in a downstream-controlled dual-overflow structure. Water Sci Technol 64(2):521–527

Meade RH, Rayol JM, Conceicao SC, Natividade JRG (1991) Backwater effect in the Amazon River Basin of Brazil. Environ Geol Water Sci 18(2):105–114

Mihnea I (2008) Danube dams—necessity or calamity? Carpathian J Earth Environ 3(1):31–38

Montz BE, Tobin GA (2011) Natural hazards: an evolving tradition in applied geography. Appl Geogr 31:1–4

Nistor I, Saatcioglu M (2012) Tsunami risk and impacts on coastlines. Nat Hazards 60(1):1

Pajici P, Andjelic L, Urovevic U, Polomcic D (2014) Evaluation of melioration area damage on the river Danube caused by the hydroelectric power plant ‘Djerdap 1’ backwater. Water Sci Technol 70(2):376–385

Piche S, Nistor I, Murty T (2014) Numerical modeling of debris impacts using the SPH method. Coast Eng Proc 1(34):19

Prowse TD, Demuth MN, Peterson M (1993) Proposed artificial river ice damming to induce flooding of a delta ecosystem. 50th Eastern Snow Conference, pp 331–338

Prowse TD, Conly FM (1998) Effects of climatic variability and flow regulation on ice-jam flooding of a northern delta. Hydrol Process 12:1589–1610

Radan SC, Radan S (2011) Recent sediments as enviromagnetic archives. A brief overview. Geo-Eco-Marina 17:103–122

Radevski I, Gorin S (2017) Floodplain analysis for different return periods of river Vardar in Tikvesh valley (Republic of Macedonia). Carpathian J Earth Environ Sci 12(1):179–187

Rahman M, Arya DS, Goel NK, Dhamy AP (2011) Design Flow and Stage Computation in the Teesta River, Bangladesh, Using Frequency Analysis and MIKE 11 Modeling. J Hydrol Eng-ASCE 16(2):176–186

Risley JC, Walder JS, Denlinger RP (2006) Usoi Dam Wave Overtopping and Flood Routing in the Bartang and Panj Rivers, Tajikistan. Nat Hazards 38(3):375–390

Romanescu G (2005) Riscul inundatiilor in amonte de lacul Izvorul Muntelui si efectul imediat asupra trasaturilor geomorfologice ale albiei. Riscuri si catastrofe 4:117–124

Romanescu G (2015) Backwater as hydrological hazard. Study case: Suhu catchment (Romania). International Multidisciplinary Scientific Geoconference SGEM 2015, Water Resources. Forest, Marine and Ocean Ecosystem. In: Proceedings of Conference, Hydrology&Water Resources vol I, 95–102

Romanescu G, Nistor I (2011) The effect of the July 2005 catastrophic inundations in the Siret River’s Lower Watershed, Romania. Nat Hazards 57(2):345–368

Romanescu G, Stoleriu C, Romanescu AM (2011) Water reservoirs and the risk of accidental flood occurrence. Case study: Stanca-Costesti reservoir and the historical floods of the Prut River in the period July–August 2008, Romania. Hydrol Process 25:2056–2070

Romanescu G, Bounegru O (2012) Ice dams and backwaters as hydrological risk phenomena—case study: the Bistrita River upstream of the Izvorul Muntelui Lake (Romania). WIT Trans Ecol Environ 159:167–178

Romanescu G, Stoleriu C (2013) An inter-basin backwater overflow (the Buhai Brook and the Ezer reservoir on the Jijia River, Romania). Hydrol Process 28(7):3118–3131

Romanescu G, Stoleriu C (2014) Anthropogenic interventions and hydrological-risk phenomena in the fluvial-maritime delta of the Danube (Romania). Ocean Coast Manage 102:123–130

Seckin G (2007) The effect of skewness on bridge backwater prediction. Can J Civil Eng 34(10):1371–1374

Seckin G, Akoz MS, Cobaner M, Haktanir T (2009) Application of ANN techniques for estimating backwater through bridge constrictions in Mississippi River Basin. J Adv Eng Softw 40(10):1039–1049

Shi J, Zhang Y, Zhang F (2011) Analysis on backwater calculation of ice jam during ice-flood period. Adv Mat Res 243–249:4458–4461

Simm DJ, Walling DE, Bates PD, Anderson MG (1997) The potential application of finite element modelling of flood plain inundation to predict patterns of overbank deposition. Hydrolog Sci J 42(6):859–875

Simpson MR, Bland R (2000) Methods for accurate estimation of net discharge in a tidal channel. IEEE J Oceanic Eng 25:437–445

Sommer T, Karpf C, Ettrich N, Haase D, Weichel T, Peetz JV, Steckel B, Eulitz K, Ullrich K (2009) Coupled modelling of subsurface water flux for an integrated flood risk management. Nat Hazards Earth Syst Sci 9:1277–1290

Stancalie G, Cracinescu V, Nertan A, Mihailescu D (2012) Contribution of satellite data to flood risk mapping in Romania. IEEE Int Geosci Remote Sens Symp (IGARSS) 2012:899–902

Stefanache M (2007) Cercetari privind evolutia unor fenomene hidrologice periculoase. Ph.D. thesis, Gheorghe Asachi Technical University of Iași, Iasi

Stoleriu CC, Romanescu G, Romanescu AM, Mihu-Pintilie A (2015) Morpho-bathymetrical conditions and the silting rate in Stanca-Costesti reservoir (Romania). Wulfenia 22(2):451–470

Sudhaus D, Seidel J, Burger K, Dostal P, Imbery F, Mayer H, Glaser R, Konold W (2008) Discharges of past flood events based on historical river profiles. Hydrol Earth Syst Sci 12:1201–1209

Sun S, Yan X, Cui P, Feng J (2011) A four-step method for optimising the normal water level of reservoirs based on a mathematical programming model—a case study for the Songyuan Backwater Dam in Jilin Province, China. Int J Environ Res Public Health 8:1049–1060

Szilagy J, Laurinyecz P (2014) Accounting for backwater effects in flow routing by the discrete linear cascade model. J Hydrol Eng 19:69–77

Tockner K, Malard F, Ward JV (2000) An extension of the flood pulse concept. Hydrol Process 14:2861–2883

US Army (1994) Ice jam floding: causes and possible solution. Engineering and Design Comphlet, 1110-2-11, Washington, D.C

US Army CRREL (1996) Drilling holes in ice to reduce ice jam potential. US Army corps of Engineers Cold Regions Research and Engineering Laboratorz, IERD Newsletter, 14 March 1996

Valverde ALA (2004) Use of flood propagation models in real time hydrologic forecast. Experiences at Segura River. BALWOIS, Ohrid, pp 1–8

Visutimeteegorn S, Likitdecharote K, Vongvisessomjai S (2007) Effects on the upstream flood inundation caused from the operation of Chao Phraya Dam. Songklanakarin J Sci Technol 29(6):1661–1674

Xia J, Teo FY, Falconer RA, Chen Q, Deng S (2016) Hydrodynamics experiments on the impacts of vehicle blockages at bridges. J Flood Risk Manage. https://doi.org/10.1111/jfr3.12228

Zelenakova M, Gaňová L, Purcz P, Satrapa L (2015) Methodology of flood risk assessment from flash floods based onhazard and vulnerability of the river basin. Nat Hazards 79(3):2055–2071

Zoppou C, O'Neill IC (1980) The validity of backwater model for dlood routing application. In: 7th Australasian Hydraulics and Fluid Mechanics Conference, Brisbane, pp 179–182

Chapter 7
Winter Phenomena (Ice Jam) on Rivers from the Romanian Upper Tisa Watershed in 2006–2017 Winter Season

Daniel Sabău, Gheorghe Şerban, Istvan Kocsis, Petrică Stroi and Răzvan Stroi

Abstract The paper proposes an analysis of the winter phenomena generated by severe persistent negative temperatures (over two months) and sudden changes in weather, as well as of the effects induced by these phenomena on the anthropogenic and natural environment. In the first chapter of the paper, a selective radiography of the main international and national publications related to the winter phenomena on the watercourses is done, with a focus on natural manifestations of ice jam type. A significant concentration of specialists and studies is noticeable in the Nordic world area or in countries with high mountain areas. Also in the first chapter, it provides information about the studied area—the upper watershed of Tisa River, in general, and the Romanian watershed of the upper Tisa, as a case study, as well as on the monitoring activity of the Romanian part of the study basin. The second chapter is dedicated to the presentation of three major components of this space—the morphological, climatic, and hydric—decisive general conditions for the winter phenomena that occurred on the watercourses. In the third part, and the most extended, the specific conditions of occurrence are analyzed at the level of detail (morphological and morphometric, weather conditions—with special regard to the synoptic, the temperature and precipitation variation in the study period), the anthropogenic conditions of the riverbed influence on the studied river sectors. It also includes an analysis

D. Sabău · I. Kocsis · P. Stroi · R. Stroi
National Administration "Romanian Waters", Someş-Tisa Water Branch,
17 Vânătorului, 400213 Cluj-Napoca, Romania
e-mail: andrei.sabau@dast.rowater.ro

I. Kocsis
e-mail: istvan.kocsis@dast.rowater.ro

P. Stroi
e-mail: petrica.stroi@sgasj.dast.rowater.ro

R. Stroi
e-mail: razvan.stroi@yahoo.com

G. Şerban (✉)
Faculty of Geography, Babeş-Bolyai University, 5-7 Clinicilor,
400006 Cluj-Napoca, Romania
e-mail: gheorghe.serban@ubbcluj.ro

M. Zelenakova (ed.), *Water Management and the Environment: Case Studies*, Water Science and Technology Library 86,
https://doi.org/10.1007/978-3-319-79014-5_7

on the rivers state, types of winter phenomena occurred during the December 1, 2016–February 14, 2017 period—where the complex winter chart was a great help in the correlation between parameters and in highlighting the phenomena severity—their evolution, monitoring, and management. At the end of this chapter, the variation of the river water levels reported to the defense levels is detailed, as of great significance in explaining the effects generated by natural occurrences. In the last part, issues related to the effects of the winter phenomena on the anthropogenic and natural environment are presented, with details on the produced damages according to the different categories of affected items. The most important aspect related to the consequences refers to damage limitation and the absence of human victims, due to the prompt, concerted, and synchronized actions of all the involved authorities and institutions.

Keywords Romanian watershed of the upper Tisa · Specific conditions of occurrence · Winter phenomena on rivers · Complex winter chart
Critical sectors · Effects of the winter phenomena on rivers

7.1 Introduction—Literature About Winter Phenomena and Their Effects

Winter phenomena, normal manifestations on watercourses from the temperate zone, happen, more or less, depending on the factors that affect them. The frequency of these rises with latitude, but also with altitude, reaching maximum values at the northern periphery, and the southern one, of the continents, in the boreal hemisphere and less in the austral one.

Thanks to the decades of experience on observations and measurements in the hydrometric network R.W.N.A. ("Romanian Waters" National Administration), it is noted that on altitude the maximum frequency of those belongs to the mountain area, the birth place of the flowing, marked by high precipitations and low thermic environments, especially at its periphery. Winter phenomena occur with a lower frequency in the hilly area as well, because of the agglomerations of ice blocks from other effluents.

A season with ice on a river can last over 100 days for the most of the rivers in Scandinavia, Russia, and Canada, and it can descend until latitudes of 42° and 30° N in North America and Asia (Bates and Billelo 1966; Rădoane et al. 2008, 2010).

From all the categories of river winter phenomena, ice needles, ice at the shore, broken ice bridge, ice bridge with water eyes, compact ice bridge, water flows over the ice bridge, ice blocks flowing, etc. (N.I.M.H.—National Institute of Meteorology and Hydrology 1966), the most dangerous seems to be, without a doubt, the one of the ice jams. These massive agglomerations of ice, formed at narrowing of water courses or at obstacles in the river bed, can block effectively the water draining, causing quick and massive floods or destructive mechanical actions generated by

Fig. 7.1 Ice jam, giant ice blocks, and the after-effects—February 2017, Vaser Valley (photos made by S.T.W.B.A.—"Someș-Tisa" Water Basin Administration, M.W.M.S.—Maramureș Water Management System)

the gigantic ice blocks, with catastrophic effects (Fig. 7.1). On the other hand, the authors are supporting the fact that ice jams are being the most hazardous river winter phenomena (Ashton 1986; Rădoane et al. 2008, 2010).

The specialized literature is full of published materials about hazards and the hydric risks and, of course, about the river winter phenomena and their effects; the references about the winter phenomena are related especially to Canada, USA, Iceland, Scandinavia, Russia, Japan, China, etc. (Derecki and Quinn 1986; Belatos 2008; Prowse and Conly 1998; Prowse and Beltaos 2002; Jasek 2003; Kowalczyk and Hicks 2003; Shen and Liu 2003; Prowse and Bonsal 2004; Morse and Hicks 2005; Korytny and Kichigina 2006; Daly 2009; Kolerski and Shen 2010; Kolerski 2014; Wang et al. 2014; Kolerski and Shen 2015; Lagadec et al. 2015; Shen 2016; Koegel et al. 2017; Kraatz et al. 2017; Zeleňáková and Zvijáková 2017).

In fact, Cold Regions Research and Engineering Laboratory (CRREL) of the US Army Corps of Engineers devotes virtual platforms and in-depth studies of the winter phenomena, ice in general, its effects and of the different measures of management and attenuation of the effects (http://www.erdc.usace.army.mil/Locations/CRREL/)—quoted by White and Eames (1999) and White et al. (2007), USACE (1984) or Government of Alberta (2013), and some famous magazines devote special editions—e.g., Hydrological Processes, Volume 19, Issue 1, January 2005, Special Issue: Canadian Geophysical Union—Hydrology Section.

Some studies refer to the causes and the climatic impact over the formation and evolution of the winter phenomena on rivers (Beltaos and Prowse 2001; Beltaos 2007; Beltaos and Burrell 2015).

Many studies are meant for analyzing the formation process of the ice jams, in some case starting form dynamic factors, e.g., the existence of waves and ice blocks (Nafziger et al. 2016; Pawlowski 2016).

Some of the publications are meant for mapping and modeling of the ice formations and the analysis of the flood risk induced by the winter phenomena, especially in ice jams (Beltaos 1993; She and Hicks 2006; Lindenschmidt et al. 2016; Boivin et al. 2017).

Other publications refer at the dynamic/mobility of the ice in ice jams under the influence of some allogenic or autogenous factors, the impact (Beltaos 1990; Lu et al. 1999; Shen et al. 2000; Liu and Shen 2004; Shen et al. 2008; Huang et al. 2016; Wang et al. 2015a, b, 2016; Zare et al. 2016).

Some publications aim for the study of the ice and the ice blocks made along the narrows and channels (Lucie et al. 2017).

The water flow and silt draining under the effect of frost and the modeling of this process is another subject chosen by the researchers (Thériault et al. 2010; Beltaos 2016; Beltaos and Burrell 2016a, b; Robb et al. 2016).

Other researchers tried even other methods of predictions of the floods generated by the winter phenomena, starting from hydro-meteorological (Gholamreza-Kashi 2016).

Some of the authors made some studies about the elements that cause the hydraulic of the river bed and, naturally, the formation and evolution of the winter phenomena on rivers (Leopold and Maddock 1953; Leopold et al. 1964; Rhodes 1977; Ichim and Rădoane 1986; Ichim et al. 1989; Rădoane et al. 2008, 2010).

The same authors support the fact that the knowledge of geometrical hydraulic is important, namely as basis for the identification of stable sections from a dynamic point of view (so necessary in the design programs), as well as for the transit of the freeze-up jams and ice along the river.

Moreover, the authors consider that under the circumstances of winter phenomena affected riverbed, the hydraulic geometry is totally disordered. The width of the riverbed is narrowing, and the depth of the riverbed decreases. Thus, the water is forced to flow more rapidly in order to continue its course (Rădoane et al. 2008, 2010).

In Romania, beside the national hydrometric network, where the regular five day between observations of the winter phenomena on rivers is the norm (intensified at hourly level in exceptional cases) (N.I.M.H. 1966 and other sources), numerous researchers conducted and are now conducting studies of these phenomena.

The history of the special research goes all the way to the period of development of the hydrometric activity, the 1960s, when some very rigorous technical instructions were elaborated. Some of them are applied even today, while others were substituted or completed with modern work methods.

From the Romanian researchers dedicated to this domain in the pioneering period, we are mentioning: Semenescu (1960), Constantinescu (1964), Ciaglic (1965), etc.

Further, other studies of scale about hazards and hydric risks were added, materialized in books or doctoral thesis, international publications, within which the area of the specialists has widened, from hydrologists, to hydrotechnical engineers, geomorphologists, and even climatologists, which contributed with works about the cause of the winter phenomena or about their effects on the river floods.

From these, some of the most representative works were written by: Miţă (1977), Mustăţea (1996), Sorocovschi et al. (2002), Colceriu (2002, 2003, 2010), Păvăleanu (2003), Romanescu (2003, 2005, 2015), Rădoane (2004), Surdeanu et al. (2005), Ştefănache (2007a, b), Rădoane et al. (2008, 2010), Giurma and Stefanache (2010), Romanescu and Bounegru (2012), Găman (2014), Cojoc et al. (2015), Romanescu et al. (2015), Romanescu and Stoleriu (2017).

7.2 The Upper Tisa Watershed and Studied Sub-basins in Regional and National Context

The upper basin of Tisa mostly overlaps with the historic Basin of Maramures—from which, only a third is located on the Romanian territory, and with the southwest part of Ukraine—Transcarpatic Region, with the sections Rahiv, Tiaciv, Hust, and Mijgiria (the largest basin in the Carpathian Mountains). This is a territory well individualized from the geographic point of view, thanks to the outer mountain formations and to the petrographic mosaic, which led to a different and complex evolution of the region (Chiş and Kosinszki 2011; Sabău et al. 2017) (Fig. 7.2).

The mentioned area, along with the outer high zone, displays a diversified landform, ranging from riversides and valley couloirs to mountains. The minimal altitude is 157 m at the exit of the Tisa River from the basin, while the highest altitude is located in Rodna Mountains, at the "Pietrosu Rodnei" peak 2303 m (Posea et al. 1980; Ardelean and Béres 2000; Boar 2005; Cocuţ 2008).

The area of the basin, excluding the basin of Tur and other affluent watercourses in the mentioned zone, is 10,354 km^2, from which 3381 km^2 (about a third) is located in Romania and 6973 km^2 (the other two thirds) in Ukraine (Chiş and Kosinszki 2011).

In the Ukrainian part (the right slope), the basin is better developed, thanks to many important right effluents (Kisva, Shopurka, Apshytsia, Teresva, Tereblia şi Rika), which create a confluence of many watercourses at the base of the mountain.

In the Romanian part of the basin (the left slope), there are only two important effluents (Vişeu and Iza), while the "Săpânţa" and "Şugătagul Mare" rivers contribute in a less measure to the development of the basin, because of the existence of the andesitic massif of Igniș Mountains in the south (Fig. 7.2).

Besides, from the two watercourses only the Vișeu River displays similar mountain characteristics like the Ukrainian rivers, because Iza collects a good percent of its discharge from the median hill zones of the basin of Maramureș. Only the left affluent, Mara, brings a share from the mountain area, though influenced by the hydrotechnical arrangements (Fig. 7.3).

In contrast to the Ukrainian part of the basin, which has a larger opening to the Panonic Basin, the Romanian part of the basin is opened only to the northwest, to west and south being closed by the neogen volcanic chain of the Oaș–Gutâi–Țibleș Mountains (Fig. 7.3). This natural organization enables cold air mass advections from

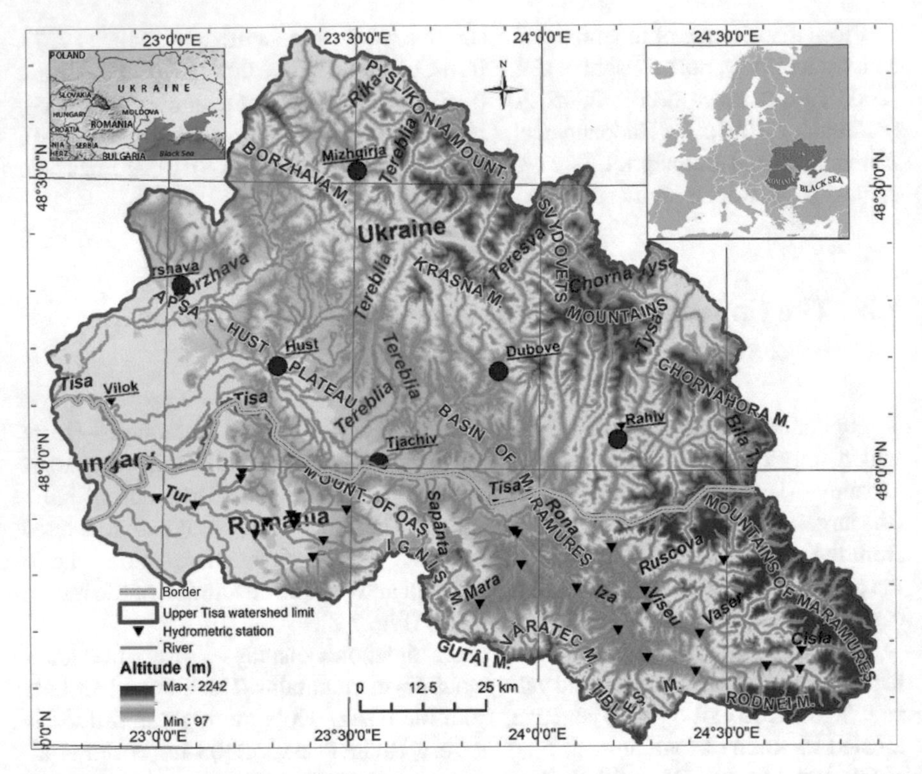

Fig. 7.2 Upper Tisa watershed on the Romanian and Ukrainian territories (adapted and completed after Lukiantes and Obodovskyi, 2015 and from Photograph source 1 & 2—see References)

north and important thermic inversions on the lower basin area which determines specific characteristics of winter flow on the two rivers.

7.3 The Hydrometric Monitoring Network and the Observation and Gauging Program

As a basin space with a strong asymmetric character (the Viseu River on the right and Iza on the left), which drains slopes with high humidity and a circularity degree almost unitary, the area of the study has been monitored from a hydrometric and pluviometric point of view since the middle of the twentieth century, with the purpose of water management and control of the risky hydric phenomena.

The Tisa collector fell under the incidence of complex monitoring only after 2000 (before only the water level was tracked at the hydrometric station of Sighetu Marmatiei), when the bases of the border cooperation between Romania and Ukraine were laid in this field.

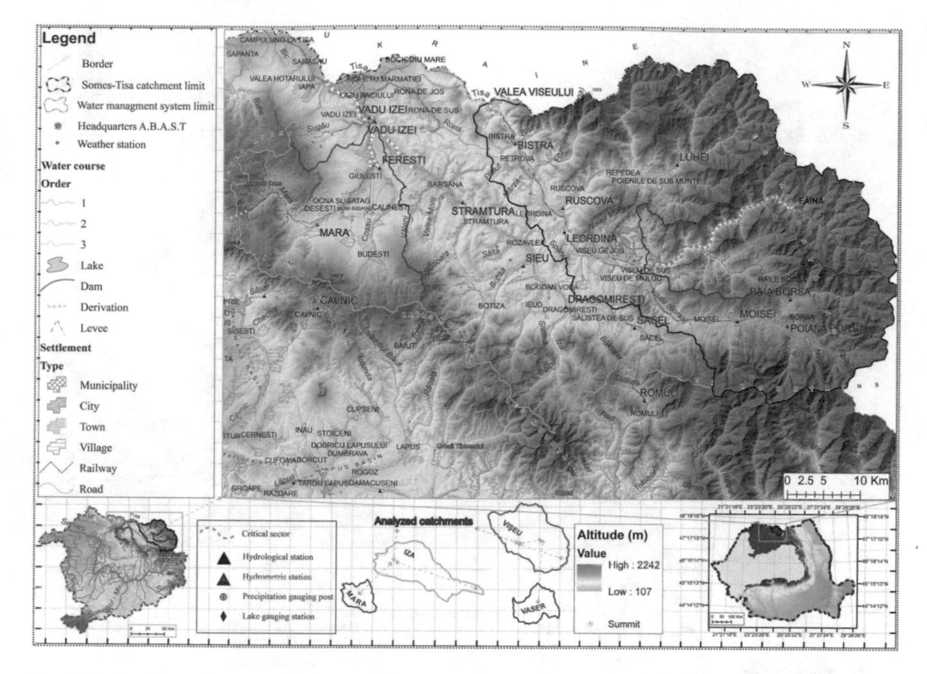

Fig. 7.3 Position of the analyzed sub-basins and of the study sectors in national context. *Source* Authors

Currently, for the Tisa collector (at the entrance on the Romanian territory and in the urban zone of Sighetu Marmatiei), and as well as the other two important rivers of Maramures, Viesu and Iza, stations for hydric and weather monitoring were organized in key areas. They were chosen either in flow formation (upper basins) or in water-gathering areas (after important confluences with different large tributaries) (Fig. 7.4).

The quality of the monitoring and the efficiency of the forecasts and of the population mobilization in emergency situations increased proportionally with general implementation of the warning levels on rivers and their afferent alert codes (Fig. 7.4).

The observation and gauging program from the river hydrometric stations targets parameters of high importance for the monitoring and management of the water resources and of dangerous hydrological and meteorological phenomena: the water level, the temperature of air and water, the rainfall, the winter phenomena, the deepening of the riverbed, the speed of the water flow, the turbidity, the water pollution, the dynamics of the riverbed, etc.

The examinations are done frequently (6.00 AM and 6.00 PM, summer time; 7.00 AM and 5.00 PM, winter time), or even more often (hourly) at the manifestation of extreme hydric events (the surpassing of the warning level) or of winter phenomena, if these are developing negatively in the area of occurrence (N.I.H.W.M.—The National Institute of Hydrology and Water Management 2013).

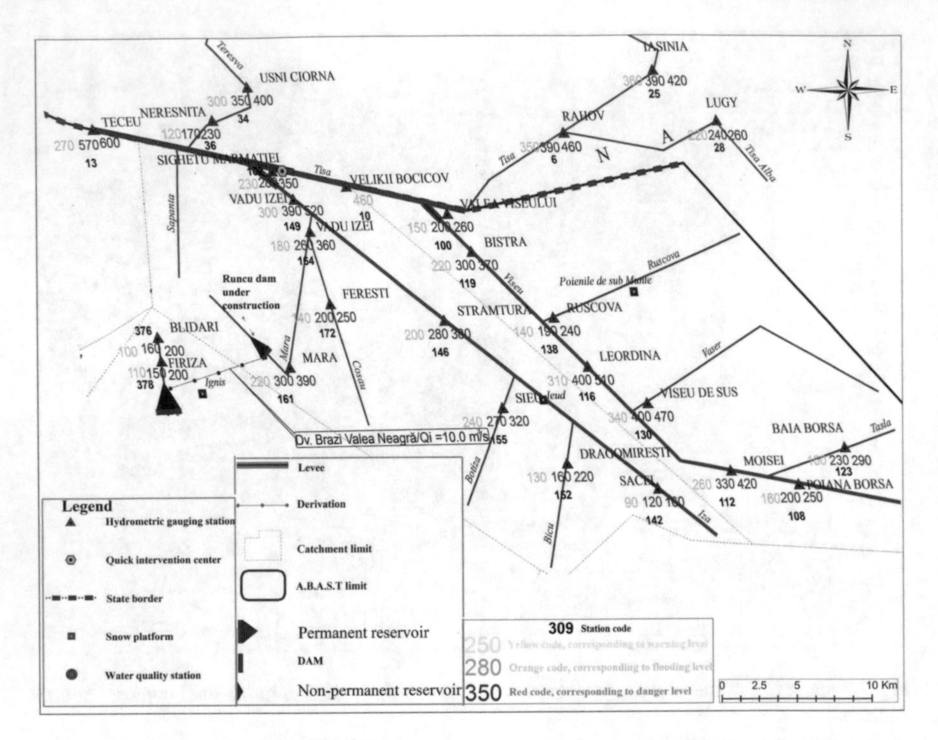

Fig. 7.4 Maximum flow management plan in the Romanian basin of upper Tisa (after S.T.B.W.A. 2016)

To ensure a more efficient gauging, automated hydrometric stations for the basin water balance were introduced on important watercourses or on the ones with increased chance of floods (with the help of DESWAT programs—Destructive Waters and WATMAN—Informational System for the Integrated Water Management). These grant the possibility of live monitoring of some parameters measured by sensors.

The measurements of liquid discharge or of river deposits in suspension are done in a less frequent manner (once every five days), being intensified at hydric events, at even hourly frequency on the smaller rivers.

The measurements of winter phenomena are done every five days, and those of submerged vegetation, every ten days.

7.4 Sources and Techniques

The text was organized using information gathered through the personal investigations of the project's staff, as well some articles, studies, and scientific reports elaborated by other consecrated researchers, or by institutions administrating the hydric environment and natural resources from Romania (S.T.W.B.A.)

The technical data regarding the hydrographical basins and the watercourses, on which analysis was based, were taken partially from the Atlas of Water Cadastre from Romania, 1992, partially from the statistics of colleagues from S.T.W.B.A., or the data were generated in PC program "GIS" (ArcGIS 10.x).

The elements regarding the winter phenomena and their production cause came from the hydrometric activity of S.T.W.B.A. The elements were either calculated in PC statistic programs (Microsoft Office 2016, SPSS) or gathered according to syntheses and studies made at the above regional institution.

The support for GIS modeling was constituted from plans 1:5.000, topographic maps 1:23.000, ortophotoplans and other satellite imagery, hazard and risk maps for floods (A.N.A.R.), coordinates and GPS files from different sources (terrain research and bibliographical). The digital mapping, the files conversion, and the modeling were made in PC specific programs (GPS Utility, Global Mapper, ArcGIS 10.x, etc.).

7.5 Major Components of the Natural Landscape Influencing the Winter Phenomena in the Upper Tisa Watershed

7.5.1 Morphological Component

The existence of magmatic and crystalline rocks gives massiveness and peripheral scale to the basin, aspect equally reverberated in the ascension of the morphologic structures (with altitudes that exceed 2000 m in the southeastern part of the region). The sedimentary component characterizes the lower zones in the inner part of the area, and along with the other petrographic formations, stamp the watercourses in the area with distinct quantitative and qualitative characteristics.

In its entirety, the relief is organized on many altitudinal and structural levels (Posea et al. 1980; Pop 2000; Boar 2005; Chiş and Kosinszki 2011) (Fig. 7.2):

- the mountain component, which closes the basin at the periphery, respectively, the mountains of Maramureş, Rodna, Ţibleş, Lăpuş, Gutâi, Igniş, Oaş, Hust, Borzhava, Gorgany, Pyslikonia, Kedryn, Krasna, Svydovets, and Chornohora;
- the piedmont structure, given by the piedmonts of Borşa, Văratec, Gutâi, and Mara-Săpânţa;
- the component of the glacis—the glacises of Săcel and Vişeu;

- the hill component, namely the interfluve between the Iza and Vişeu Rivers;
- the plateau structure—the Hust-Apşa table land;
- the basin component, developed frequently under the form of larger basins in confluence areas—Borşa, Vişeu, Ruscova, Bârsana, Vadu Izei, Mara, Rona, Sighet, Yasynia, Chorna, Synevyr Kolochava, and Studeniy Mizhgiria;
- the valley couloirs—Vişeu, Iza, Tisa, Shopurka, Apshytsia, Teresna, Tereblia, and Rika;
- gorges—Vişeu, Surduc and Tisa;
- the larger riversides—the riverside of Tisa.

The connection of the Maramureș Basin with the other neighboring regions, in the Romanian sector, is made through the altitude passes: Huta 587 m to Oaş Basin, Gutâi 987 m to Baia Mare Basin, Neteda 1040 m to Lăpuş Basin, Şetref 818 m to Someşan, and Prislop table land 1416 m to Obcinele Bucovinei Mountains.

By its traits and by its defining elements (the mountain basin character, slope orientation, slope, depth of the fragmentation, drainage density, etc.), the relief engrains local specificities, with important effects on the climate and hydrological phenomena and processes (Ujvári 1972; Sabău et al. 2017).

7.5.2 The Climatic Component

The relief organization generates the climatic specifics as well, among other invasions of cold air masses with a scandinavian-baltic origin along the basin corridor and important thermic inversions in the core area of Maramures Basin, etc., which give specific characteristics of winter flowing on the two collecting rivers Viseu and Iza and on their main tributaries.

Regarding the average yearly temperatures of the air (between the 1961s and the 2000s), the Basin of Maramures and the afferent mountain zone fit in the interval 9–0 °C (Climate of Romania 2008, p. 135).

Referring to January, for the area of study, the same source highlights the following:

- the average temperature of the soil surface varies between −5 and 1 °C; the absolute extreme minimum was recorded at Sighetul Marmatiei (−33.5 °C), on January 25, 1963;
- the average air temperature in January varies between −3 and 8 °C, and the average of daily minimum temperature of the air varies between −6 and −12 °C (Fig. 7.5a, b);
- the yearly average number of days with frost ($t_{min.} \leq 0$ °C) varies between 100 in the lower zone of the basin and 180 in the high peripheral zone;
- the yearly average number of frosty nights ($t_{min.} \leq -10$ °C) varies between 20 in the lower zone and 70 in the high one;
- the yearly average number of winter days ($t_{min.} \leq -10$ °C) varies between 30 in the lower zone and 110 in the high zone for south and east.

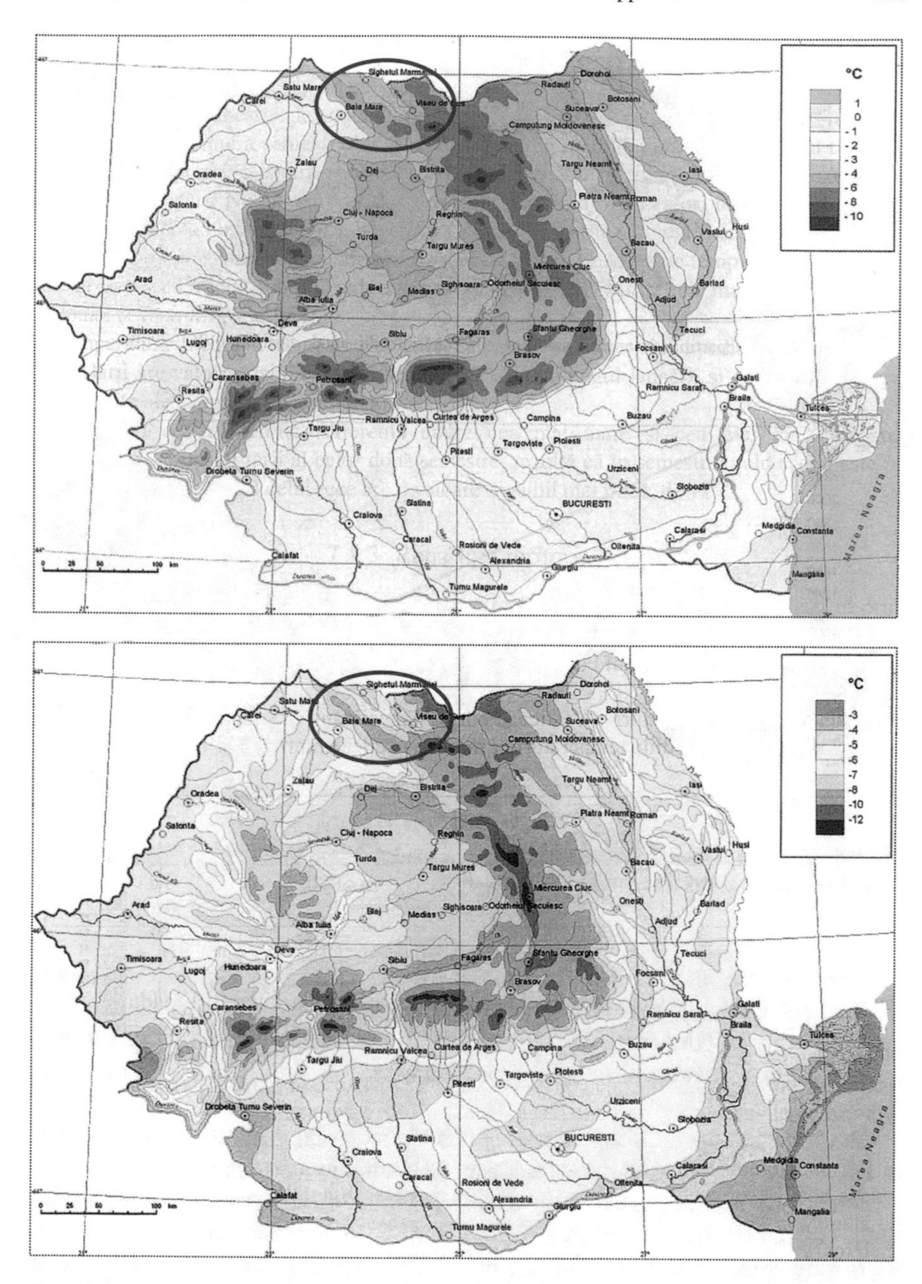

Fig. 7.5 Variation of the average air temperature (above) and the average of the minimum daily temperature of the air (beneath) in January between the 1961s and the 2000s (after Climate of Romania 2008)

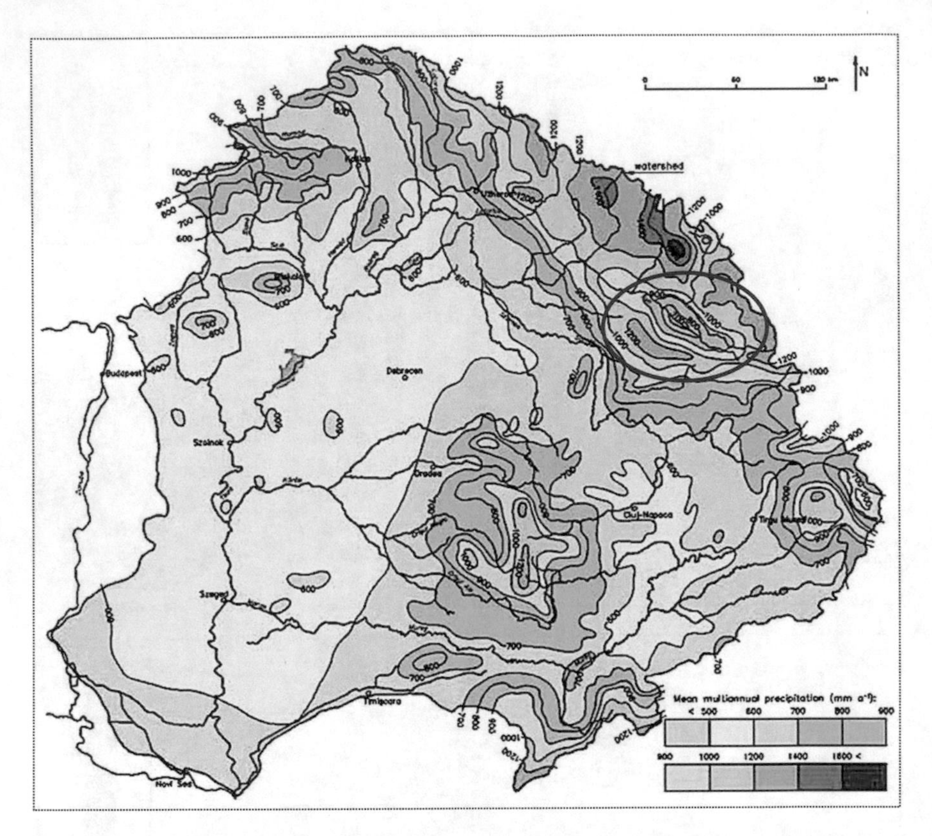

Fig. 7.6 Multiannual isohyets of the average precipitation in the Basin of Tisa (after Sub-Basin Level Flood Action Plan Tisza River Basin 2009)

These thermic particularities occur on the background of frequent cold air invasions brought by ridges of the Siberian Anticyclone, sometimes in junction with the one of the Azoric Anticyclone and of the ones brought by the thalweg of Islandice Low Pressure, with the route over the Baltic Sea.

The values of the multiannual average precipitation are high and rising with altitude: On the core area of Maramures Basin, the multiannual averages vary around the value of 700 mm, rising significantly at over 1200 mm in the massive of Rodna, Ţibleş, Igniş, Gutâi, Oaş, Hust, Borzhava, Krdryn, Krasna, and Pyslikoniaiar and approaching or surpassing 1400 mm in the mountains of Maramureş, Chornohora, Svydovets, and Gorgany (Ujvári 1972; Cocuţ 2008; Chiş and Kosinszki 2011) (Fig. 7.6).

The values are confirmed also in the reference work for the Romanian territory (Climate of Romania 2008, p. 262).

The surplus of precipitation, extremely obvious in the upper basin of Tisa, generates an equally maximum flow, with multiple extreme hydric events and with frequent exceeding of the warning levels on watercourses.

7.5.3 The Hydric Component

Tisa, the collector river with a length of 209 km on the Ukrainian territory, is the sum of the Black Tisa (Chorna Tysa) and the White Tisa (Bila Tisa), united at 4-km north from Rakhiv (Fig. 7.2).

The Black Tisa has the spring on the northeastern slope of the Svydovets Mountains (the Massive of Okoly) at an altitude of 1400 m. It has a length of 49 km, and a hydrographical basin unfolded on a surface of 567 km^2 (entirely in the mountain space), which defines the hydrological regime and the aspect of the valley, deep and sinuous. The width of the riverbed oscillates between 10 and 25–50 m, and the depth of it varies between 0.5 and 2 m in the summertime and 4–6 m during the high-water period. The speed of the current is 1 and 1.5 m per second in the middle of the estival season, and the medium discharge is at about 12.3 m^3/s in the Bilyn section. Near Yasinia, it receives an important left affluent, the Lazeschyna River (Yatsyk and Byshovets 1991; Paparyha et al. 2011).

The White Tisa has the spring on the western slope of the Chornohora Massif in the region of Korbul, at an altitude of 1600 m, and it follows an east–west direction, separating the Chronohora Massif from the Rakhiv Massif. It is a typical mountain river with medium high slopes (10 m/km) and a deep narrow valley, a little sinuous, with steep and wooded slopes, which frequently presents a canyon aspect. The river has 19 km in length and a 489 km^2 surface of the basin (Yatsyk and Byshovets 1991). The medium speed of the water is 2–3 m/s, and the medium discharge is app. 13.5 m^3/s, in the section of Roztoky (Paparyha et al. 2011).

The most important right effluents of Tisa are Kisva, Shopurka, Apshytsia, Teresva, Tereblia, and Rika. The left ones are Vişeu, Iza, and Săpânţa (Ujvári 1972; Bashta and Potish 2007; Chiş and Kosinszki 2011) (Fig. 7.2).

On the Romanian territory of the Maramureş basin and of the surrounding mountain area, the main autochthonous watercourses collected by Tisa are Vişeu with a surface of 1581 km^2 and a total length of 82 km, Iza with a surface of 1293 km^2 and a total length of 80 km. The other left effluents that flow on a SE-NV direction, Săpânţa and Şugătagul Mare, are less important (Fig. 7.3).

In this area, the rivers are characterized by a medium slope with values between 0.2 and 8.9‰ and a sinuous coefficient between 1.04 and 2.16. The hydrographical basins with surfaces larger than 100 km^2 are eleven. Furthermore, the medium altitudes are relatively high, a fact that indicates (theoretically, along with the resistance to erosion) a high hydro-energetic potential in the mountains of Rodna and Maramures, Gutîi and Ţibleş (The Atlas of Water Cadastre of Romania 1992).

The density of hydrographical network in the Romanian basin has values between 0.5 km/km^2 in the lower part of the basin and 0.8 km/km^2 in the mountain zone.

For the Tisa River, at the exit from the Romanian territory, it was calculated a multiannual medium discharge of 130 m^3/s (specific discharge of 20.2 1/s.km^2), with significant hydrologic contribution from the Romanian territory because of the rivers of Vişeu (33.9 m^3/s) and Iza (16.6 m^3/s). The Tisa River has a specific discharge three times higher than the Someş River, even though the surface of Tisa's basin is half of Someş, because of the high recorded precipitations in the reception basin (S.T.B.W.A. 2015).

The height of the flow reveals the existence of abundant liquid flow in the studied space, with values that exceed 1000 mm, confirmed also by Flood Protection Expert Group, Hungary, Romania, Slovakia, Serbia, Ukraine, 2009, and by the GIS models made afterward by S.T.B.W.A. (2015) and the other authors of this study (Fig. 7.5).

Tisa and its effluents in the study zone have a Carpathian-type hydrologic regime, having the maximum volume of the flow in the month of April and the minimum volume in winter (Ujvári 1972; Sorocovschi and Şerban 2012).

The Carpathian regime type (TC) is characterized by vertical zoning of the hydric regime elements. Thus, in winter, along with the growth of the altitude, the period with low waters is longer and the floods' frequency decreases. In spring, the period of the nivo-pluvial waters rises simultaneously with the delay of the snow melting period ending. At altitudes that exceed 1200–1400 m, the high waters of spring are followed directly by the pluvio-nival waters of summer, which last 3–4 months.

The Western Carpathian subtype (CW) is specific, on a limited area in the Basin of Mara, for the volcanic mountains of Oaş-Igniş. The main characteristic of this subtype regime is relatively early beginning of the high spring waters, which last 1–2 months (May–April). Then, the floods from the beginning of summer come, which transform in high waters. The hydric phases continue with summer low waters and fall floods, which are having a frequency of about 30–40%. In winter, the genesis of the catastrophic nivo-pluvial floods is possible, determined by sudden invasions of warm air. At altitudes beyond 1000 m, the minimum flow happens in winter and in the lower regions, in summer or fall.

The Transylvanian Carpathian (TC) subtype is specific to rivers that have the source under altitudes of 1600–1800 m, on the western slopes of the Oriental Carpathians, as well to a limited area in the Mountains of Maramureş. It has similarities with the Western Carpathian subtype, with the only discrepancy that in winter time the low waters predominate. The nivo-pluvial floods from this period have a frequency of only 10–20%. The water supply type of the rivers is pluvio-nival and moderate subterranean (Ujvári 1972; Sorocovschi and Şerban 2012) (Fig. 7.7).

7.6 The Specific Conditions of Winter Phenomenon Developing Between December 1, 2016, and February 14, 2017

The watercourses, which recorded the most severe winter phenomenon, followed by unusual events, were the Tisa river collector, its affluent Iza and less the Viseu River, like other effluents of them: Vaser, effluent of Viseu—the most severe case, Mara affluent of Iza and Cosău effluent of Mara (Figs. 7.3 and 7.8).

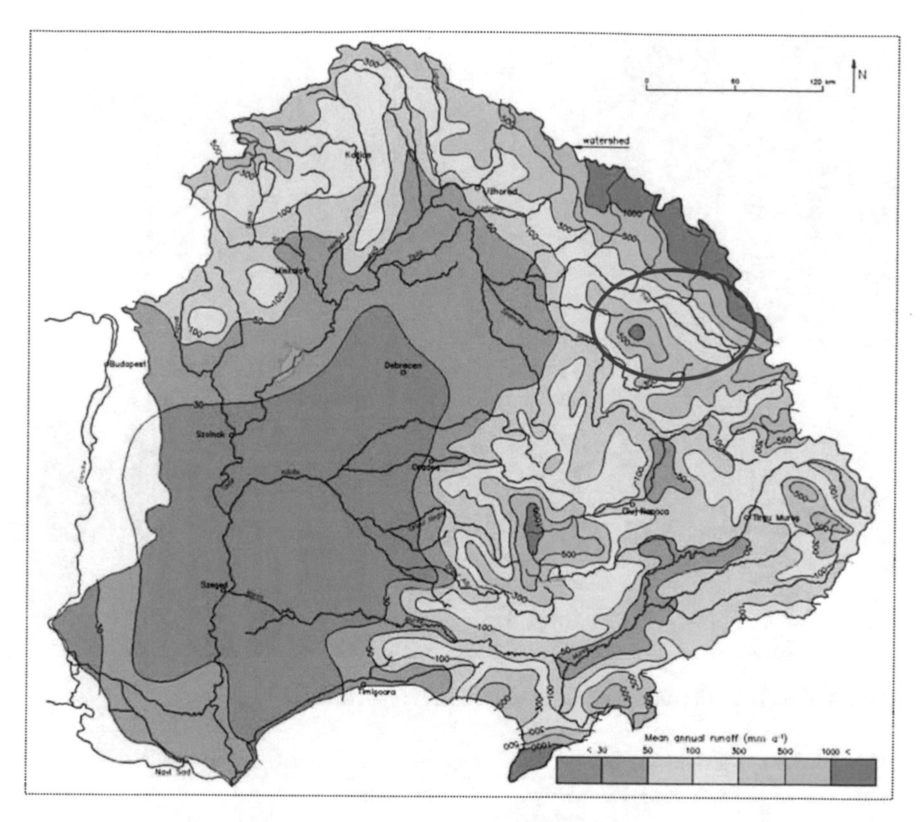

Fig. 7.7 Multiannual values of the medium flow layer (Y—mm) in the basin of Tisa (after Sub-Basin Level Flood Action Plan Tisza River Basin, 2009)

Their position, as well other characteristics and special conditions in which were these watercourses, generated particularities for each one of them, helping some critical sectors to appear.

7.6.1 The Morphologic and Morphometric Conditions

One of the most targeted sectors and spaces of produced phenomenon was the confluence ones, the low zones from the connection of the watercourses, where the ice blocks that came from upstream were massively crowded. In this manner are all the cases mentioned above, less the Viseu River (Fig. 7.8).

Another category of sectors with problems was the gorge-type ones, where the variation between the narrowing sectors and the stretching ones created optimal conditions for the narrowing of the ice blocks. In this situation is also, firstly, the Vaser

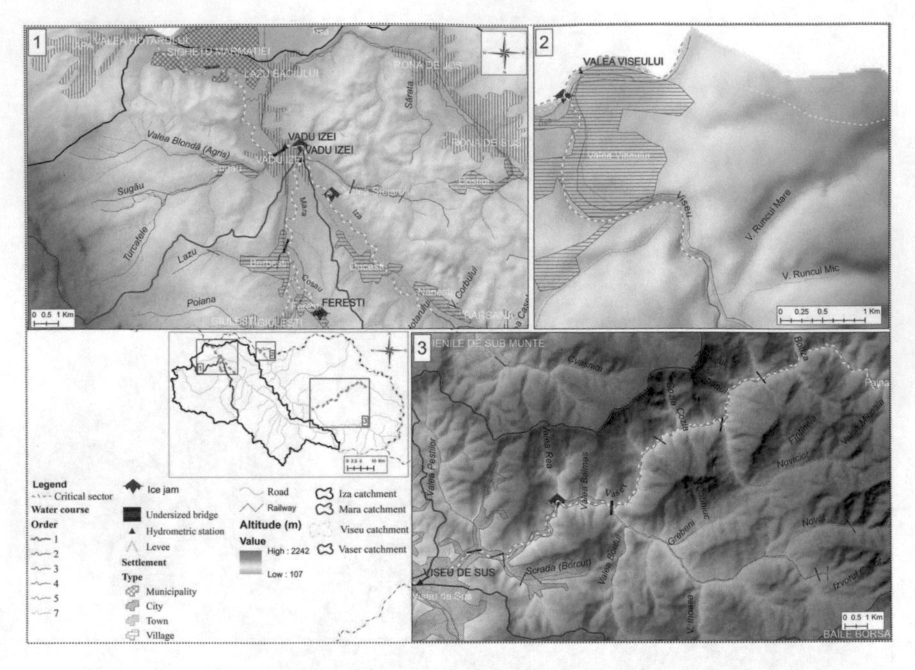

Fig. 7.8 Detail of critic sectors on the case studies. *Source* Authors

Table 7.1 Length and the sinuosity of the studied courses on critical sectors

Water course	Length (km)	Sinuosity coefficient	Critical sector
Tisa	3.11	1.28	Entry in Romania—confluence with Vișeu river
Vaser	32.38	1.41	Vișeul de Sus–Făina
Iza	22.52	1.3	Sighetu Marmației– Nănești
Mara	8.76	1.25	Vadul Izei–Giulești
Vișeu	3.61	1.48	Valea Vișeului—Upstream of Valea Vișeului
Cosău	2.2	1.24	Confluence with Mara River–Ferești

River, along with its collector the Viseu River, on the sector before the confluence with Tisa (Fig. 7.8).

The sinuosity of the watercourses was another element, which favoured the chaotic movement and the agglomeration of the ice blocks. No matter that it was of topographic or hydraulic origin it combined with narrowing's of the flowing sectors due to the anthropic factor after other riverine improvements (Table 7.1; Fig. 7.8).

From all analyzed watercourses, as can be seen in Table 7.1, the Vaser and it collector Viseu Rivers are outstanding with the biggest values of the sinuosity, in both cases mostly of topographic origin. Lower values, but spatially important, are also recorded on Iza and Mara rivers in the lower confluence zone near the city of Sighetul Marmației.

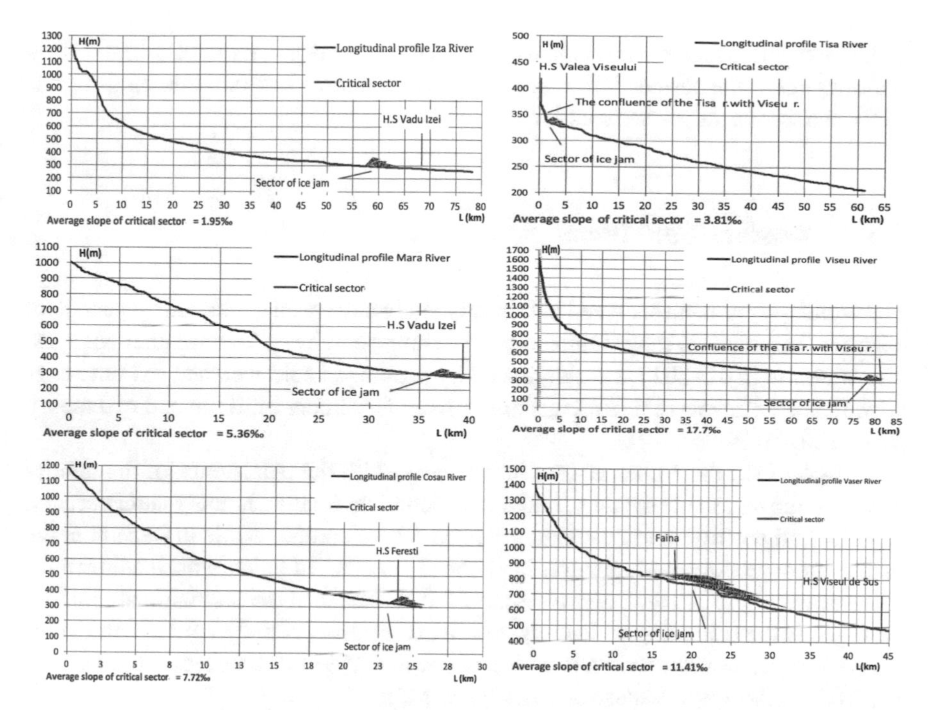

Fig. 7.9 Longitudinal profiles of the analyzed watercourses and the critical sectors, affected by the flowing and the agglomerations of ice. *Source* Authors

An element, again, extremely important is the aspect of the longitudinal profiles of the analyzed watercourses and their medium slope (Fig. 7.9).

Excepting the Tisa collector, which enter the Romanian territory only at an altitude of 400 m, all the other rivers have their source at altitudes that exceed 1000 m, the Viseu River even beyond 1600 m, and the Vaser River at altitudes beyond 1400 m. As shown in Fig. 7.9, all ice agglomerations of ice jam type, followed by the critical sectors of induced effects manifestation, were produced on lower courses of the analyzed rivers, except Tisa, where the ice jam and its effects have targeted narrowing the area at the confluence with the Viseu River. Also in the case of Vaser River, the ice jam was formed in the medium zone of the profile, but the effect was felt until the shedding.

Three of the cases of ice agglomerations were stronger in the upstream sector (Tisa, Mara, and on a less portion the Viseu River), and in the other three cases, the effects manifested mostly downstream (Iza, Cosăul, and especially Vaser).

An important contribution in the manifestation of the phenomena had, also, the slope, which, associated with other morphological and morphometric elements, favored the congestion and stagnation of the ice floes (Fig. 7.9).

In most of the cases, high values of the slope can be seen in the upper sectors of the rivers, where the mountain stage stands out, while in the lower sectors its values are decreasing a lot.

On the critical sectors, the slope was different, form river to river, very high values standing out in the case of Vaser and Viseu rivers (above 11‰), while on the other watercourses the values were under 8‰.

7.6.2 Weather Conditions

The winter phenomena analyzed in this very study have been generated by persistent negative temperatures (approximately two months) and by sudden weather changes. The winter of 2016–2017 was dominated by prolonged cold weather; extreme negative temperatures were recorded, starting from December until the end of January (Lucza 2017).

In December, the winter has set in normally, with the persistence of the anticyclone regime and stable weather, but in the upcoming months, January and February, negative and positive thermic anomalies have been recorded. While in December the medium temperature in Maramureş has been of −4.6 °C, in January the medium temperature was −7.5 °C, with approximately 3.0 °C less than the multiannual monthly average. At the beginning of January, the weather was extremely cold and the minimum temperature value recorded on interval was −25.9 °C (recorded on the January 10, 2017, at the Târgu Lăpuş meteorological station).

Even though the quantities of fallen precipitation in the period between December 2016 and January 2017 did not overcome the multiannual average values, because of the negative temperatures, the precipitation was almost in form of snow. The thickness of the snow layer and the water supplies accumulated by it have risen in the whole period, so as until on January 31 the water equivalent from the snow layer in the basin of Viseu was of 96.46 mil. m^3 and of 71.41 mil. m^3 in the basin of Iza.

The persistence of the negative temperatures has driven to the formation of winter phenomena on watercourses, the thickness of the ice rising directly proportional with the intensity of the frost. On the watercourses of Maramures, the first winter phenomenon appeared on the December 4, 2016, and until the beginning of January, the ice bridge has been installed on all the watercourses of the mountain zone, with measured thicknesses up to 38–40 cm in the hydrometric stations section.

After the severe frosting from January, in the first days of February, under the circumstances of an advection of warm and damp air, the quick melting of the snow and the heating of the air started. Between the 2 and 3 February on the southern slope of the Gutâi mountains, at the Cavnic pluviometric station there were recorded quantities of liquid precipitation with values of 90 mm. Thus, the total values between 2 and 6 February were 170 mm.

Though the majority of the fallen precipitation in areas with the altitude below 1300 m was in liquid form, the thick layer of the snow acted as a sponge. Thus, the flowing and gathering of the water in watercourses was slowed. However, the sudden warming, which persisted for a few days, with positive minimum temperatures associated with high precipitation, caused the melting of wet snow. On February 3,

06 AM local time, the maximum temperature of 5.2 °C was recorded at the Sighetu Marmaţiei hydrological station.

Under these circumstances, the abundant nivo-pluvial input has determined the emergence of floods on the upper sections of watercourses. As a result, the level fluctuations destroyed the ice formations accumulated on watercourses, mobilizing and dislocating them downstream, where ice jams were formed at narrower sections. The water supplies of the rivers being supplemented also from the low precipitation fallen between the 5 and 6 February.

7.6.2.1 Synoptic Conditions

In December, the ridge of Azoric and Siberian anticyclones were united all over the south of Europe forming a high-pressure belt which persists, having more southern positions while descending to lower latitudes of the polar front during the cold season (Fig. 7.10).

The winter started normally with the persistence of the anticyclonic regime and stable weather in the most part of the Balkans and Central Europe. The short invasion

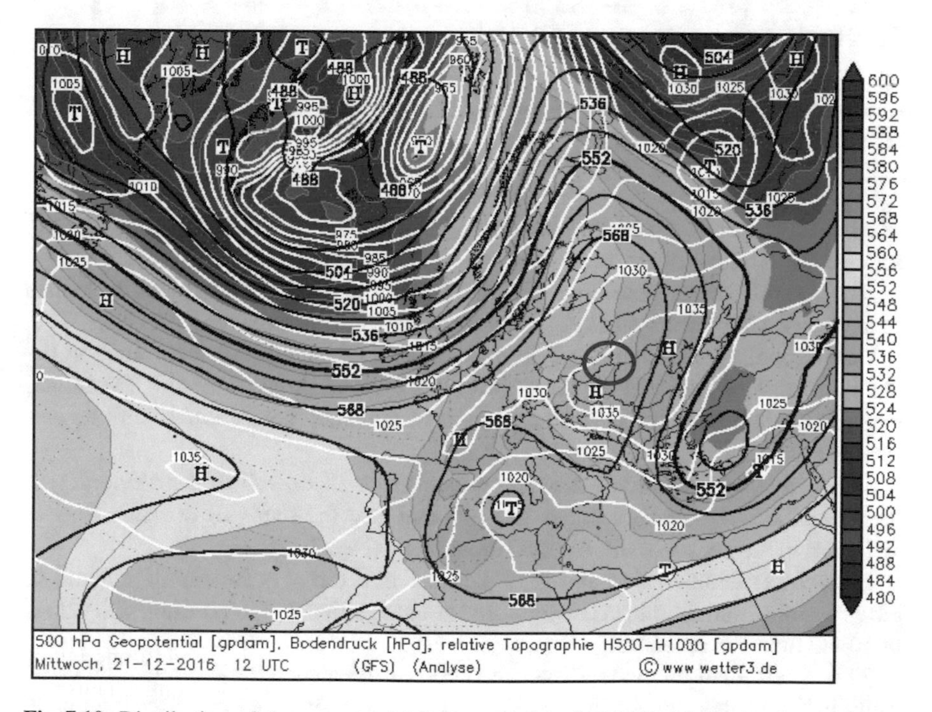

Fig. 7.10 Distribution of the geopotential field at a level of 500 hPa (damgp), of ground-level pressure field (hPa), and of relative topography 500–1000 hPa: December 21st, 2016 at 12.00 UTC. *Source* www.wetter3.de

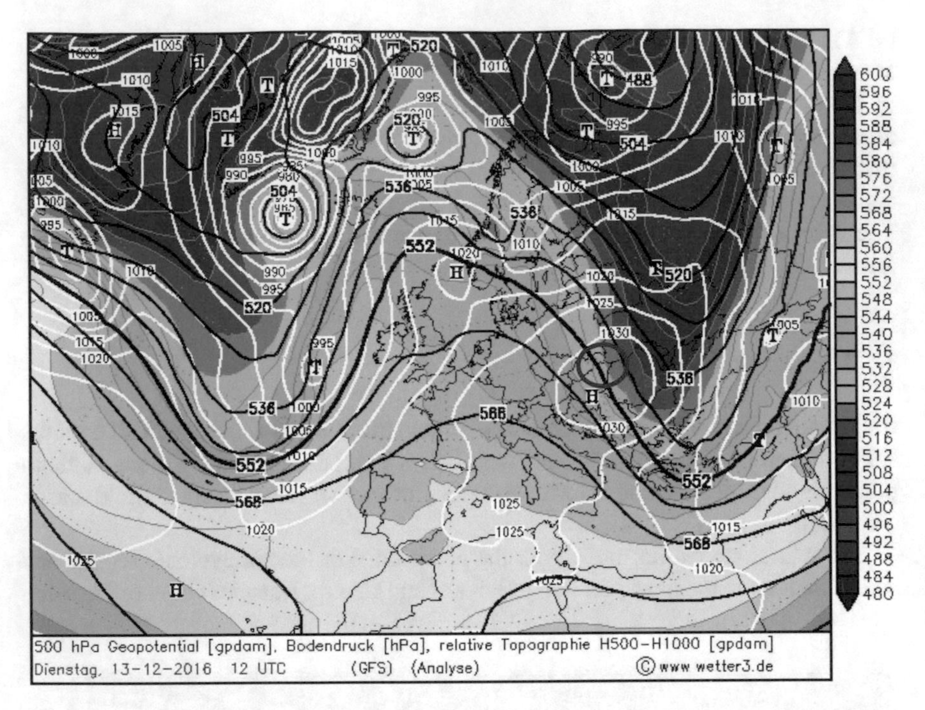

Fig. 7.11 Distribution of the geopotential field at a level of 500 hPa (damgp), of ground-level pressure field (hPa), and of relative topography 500–1000 hPa: December 17th at 12.00 UTC. *Source* www.wetter3.de

of a moist air mass, then of cold air waves, happened at the beginning of the month, followed by a temperature drop between December 12th and 17th (Fig. 7.11).

The period between December 18th and 25th was characterized by stability, thanks to the dense and cold air mass persistence, which generated dryness.

Starting with the December 29th, the invasion of the cold air continued at the medium troposphere level on the northwest direction. Alongside with the frontal passage, slight precipitation, dispersed snow or rain, and mostly, a new cooling was registered.

In the month of January, the baric gradients rose as a result of the deepening of the Icelandic Low Pressure under 985 hPa and also the rising of pressure to over 1040 hPa in the centre of the Azoric Anticyclone (Fig. 7.12).

Between December 26th and 29th, because of a direct polar circulation and of advection of a more humid arctic maritime air, insignificant precipitation with values of 10–20 mm was recorded.

This determined an evident intensification of the atmospheric circulation especially in the northern half of the continent. In the southeastern part of the continent, as a result of the anticyclonic belt emergence through the blending of the Azoric Anticyclone with the Siberian one, as well as of weakening of the cyclonic activity in the Mediterranean basin, the circulation of the air is less intense. Under these

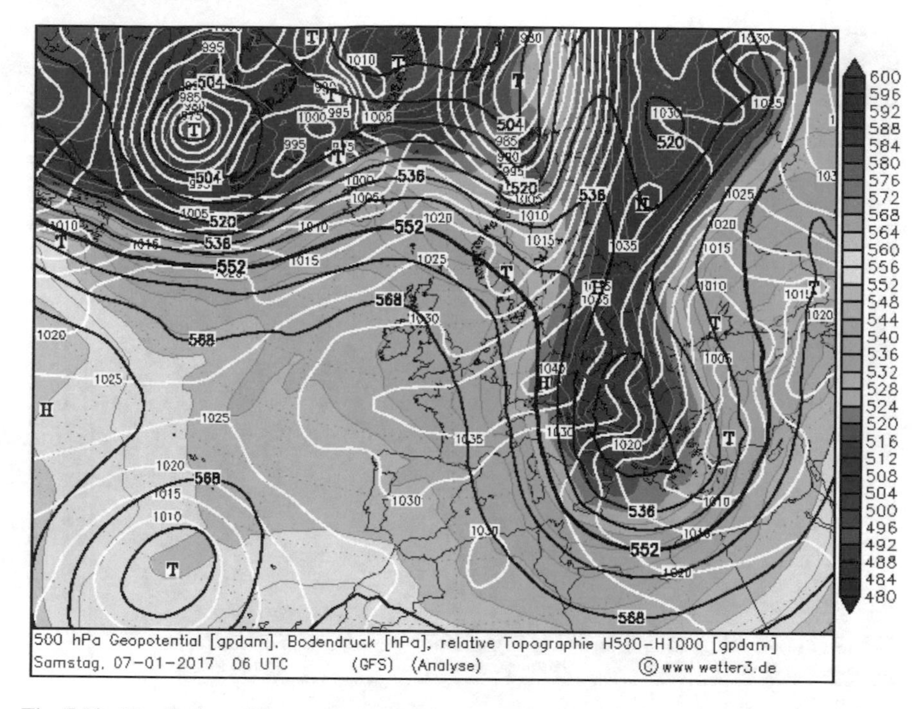

Fig. 7.12 Distribution of the geopotential field at a level of 500 hPa (damgp), of ground pressure field (hPa), and of relative topography 500–1000 hPa: January 7th 2017 at 06.00 UTC. *Source* www.wetter3.de

conditions, the anticyclonic regime was rarely disrupted by short invasions from the northern sector, furthered by the fluctuations of the southern waves of the jet stream (Fig. 7.13).

In the first part of January, almost the whole continent was under the influence of an anticyclonic field at the ground level, whose ridge stretched from the Biscaya Bay to the Carpathians, on a west–east direction (Fig. 7.13). While the eastern zones, from the front part, were dominated by a mass of cold air, the central zones and the western part of the continent were affected by the relatively warm air. In this context, because the Basin of Maramureş is located in the east, it was affected by a frosty weather.

Meanwhile, due to a polar direct circulation, the cold mobile cyclone has moved from Greenland, along the Northern Sea, to the Central European Plain and to the Alps, consequently to the east. Hence, the decreasing of the air pressure at ground level, followed by the strengthening of the baric gradient in the geopotential zone from the medium layers of the troposphere, occurred on the Romanian territory, too.

The continuous advection of the polar continental air mass with a strong northern circulation caused the terminating of the cyclone circulation over the Balkans Mountains and the Adriatic Sea.

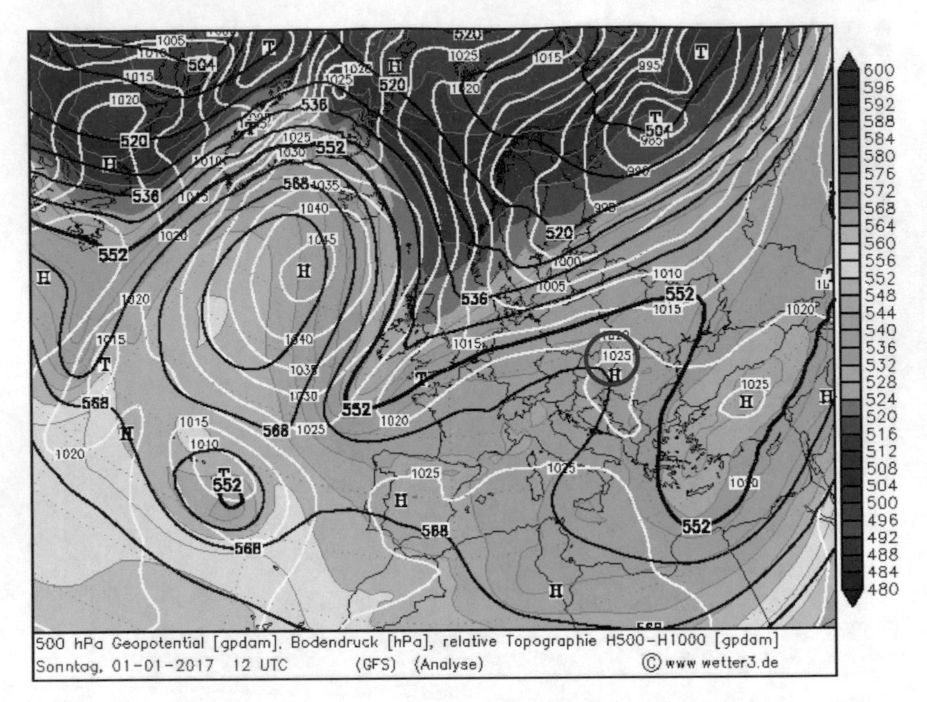

Fig. 7.13 Distribution of the geopotential field at a level of 500 hPa (damgp), of ground pressure field (hPa), and of relative topography 500–1000 hPa: January 1st, 2017 at 12.00 UTC. *Source* www.wetter3.de

The afterward deforming of the polar jet stream, along with the growth of the amplitude of the southern waves, determined, beginning with 6 January, an intensive advection of the polar continental air mass, first in Central Europe, Pannonian Plain, the Balkans Mountains and the Adriatic Sea, then to the northeastern coast of Africa and the east of Mediterranean Sea (Fig. 7.14).

So, the extreme temperatures registered between January 1st and 10th were generated by this synoptic situation, when the lowest temperature of −25.9 °C was recorded (the interval of consistent extension of the ice thickness on watercourses from the basin of Maramureş).

The cyclone of high altitude already formed on the 6th of January, with very low values of the geopotential at the isobaric surface of 500 hPa and relative of 1000/500 hPa, the very cold relative of 1000/850 hPa, indicates a mass of continental arctic air (Fig. 7.15).

In the following days, until the 8th of January, the cold air continuous advection from north led to the formation of a "cold air lake" over the basin of Maramureş. At the same time, a subsidence regime dominated the inferior layer of the troposphere at the periphery of the anticyclone, which stretched out on the most part of the continent.

Until the 12th of January, the Maramureş is under the influence of polar air masses between the cyclonic circulations of altitude, also the waves of humid air at the

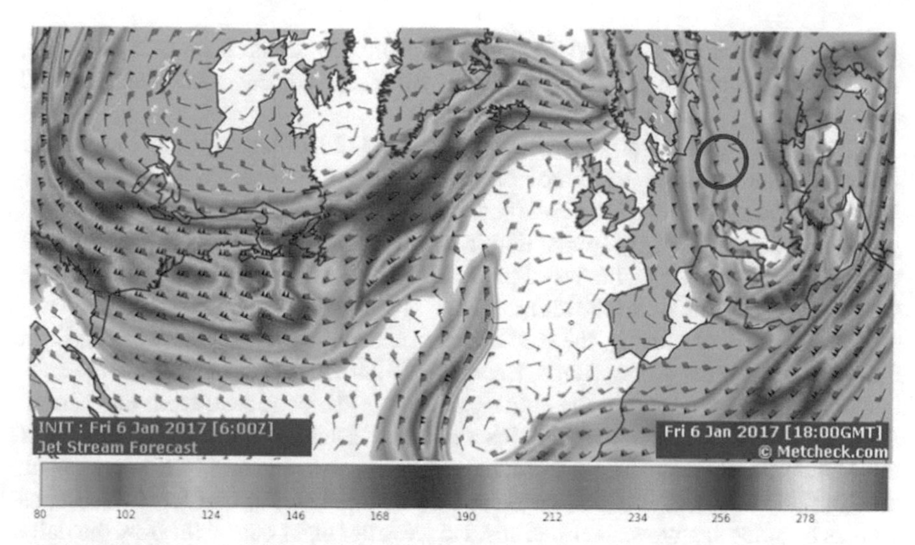

Fig. 7.14 Distribution of the jet stream at the level of 500 hPa: January 6th, 2017, at 18.00 GMT. *Source* www.metcheck.com/WEATHER/jetstream_archive.asp

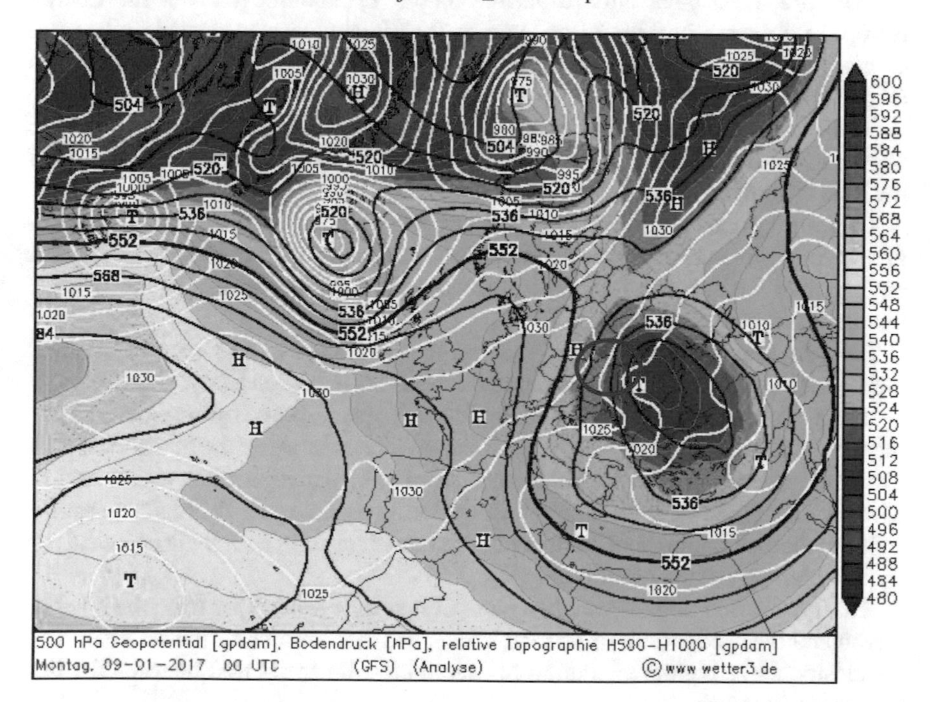

Fig. 7.15 Distribution of the geopotential field at a level of 500 hPa (damgp), of ground pressure field (hPa), and of relative topography 500–1000 hPa: January 9th, 2017 at 00.00 UTC. *Source* www.wetter3.de

periphery of the surface anticyclone, coming from the central parts of the Europe, which brought occasional snow. Some snow was recorded occasionally at night and discontinuously, in the mountain area, too.

The transformation of air masses occurred on the 12th of January, after the gradual decrease of the pressure, under the influence of a set of fronts from northwest in the interior of the cyclone, which developed on the most part of the continent. After the passing of the atmospheric fronts, dominated by a higher pressure of the air at the terrestrial surface, the weather was characterized by the fog phenomenon, along with the progressive decrease of the temperature.

Until the 24th of January, the cold anticyclonic regime predominated, with short forays of the low disturbing fronts. Between the 24th and 27th of January, the anticyclonic regime persisted in the entire east part of the continent, from the Central European Plain to the Ural Mountains and to the south, to the surface cyclone of the Mediterranean zone. The maximum barometric was centered in Maramures on the 26th of January, and it was characterized by a nebulosity and fog decrease, as well as by some sunny weather periods. Hence, after eight days with frost the daily temperature surpassed 0 °C in most of the settlements of Maramures.

Due to a direct polar circulation, the mobile cyclone moved over the Central Europe to the east. Until the 31 of January, the Maramures was located in the mass of cold air of the ridge, where the high pressure and the cold weather persisted (Fig. 7.16).

The last day of January, was marked by the shifting of the cyclone and of its cold air waves from the Central Europe to the northeast (with its periphery it disturbed the base layer of the thermic inversion).

On a background of tropical circulation, the geopotential increase and the advection of the warmer air from the south led to weather instability in Maramures at the beginning of the month, with temperatures significantly higher than the average for the first ten days of February.

Nevertheless, the persistence of the cyclone, which dominated the weather from the western part of the Europe, generated a pressure reduction over the Balkans.

The incursion of cold air from the Genova Bay was followed by the emerging of a mobile cyclone in the zone of cyclogenesis, both happening on the 5th of February. The cyclone with the frontal zone transferred over the Apennines to the east of Mediterranean, fed with humidity, followed afterward a trans-Balkanic trajectory. This fact caused raining and snowing in the mountain zone with higher intensity in the southern part of the Maramureş (Fig. 7.17).

This type of circulation brought a mass of maritime tropical air from the Atlantic Ocean and the Mediterranean Sea, which caused the warming of the weather, high nebulosity, and a high precipitation in Maramures, for the period between the 2nd and 6th of February.

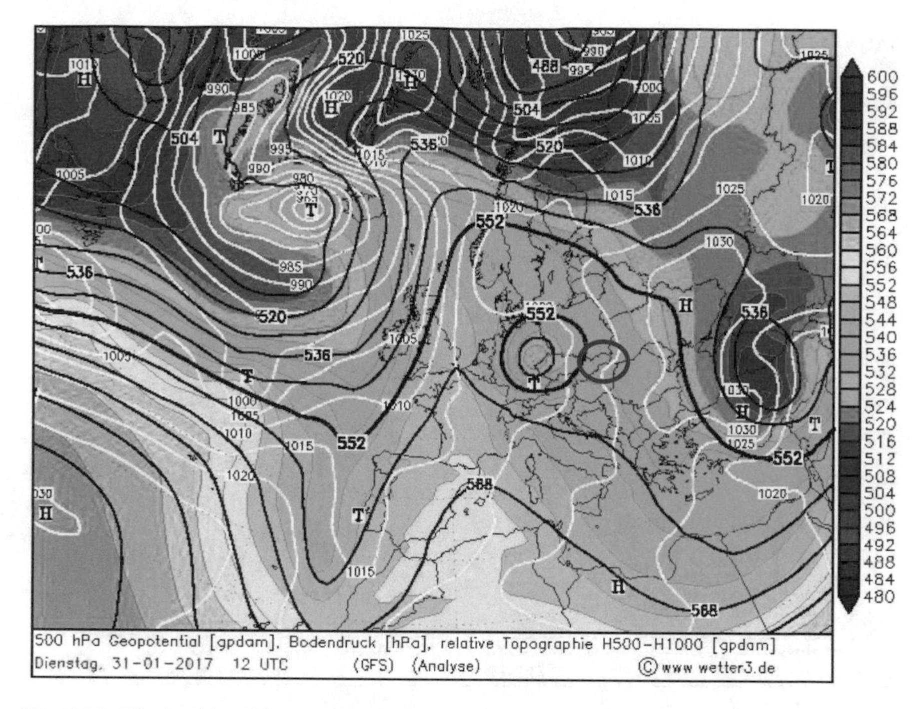

Fig. 7.16 Distribution of the geopotential field at a level of 500 hPa (damgp), of ground pressure field (hPa), and of relative topography 500–1000 hPa: January 31st, 2017 at 12.00 UTC. *Source* www.wetter3.de

7.6.2.2 The Evolution of Air Temperature and Rainfall

According to the synoptic situation detailed above, the winter of 2016–2017 was unusual for the last decades because of its severity and the manifestation way of the meteorological parameters.

The records made by S.T.W.B.A. through the network of hydrometric stations and at the frequency previously presented as well show this aspect. These recordings have made possible the realization of a *complex winter chart*, which allows the conditional explanation of the production and manifestation of the analyzed phenomena.

The sitting of the hydrometric stations is not, necessarily, in the area of the flow formation but, rather, in the area of the accumulation of it, in key sections, which allow the control and effective management of special circumstances, occurred due to hydrological extreme events. That's why there is a certain quantitative/value difference between what was actually measured in their section and what was determined by different methods or gauging campaigns on basin surfaces.

The air temperature, measured even hourly during the manifestation periods of the dangerous hydric phenomena, was rarely positive, and this for very brief periods; periods identified as the same with those with rainfall in the area of study, which gave, besides, oscillations that are more than obvious, on the chart (Fig. 7.18).

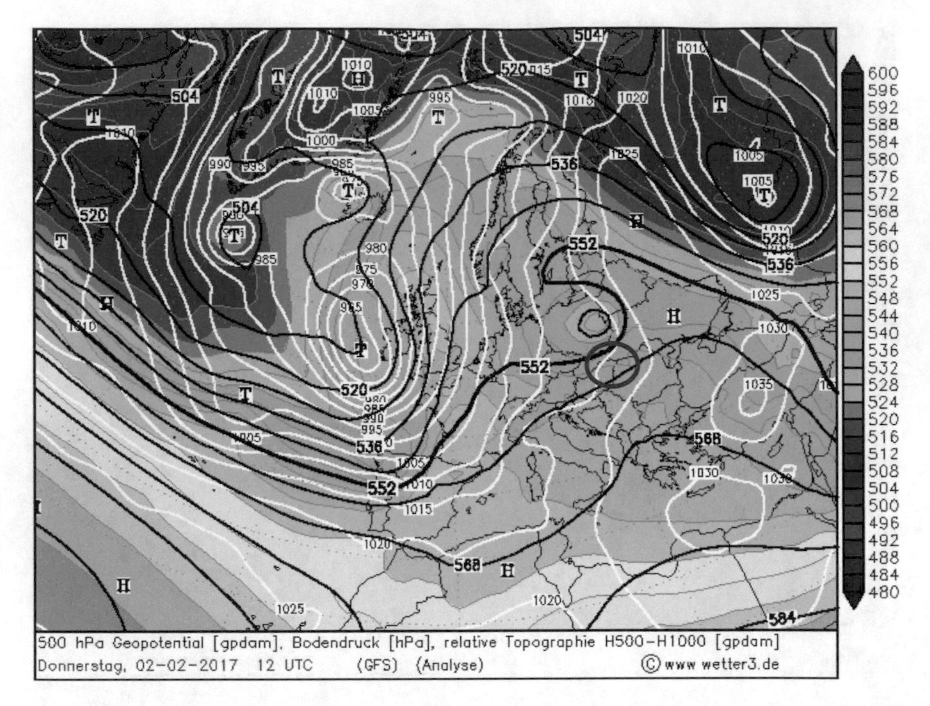

Fig. 7.17 Distribution of the geopotential field at a level of 500 hPa (damgp), of ground pressure field (hPa), and of relative topography 500–1000 hPa: January 31, 2017 at 12.00 UTC. *Source* www.wetter3.de

The thermic values varied between 3 °C the maximum and −19 °C the minimum, and there were recorded multiple days with strong negative values in a row: −3 … −14 °C, in the first period of the analysis interval and even −10 … −19 °C in the middle of the interval. To the end of it, the thermic values raised very much staying positive even for more than a week (February 1st to 11th, 2017). Then, a short period of time of low temperatures took over again: under −5 °C.

These thermic characteristics were very favorable for the increase of ice thickness on watercourses, with a strong impact on the phenomenon of liquid flow, as it will be seen in one of the following chapters.

The registered rainfall quantities were not important—a maximum of 14 mm in the week preceding the formation of the ice jams (predominately rain).

Throughout the entire analyzed period they were, however, mostly in the form of snow, which facilitated the accumulation of a substantial layer (Fig. 7.18), especially on slopes, with a subsequent effect on the flow. Important was, however, the frequency nearly regular, identifying six periods with rainfall at approximately equal intervals, except for the last decade of January, when they were quasi-absent.

As expected, the quantities of solid precipitation, but also liquid, were almost double in the case of the Vaser basin—hydrometric station of Viseul de Sus, on the slopes with western exposure of the Mountains Maramures (compared with the hearth of the basin, where the other gauging stations are located).

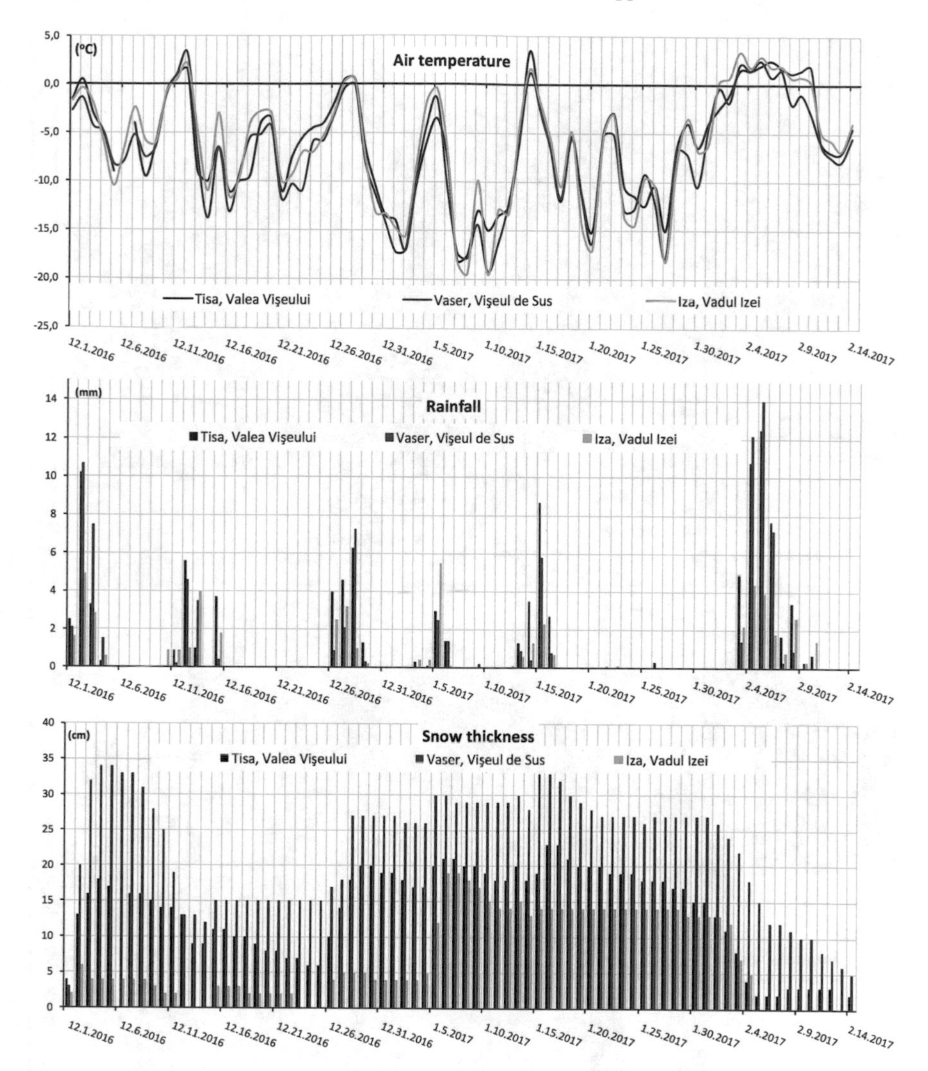

Fig. 7.18 Complex chart of the winter I—air temperature, rainfall, and snow thickness at the hydrometric stations on the studied rivers (*data source* S.T.B.W.A. 2017)

7.6.2.3 The Average Thickness of the Snow Layer and the Water Supply in the Basin

These two essential elements in the process of winter hydrological forecasts elaboration have recorded a very interesting spatial distribution with relevance in the formation of extreme hydric studied events, in the basin of Maramures (Fig. 7.19).

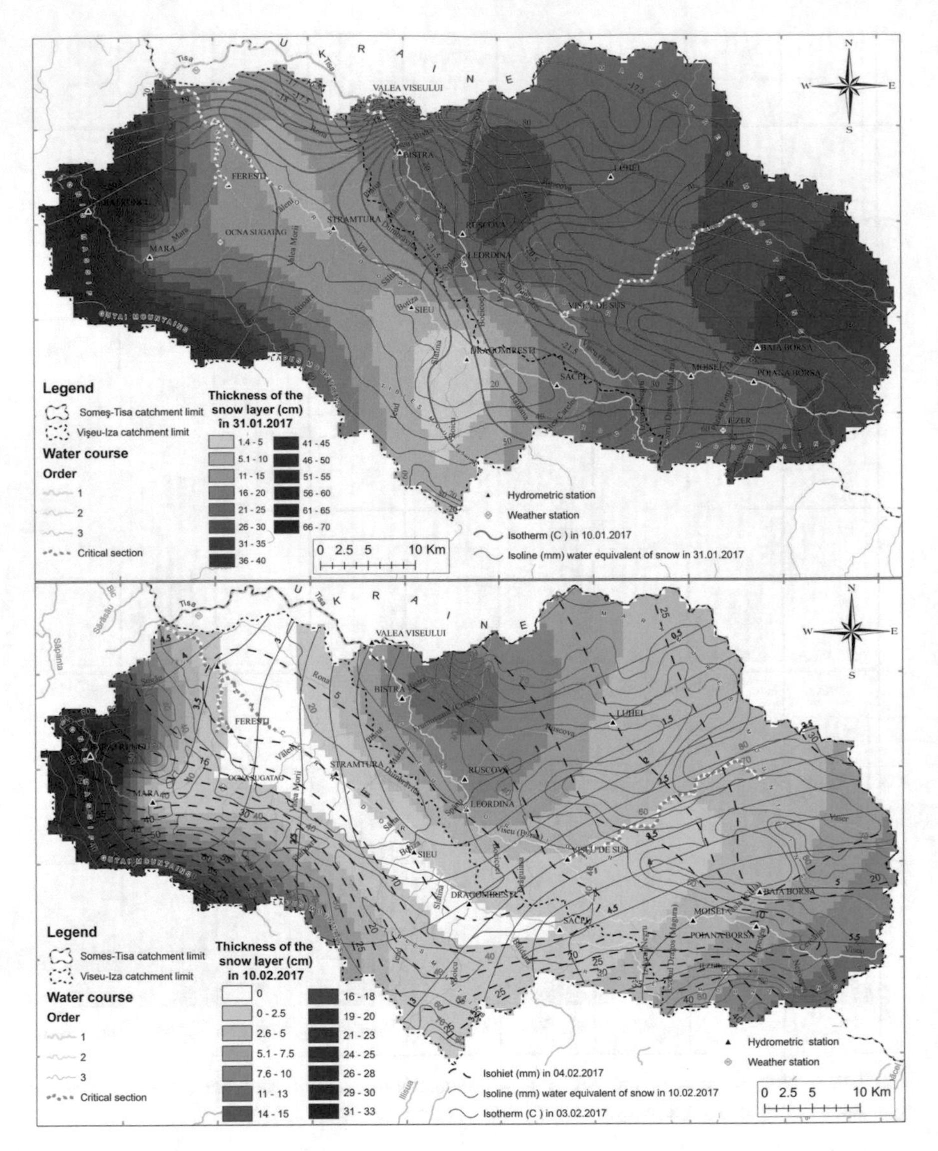

Fig. 7.19 Average thickness of the snow layer and the water supply in the basin. Up, January 31st, 2017; Down, February 10th, 2017 (*data source* S.T.B.W.A. 2017)

Even before the consistent warming period (February 1st to 11th, 2017), doubled by liquid precipitation, as well as after this, the hearth of basin of Maramureș Basin, individualized of the middle and lower course of the Iza River, was covered by a *layer of low snow* (under 15 cm) on January 31st, 2017, or devoid of this bedspread on February 10th, 2017 (Fig. 7.19).

On the contrary, the eastern slopes of the mountains of Oaş and Igniş and the northern ones of the Gutâi Mountains and the western slope of the Maramureş Mountains stored significant quantities of snow (between 25 and 70 cm before warming and 5–33 cm after it).

Besides, the group of the Gutâi Mountains is recognized as a pole of rainfall in Romania with values around 1200 mm, after the Vladeasa Mountains (Apuseni Mountains, where at the Station Stâna de Vale the quantities range up to 1500 mm), Maramureş Mountains, Rodnei Mountains, and the Făgăraș Mountains.

Due to the thermal oscillations and repeated precipitation, as described above, on **January 31st, 2017**, *the water supply in the snow layer* had very high values, which ranged from 80 to 90 mm in Maramures Mountains, 70–80 mm in the Gutâi Mountains to values much more modest between 15 and 30 mm on the hearth of the basin, along the axis of the Iza river, in particular (Fig. 7.19).

After the warming period and the period of constant liquid precipitation in the interval 02-05.02.2017 (80 mm in the Gutâi Mountains and 40 mm in Rodna Mountains—Fig. 7.18) on **February 10th, 2017**, the *water reserve in the layer of snow* decreased slightly just on the hearth of the basin, while in the mountain area remained fairly high (70–80 mm in the mountains, Lapus and Tibles, 70 and 80–90 mm in the north and, respectively, south of the Maramureş Mountains—Fig. 7.19).

As it will be observed in the next chapter, these large amounts of water stored on the slopes had important effects on the flowing regime of the rivers and on the mobilization of the ice floes downstream.

7.6.3 The Anthropogenic Conditions Influencing the Riverbed on the Studied Sectors

The areas of ice jams formation are heavily populated and with a natural landscape influenced by human intervention (including the watercourses landscape), through systems of damming, defense, and consolidation of banks and overpasses like bridges and walking boards (Fig. 7.8).

Moreover, there are natural obstacles caused by the river topographic sinuosity, narrowing's of the minor and major riverbeds, thresholds, and rapids.

In the case of rivers Tisa and Viseu, it is noted only a narrowing generated by the railway bridge which crosses the tributary, less than a hundred meters from their confluence, however, without a major impact on the occurred phenomena, as its hydraulic calibration is correctly made.

The effect of self-resistance to the advance of the ice and double/auto jam is, however, the one that generated the blocking of the two valleys and the special problems of mechanical and flooding nature. The hydraulic capacity and the channel roughness of Tisa after the reception of Viseu were not sufficient and adapted to mechanical flow massive and abundant floes, which led to agglomeration and the emergence of significant amounts of water behind the jams. As we will see in the next section, the overflowing water volume created the greatest problems and induced the state of alert.

What's more, for the Vaser River, eight sections of overpassing can be counted (footboards and bridges), probably undersized for this type of phenomenon, which is historical for this watercourse. Another possible factor was the narrow railroad embankment, route of importance for logging and tourism, which accompanies the river till its middle course and which sometimes contributes to the narrowing of the flowing section.

Going forward, without a doubt, the most complex area of manifestation of the winter phenomena on watercourses is the confluences and the vulnerable and low sectors of the rivers Tisa, Iza, Mara, and Cosau (Fig. 7.8). Numerous sections of embankments, defenses and consolidation of banks, sections of riverbed overpassing, confluences with backwater zone, and narrowing of watercourses due to sectors of quasi-gorge are found here. There are three sections of overpassing-type bridges, which can pose problems of flowing at high levels of the flow and high density of ice blocks: in the town of Berbești on the Mara River, in the town of Vadul Iza on Iza River, and at the entrance of the Iza River in the municipality Sighetul Marmatiei.

Also, this area is intensely populated, and homesteads reach, sometimes, until in the immediate vicinity of the minor riverbed, so the infusion of different wastes to the riverbed is not ruled out.

7.6.4 The Condition of Rivers, the Types of Winter Phenomena, and Their Evolution, Monitoring, and Management

The stark negative evolution of air temperature was bound to leave significant traces on the watercourses (Fig. 7.20).

In this regard, all studied courses developed the winter phenomena, starting with ice at bank or spongiously ice (possibly other early or transition forms for the advanced phenomena), along with the temporal onset of negative temperatures (Vaser) or three days after it (Mara, Iza, and Tisa). This situation is understandable, as the Vaser Valley belongs entirely to the high mountain area (Fig. 7.3), where the supercooled air had descending character along the homonym canyon till the exit from the mountain area, contributing to the accelerated cooling of the river water. Other watercourses cross the much-enlarged corridors and small basins to the hydrometric stations where parameters of interest were monitored, areas where the supercooled air installed and made its effect with a slight delay.

After the second decade of the month of December, the ice gained consistency, the observed thicknesses exceeding 15 cm, in some cases (Fig. 7.20), thickness from which, in the case of the compact ice, the access of observers on the bridge of ice is permissible (N.I.M.H. București 1996; N.I.H.W.M. București 2013).

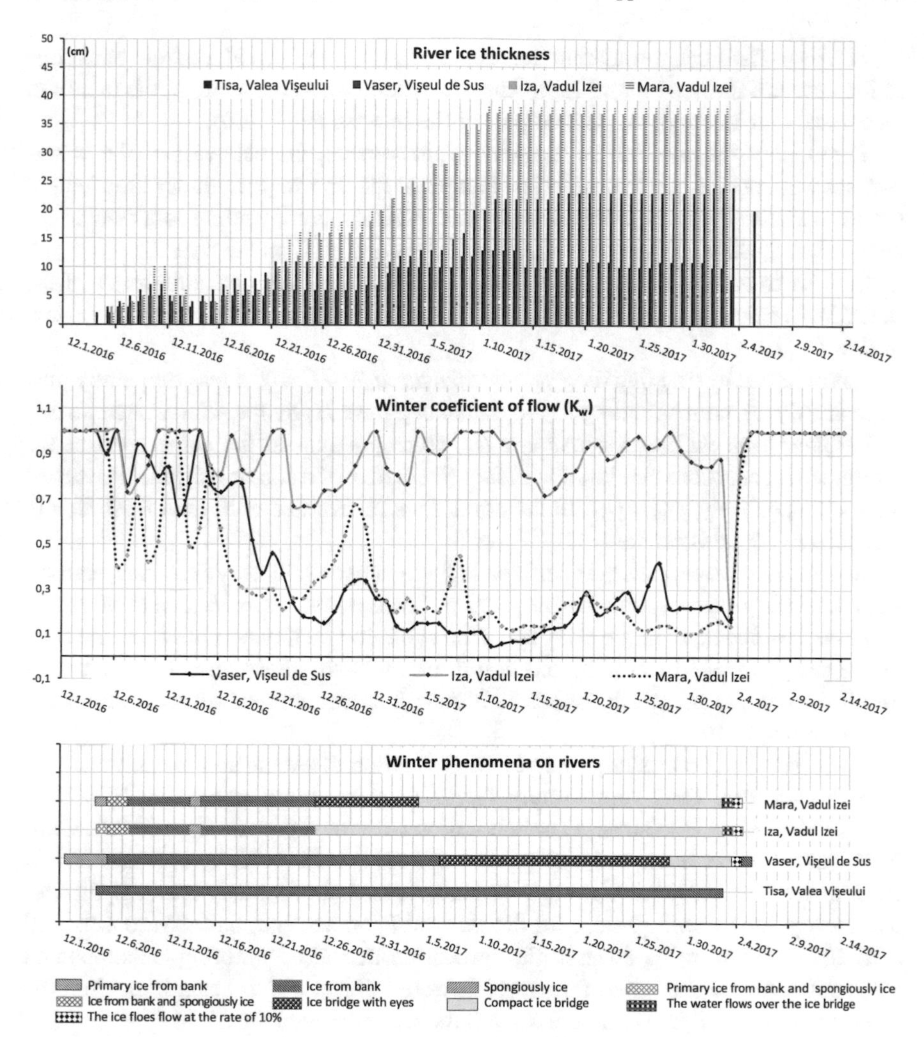

Fig. 7.20 Complex chart of winter II—the ice thickness, winter coefficient of flow, and the types of winter phenomena at the gauging stations on the studied rivers (*data source* S.T.B.W.A. 2017)

Besides, on the same temporal coordinates the transition to an evolved form of winter phenomenon was made as well—the ice bridge, which installed relatively quickly on the rivers Iza and Mara, due to the relatively low flow levels and the persistence of the very low temperatures on the hearth of Maramures Basin. In a first phase, the Vaser River has developed a form of transition to the compact ice bridge, respectively, *the bridge of ice with stitches of water*, due to the high speed of water flow induced by a corresponding slope and the numerous rapids, taking into account the mountain pass sector it crosses.

The river Tisa kept, throughout the period of study, the form of ice from bank, due to the constant flow, the ice with thicknesses around the value 5 cm, in the month of December and 10 cm in the second part of the interval (Fig. 7.20).

Throughout the duration of the winter phenomena, the behavior of the watercourses was specific, with substantial changes in the flow regime, under the effect of ice formations (Fig. 7.20). The flow regime was disrupted, leading to a flow influenced even to 80% on the rivers Vaser and Mara and 30% on the river Iza.

In order to capture the correct degree of influence, the regional authorities of the S.T.B.W.A. calculated and monitored the evolution of the so-called *winter coefficient of flow (K_w)*.

It is obtained by dividing the *effective measured flow Qm* (not less frequently than once at 5 days), corresponding to the *influenced level H_i* by the winter phenomena, to the *flow from the lymnimetric key Q_{lc}* corresponding to the same level at the hydrometric rod, but in the situation of free riverbed. In the conditions of free riverbed $K_\mathrm{i} = 1$, respectively, $Q_\mathrm{m} = Q_\mathrm{lc}$. In the conditions of riverbed with total frost $K_\mathrm{i} = 0$, respectively, there is no flow.

If in the first five pentads the coefficient of winter had oscillatory evolution, starting with the sixth pentad he remained at very low levels on the rivers Vaser and Mara and continued to fluctuate on the river Iza. This evolution continued until after the time of the period of rapid warming and liquid rainfall start, after which the coefficient climbed, again, to unit, as in the period before the production of winter phenomena, in November, 2016.

In the last decade of study interval, the evolution of events and phenomena will take another turn. Thus, on February 1st, 2017, the national authorities R.W.N.A. and N.I.H.W.M. of Bucharest, in collaboration with the regional—S.T.B.W.A. Cluj-Napoca, and with national and regional authorities in the field of Meteorology N.M.A. Bucharest and R.M.C. Transilvania Nord have issued, for the study area, a hydrological code yellow warning, valid for the interval February 2nd, 6.00 PM to 4th, 4.00 PM, 2017. The warning covered the forecasted period of warming, accompanied by precipitation in the liquid form, which were to have a significant impact on the overall condition of the watercourses and the winter phenomena.

On February 2nd, 2017, at 10.00, another meteorological code yellow alert was received, N.M.A. Bucharest for the interval February 3rd, 3.00 AM to February 4th, 6.00 AM, 2017, related to increasing temperatures and rainfall, and the same day at 22.54, a code yellow warning for dangerous weather phenomena was received from the N.M.A. to R.M.C. Transilvania Nord for the interval February 2nd, 11.00 PM to 3rd, 3.00 AM, 2017.

On the February 3rd, 2017 at 8.10 AM, a hydrological code orange warning was issued for immediate phenomena on the rivers in the area, from the R.W.N.A. and N.I.H.W.M. Bucharest, for the interval February 3rd, 8.20 AM to February 3rd, 1.00 PM and 6.00 PM, 2017, we received a new hydrological code orange warning also from the R.W.N.A. to N.I.H.W.M. Bucharest, for the interval February 3rd, 2.00 PM to February 5th, 10.00 AM, 2017.

On the February 5th, 9.45 AM, 2017, the last hydrological code orange warning was received from the R.W.N.A. and N.I.H.W.M. Bucharest, for the interval February

5th, 10.00 AM to 7th, 4.00 PM, 2017. In the area, following the sudden warming and with the arrival of the rains, the situation worsened, confirming the forecasts, and forming ice jam/river blockades, combined with the flash floods produced by the rapid warming and the fallen liquid precipitation, as follows (Fig. 7.8):

- on the Cosau River, upstream from the Feresti hydrometric station, on a length of approximately 200 m;
- at the confluence of Tisa with the Viseu River in the village of Valea Viseului, on a length of 500 m, with a thickness of 3 m of the floes layer, both on the Viseu River as on the Tisa River, followed by a quick level rise, because of the constant discharges of the two rivers;
- on the Iza River, downstream and upstream from the confluence with Valea Băii, between the settlements of Barsana and Stramtura, on a length of 1 km approximately;
- on the Iza River, upstream of the villages of Vadu Izei and Nănești, with a length of the ice jam of 7.5 km;
- on the Ruscova River, in the town of Ruscova, with a length of the ice jam of 250 m, and a thickness of 2.45 m;
- on the Vaser River, ice jams, starting at the hydrometric station of Viseu de Sus up to the settlement of Faina, thus on a length of 32 km.

Through the Dispatcher Maramures W.M.S. and the Operational Centre of the Inspectorate for Emergency Situations (O.C.I.E.S.) “Gheorghe Pop de Băseşti” of the county of Maramureș, warnings and forecasts were sent to all the settlements in vulnerable areas. As a result, in the period February 3rd to 5th, 2017, for unlocking/clearing of the watercourses of ice blocks and ice jams, formed in exposed sections referred to in the previous chapters, the Maramureș W.M.S. intervened mechanically (an excavator, a backhoe, a dumper of 24 t, the intervention vehicles, pyrotechnic material) and through qualified personnel (2 mechanics, 5 drivers, 1 pyro-technician).

7.6.5 Variation of the Water Levels on the Rivers, Reported to the Official Defense Levels

During the manifestation of the winter phenomena, the water levels on the major rivers had *variations of small amplitudes* (generally less than 50 cm), specific to winter season, with two different exceptions from the points of view (Fig. 7.21).

The first of exceptions was the Vaser River, which has registered important variations of the water level (over 150 cm) starting from the third decade of the month of December, 2016. The explanation is related to the mountain areas specific features of this watercourse—the gorge area, with a high slope and a higher speed of turbulent flowing, which caused an intense dynamic in the riverbed, with the mobilization of large amounts of ice blocks and supercooled water. The result was their accumulation

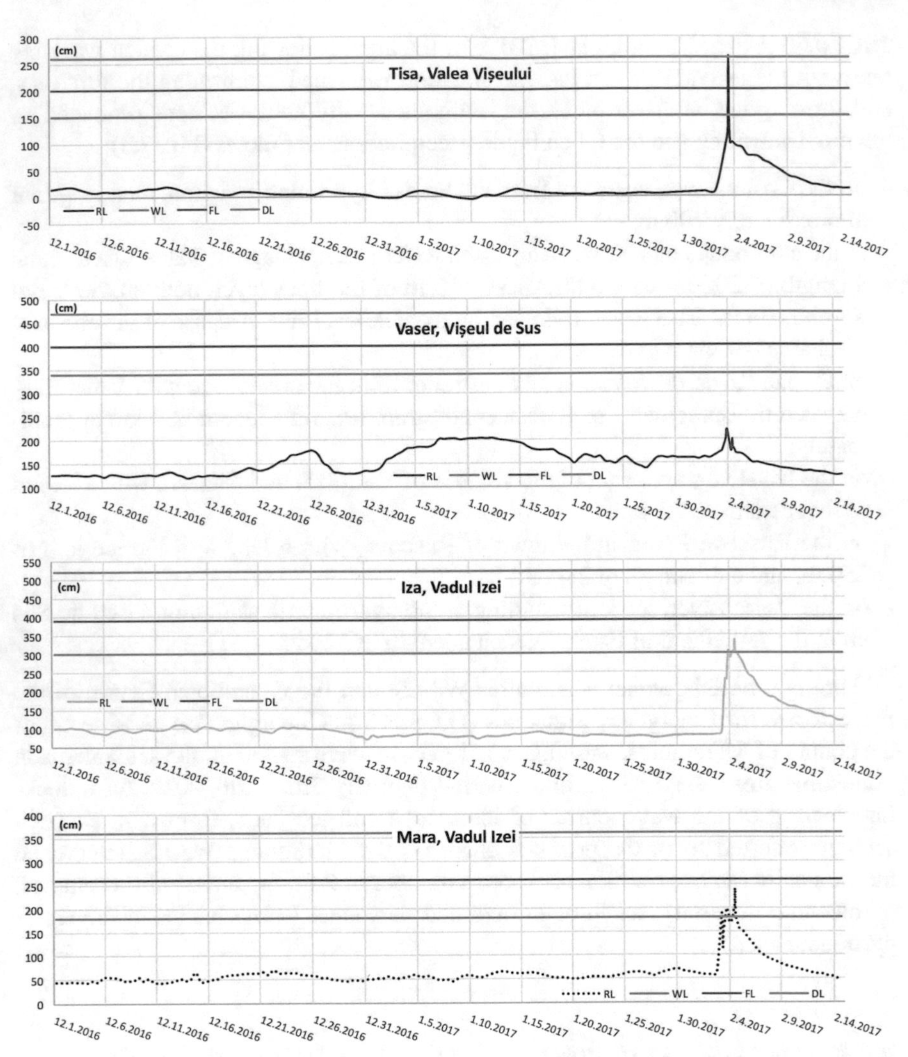

Fig. 7.21 Hydrographs of the measured water levels and the defense levels at the hydrometric stations on the studied rivers. RL—recorded level; WL—warning level, corresponding to yellow code warning; FL—flooding level, corresponding to orange code warning; DL—danger level, corresponding to red code warning (*data source* S.T.B.W.A. 2017)

much earlier than on other rivers and the development of an ice jam of impressive length (32 km).

The second exception refers to the time interval related to February, 2017, when the fast warming of the air occurred, accompanied by liquid precipitation and the watercourses unleash following the emerging flash floods. The amplitude of water levels, between the moment of maximum, reached immediately before the breaking of the ice dams and the moment of pre-warming, was extremely high on the rivers: Iza at the Vadul Izei hydrometric station—271 cm; Tisa at the Valea Vişeului hydrometric station—260 cm; Mara at the Vadul Izei hydrometric station—185 cm; and much lower on the Vaser River at the Vișeul de Sus hydrometric station—60 cm (Fig. 7.21).

Under the effect of watershed conditions—the extreme negative temperatures—the share of water from the slope was reduced to a minimum value, leaving the underground springs to sustain the rivers' flows. Also, an important storage of water in the mass of ice on riverbeds occurred.

However, if we consider these last two ideas, as well as the variation of *winter coefficient of flow* presented above, we find, in fact, that the liquid flow was the hydric parameter most affected by the phenomena, while the water level, influenced by the ice pressure at the hydrometric rod, recorded a different variation and was more affected by the water storages behind of ice dams.

Analyzing Fig. 7.21, we notice, for the rivers Tisa, Iza, and Mara, a very good connection between the small water-level variations from the majority of the studied period and the amplitude of the flash flood from the end of this period. The maximum water level always exceeded the warning level (WL) for two of these rivers, or even the danger level (DL) for the Tisa River, at the Valea Vişeului hydrometric station. On the contrary, where visible variations of the water level were noticeable throughout the period with phenomena, the amplitude of the flash flood from the end of this period was extremely small, failing to reach even the warning level (WL).

7.7 Effects of the Occurred Winter Phenomena

The ice jam effects, in general, of the winter phenomena on rivers, are difficult to manage, taking into account their frequently mechanical character, but, also, of a different type (Fig. 7.22).

The damage can be much severe than in the case of "ordinary" floods, because of the high density of the transported solid bodies (blocks of ice, other dislocated, and carried away objects), which have an amplified hitting power due to their weight associated with the movement velocity.

The mechanical effects are very commonly associated with floods, whereas, generally, behind the ice jams significant amounts of water accumulate and, by the sudden breaking of the ice dams, this quantity of water represents a devastating water front

Fig. 7.22 Mechanical and related effects induced by the agglomerations of ice on households and mobile properties—February, 2017, the Vaser Valley (photos by S.T.W.B.A.—Maramureş W.M.S.)

with solid objects, which move with very high speed and destroy everything in their path.

7.7.1 Effects on the Anthropogenic Environment

This type of effects was one of the most visible, given the human nature and the location of the hearth of localities, the objectives with anthropogenic origin or the properties in the watercourses action area. In our case, the value of damage exceeded 1255,588 lei (275,347 EUR—Table 7.2).

In a relatively poor area, devoid of large investments and major activities, the value of damage quantified in Lei and converted in EUR is quite high, taking account of the local dimension of the manifestation areas and the types of phenomena and hydric events, which occurred.

On the other hand, due to the prompt, concerted, and synchronized action of the R.W.N.A. and N.I.H.W.M. from Bucharest, S.T.W.B.A. from Cluj-Napoca, N.M.A. from Bucharest, and R.M.C. Transilvania Nord from Cluj-Napoca, Dispatcher of Maramures W.M.S. and the "Gheorghe Pop de Băseşti" Operational Centre of the Inspectorate for Emergency Situations (O.C.I.E.S.) of Maramureș county, local authorities (mayors' offices), of performed warnings, forecasts, and interventions, the damage was considerably limited and, most importantly, there were no human victims.

Making a more detailed analysis on each occurrence area and on each affected locality, important differences between them come into prominence (Table 7.3).

Table 7.2 Centralization of the winter phenomena effects on the analyzed rivers (data source S.T.B.W.A. 2017)

Name	Unity	Measure unity	Value	
			Lei	EUR
Bridges	2	–	782,620	171,627
Agrarian terrains	312	ha	312,000	68,420
Household annexes	8	–	55,370	12,143
Houses	10	–	39,368	8633
Street networks	1.1	km	25,500	5592
Decks	2	–	12,930	2836
Fodder deposits	48	t	9600	2105
Agrarian roads	0.8	km	7200	1579
Shores erosion	300	m	6000	1316
Pastures	5	ha	5000	1096
Total	–	–	1,255,588	275,347

Table 7.3 Detailed situation according to localities of the affected objectives types and of the damage caused by the winter phenomena (data source S.T.B.W.A. 2017)

Crt. No.	Locality	Affected objectives					Affected by
		Name	Unity	Measure unity	Value		
					Lei	EUR	
1.	Șieu Village	Agrarian terrains	7	ha	7000	1535	Ice jams, abundant precipitations, disposal of water from the existing snow layer
		Household annexes	1	–	7350	1612	
		Shores erosion	300	m	6000	1316	
	Total	–	–	–	20,350	4463	
2.	Sighetu Marmației Municipality	Bridges	1	–	775,560	170,079	Ice jams, abundant precipitations, disposal of water from the existing snow layer
	Total	–	–	–	775,560	170,079	
3.	Vadu Izei Village	Agrarian terrains	5	ha	5000	1096	Ice jams, abundant precipitations, disposal of water from the existing snow layer
		Pastures	5	ha	5000	1096	
	Total	–	–	–	10,000	2193	

(continued)

Table 7.3 (continued)

Crt. No.	Locality	Affected objectives					Affected by
		Name	Unity	Measure unity	Value		
					Lei	EUR	
4.	Vișeu de Sus City	Houses	5	–	34,200	7500	Ice jams, abundant precipitations, disposal of water from the existing snow layer, leaks from slopes
		Household annexes	7	–	48,020	10,531	
		Decks	2	–	12,930	2836	
		Bridges	1	–	7060	1548	
		Street networks	0.6	km	21,000	4605	
	Total	–	–	–	123,210	27,020	
5.	Oncești Village	Houses	5	–	5168	1133	Ice jams, abundant precipitations, disposal of water from the existing snow layer
		Agrarian terrains	300	ha	300,000	65,789	
		Agrarian roads	0.8	km	7200	1579	
		Street networks	0.5	km	4500	987	
		Fodder deposits	48	t	9600	2105	
	Total	–	–	–	326,468	71,594	
	General Total	–	–	–	1,255,588	275,348	

The highest damage was registered in the Sighetu Marmației Municipality, through the damage of the bridge from the southern entrance in the city, caused by the winter phenomena occurred on the Iza River, after the confluence with Mara River (Fig. 7.8).

Significant damage, reported to the number of inhabitants and to the size of the settlement, registered also the Onceşti village, as a result of the effects of the phenomena from Iza River (300 ha of agricultural land—already quite rare and low productive in the area due both to soil and climate factors).

Also, Vișeul de Sus city has been seriously affected by natural events from the Vaser River, and, in addition, by the mud and ice torrential overflow from the slopes. Here stood out several houses, household annexes, street networks, etc. (Table 7.3).

7.7.2 *Effects on the Natural Environment*

Protected areas' habitats present different flooding reactions, due to their specificity. Periodic floods are critical to maintaining the ecological integrity and biological productivity of the river floodplains (Rasmussen 1996; Poff et al. 1997; Jurajda and Reichard 2006). In the river system, massive floods are the main cause of variability and environmental disruption (Michener and Haeuber 1998; Serban et al. 2012).

An important differentiation occurs between floods with lateral invasion and those of erosive type that have different effects on fish populations. Erosive floods are characterized by a fast-moving and turbulent water with power of drawing and moving the riverbed-related components, often with dramatically destructive impact on the riverbed's natural habitats and riparian areas (Mathews 1998; Jurajda and Reichard 2006; Serban et al. 2016).

An important flood effect is the significant reduction in the abundance of fish species, in some cases, up to extinction, between the pre-flood (before flood) and post-flood (after flood) moments (Jurajda et al. 1998).

Tangent or overlapping the analyzed area, three environmental zones with special status can be identified, namely *The Upper Tisa and the "Pădurea Ronișoara" Protected Area* (http://eeagrants-tisa.ro/rosci-0251-tisa-superioara/), and two parks, namely *The Maramureș Mountains Natural Park* (https://www.muntiimaramuresului.ro/index.php/ro/) and *The Rodna Mountains National Park* (https://www.parcrodna.ro/) (Fig. 7.23).

In the Upper Tisa Natura, 2000 protected areas, from east to west, were identified the following types of fish species and their habitats sensitive to winter river phenomena and floods (Fig. 7.24—underlined species are representative and are found on the standard forms of the studied sites) (Serban et al. 2016):

- in Sect. 7.1, the A Habitat, Vișeu Valley, based on electrofishing, were found *Alburnoides bipunctatus* (spirlin), *Barbus Barbus* (barbel), *Barbatula barbatula* (stone loach); the habitat is one of *Hucho hucho* (huchen), with submerged rocks in the riverbed and depth as well, with river sides where *Rubus vitis idaea* (cran-

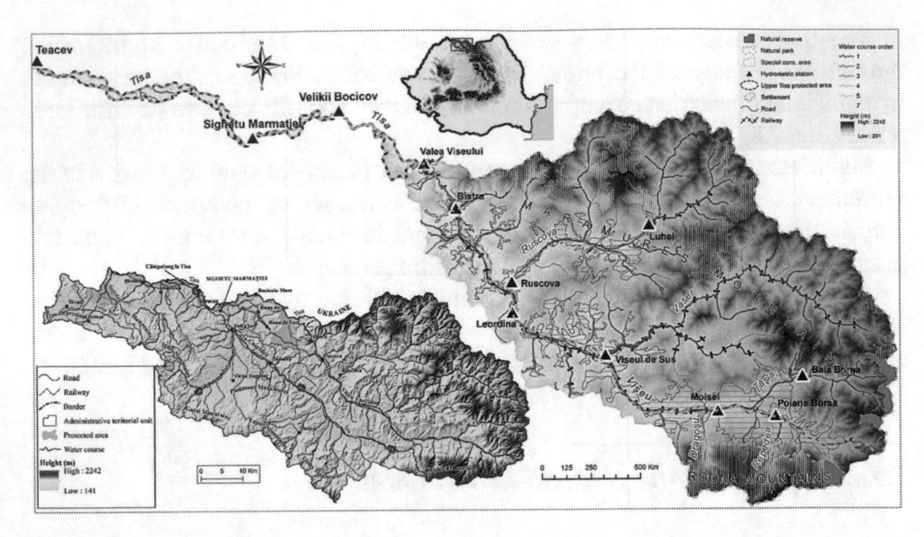

Fig. 7.23 Protected areas from the Romanian Upper Tisa watershed. *Source* Authors

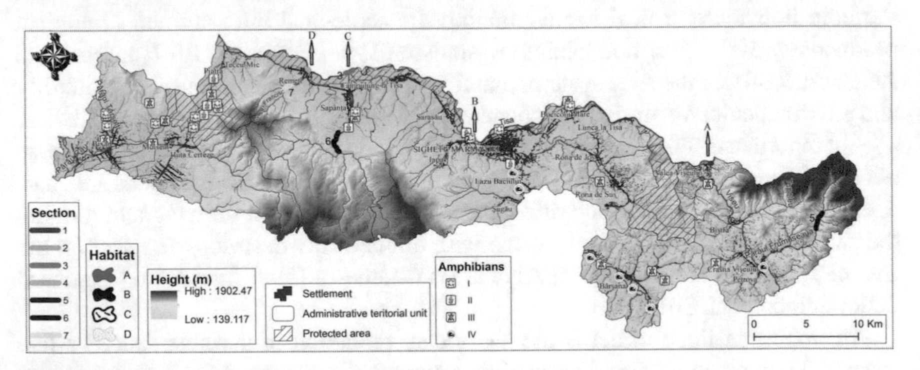

Fig. 7.24 Species and habitats identified in the administrative territorial units of Pricop—Huta-Certeze and Upper Tisa Natura 2000 protected areas vulnerable to floods and winter phenomena (by different sources and own research). **I** Triturus cristatus (Northern crested Newt); **II** Triturus montandoni (Carpathian Newt); **III** Bombina variegata (Yellow-bellied toad); **IV** Lutra lutra (Eurasian otter) (after Serban et al. 2016)

berry), *Salix fragilis* (crack willow), *Alnus incana* (white alder), *Alnus glutinosa* (black Alder) were discovered; the *white-throated dipper* (cinclus cinclus) was also located there (Oprea and Irimia 2015);

- in Sect. 7.2, the B Habitat, downstream of Sighetu Marmaţiei, electrofishing based, *Leuciscus cephalus* (european chub), *A. bipunctatus* (spirlin), *B. Barbus* (barbel), *Cottus poecilopus* (alpine bullhead) were discovered; habitat is one of floodplain forest, with *Populus alba* (white poplar), *Fraxinus augustifolia* (narrow-leaved ash), *Clematis vitalba* (old man's beard), *Humulus lupulus* (hops), *S. fragilis* (crack willow), *Salix alba* (white willow), *Acer negundo* (maple), and heavily herbaceous river sides (Oprea and Irimia 2015);

- in Sect. 7.3, the C Habitat, located in Săpânța village, based on electrofishing, *L. cephalus* (european chub), *Phoxinus phoxinus* (common minnow), *Alburnus alburnus* (bleak), *Barbus peloponnesius petenyi* (mediteranean barble), *Orthrias barbatulus* (stone loach), *Cottus gobio* (european bullhead) were discovered (Harka et al. 1999); the habitat is one of floodplain forest, with *Populus* (poplar), *F.augustifolia* (narrow-leaved ash), *C. vitalba* (old man's beard), *H. lupulus* (hops), *S. fragilis* (crack willow), *S. alba* (white willow), *A. negundo* (maple), with heavily herbaceous river banks;
- in Sect. 7.4, the D Habitat, in the Remeţi village, based on electrofishing, *B. Barbus* (barbel), *L. cephalus* (european chub), *Leuciscus souffia* (western vairone) *A. alburnus* (bleak), *B. barbatula* (stone loach), *P. phoxinus* (common minnow), *Pseudorasbora parva* (stone moroko) were discovered; the habitat is the confluence one, dammed area with floodplain forest, *S. fragilis* (crack willow), *S. alba* (white willow), *P. alba* (poplar), *Robinia pseudoacacia* (black locust), *F. augustifolia* (narrow-leaved ash) (Oprea and Irimia 2015);
- in Sects. 7.5, 7.6, and 7.7, on the Frumușeaua, Valea lui Francisc and Săpânța valleys, upstream from Crasna Vișeului, Săpânța, and Remeţi villages, based on electrofishing, *Eudontomyzon danfordi* (Carpathian brook lamprey), was observed, particularly sensitive to the impact of anthropogenic activities (Telcean and Cupsa 2011).

Regarding the **amphibian populations**, their behavior during flood is less known, although it is obvious that the phenomenon plays a vital role in the abundance and spread of such species (Tockner et al. 2006; Kupferberg et al. 2012; Ocock et al. 2014). ***Amphibians*** are very sensitive to environmental and habitat changes, primarily to changes in rainfall and flow regime, but also temperature, turbidity, water pollution with pesticides or other chemicals, or pollution with plant and animal residues (including sawdust) (Frogs in wetlands 2015). Distribution of frogs and some other amphibians related to the study area is shown in Fig. 7.24.

The existence of the Eurasian otter (Lutra lutra) in a habitat depends largely on the permanence of water and the availability of adequate food resources (Melquist and Hornocker 1983; Kruuk 1995; Prenda et al. 2001). Also, the existence of the Eurasian otter is influenced by the surrounding areas characteristics, water pollution and human-induced disturbance factor (Bas et al. 1984; Delibes et al. 1991; Prenda et al. 2001). Ardelean and Béres (2000) showed: "Our research has proven that Eurasian otter presence starting from Tisa's riverside coppice to mountain streams with a lot trout (Iza, Vișeu, Mara, Săpânța, Rica etc.)".

The winter phenomena occurred in the study area had a lower and limited impact on these three protected areas, taking account the cold season, when the biotic activity in the ecosystems is very low.

Also, the space of the phenomena manifestation is limited to the periphery of the protected areas—just the Vaser Valley with the related phenomena go deep into the area of Maramureş Mountains Natural Park, as well as to the riverside habitats exposed to the ice blocks dynamics and generated swift flash floods (Figs. 7.8 and 7.23).

7.8 Conclusions

The winter phenomena occur in a more or less intense/frequent manner, depending on the factors which influence them. From the experience of decades of observation and gauging in the R.W.N.A. hydrometric network, it is found that on altitude, the maximum frequency of them belongs to the mountain area, the space of flow genesis, characterized by high quantities of rainfall and low average temperatures, especially at its periphery. With a lower frequency, the winter phenomena appear in the hilly area as well, on the background of the agglomeration of floes from the various tributaries.

The upper basin of Tisa mostly overlaps with the historic Basin of Maramures—from which, only a third is located on the Romanian territory. In contrast to the Ukrainian part of the basin, which has a larger opening to the Panonic Basin, the Romanian part of the basin is opened only to the northwest, to west and south being closed by the neogen volcanic chain of the Oaș–Gutâi–Țibleș Mountains. This natural organization enables cold air mass advections from north and important thermic inversions on the lower basin area which determines specific characteristics of winter flow on the two rivers.

Morphological, climatic, and hydric conditions are favorable to the development of the winter phenomena on the watercourses from the Romanian sector of the basin, according to the literature and to observations and measurements made by the responsible institutions.

The watercourses, which recorded the most severe winter phenomena, followed by unusual events, were: the Tisa river collector, its effluent Iza, and with less impact the Viseu river and other tributaries of them as well: Vaser, tributary of Viseu—the most severe case; Mara, tributary of Iza, and Cosău, tributary of Mara. Their position, as well other characteristics and special circumstances under which these watercourses, were generated particularities for each of them, determining the emergence of several critical sectors.

One of the most by the occurred phenomenon targeted sectors and spaces was the confluence ones, the low areas at the connection of the watercourses, where the ice blocks that came from upstream were massively crowded. Another category of sectors with problems was the gorge-type ones, where the alternation between the narrowing sectors and the widening ones created optimal conditions for ice blocks agglomeration. The sinuosity of the watercourses was another element, which favoured the chaotic movement and the agglomeration of the ice blocks, whether of topographic or hydraulic origin. An element, again, extremely important is the aspect of the longitudinal profiles of the analyzed watercourses and their average slope, which combined with other morphological and morphometric elements favoured the congestion and stagnation of ice floes.

The winter phenomena analyzed in the study were generated by persistent negative temperatures (for approximately two months) and by sudden weather changes. The winter of 2016–2017 was dominated by prolonged cold weather; extreme negative temperatures were recorded, starting from December until the end of January. According to the detailed presented synoptic situation, the winter of 2016–2017 was an unusual one for the last decades by its severity and manifestation way of the meteorological parameters.

The records made by S.T.W.A. through the network of hydrometric stations have made possible the construction of a complex winter chart, which allows the conditional explanation of the phenomena occurrence and manifestation which are the subject of analysis of this material.

The thermic values varied between 3 °C the maximum and −19 °C the minimum, and strong negative values were registered for several days in a row: −3… −14°C, in the first period of the analyzed interval and −10… −19 °C in the middle of the interval. To the end of the period, the thermic values rose very much remaining positive for even more than a week (February 1st to 11th, 2017), followed, again, by a shorter period of time of low temperatures: under −5 °C.

The registered quantities of rainfall were not important—a maximum of 14 mm in the week preceding the formation of the ice jams, when the raining predominated. Throughout the entire analyzed period they were, however, mostly in the snow form, which facilitated the accumulation of a substantial layer, especially on slopes, with a subsequent effect on the flow. Important was, nonetheless, the nearly regular frequency, identifying six periods with rainfall at approximately equal time intervals, except for the last decade of January, when they were quasi-absent.

The areas of ice jams formation are densely populated and with a natural landscape influenced by human intervention, including the influence of watercourses landscape through embankment systems, defense improvements, and consolidation of banks overpasses of bridges and footbridges type. To these are added the natural obstacles caused by the sinuosity of rivers, narrowing of the minor and major riverbeds, thresholds, and rapids.

All studied watercourses developed winter phenomena, starting with ice at bank or spongiously ice (possibly other early or transition forms for the advanced phenomena), along with the temporal onset of negative temperatures (Vaser) or three days after it (Mara, Iza, and Tisa). After the second decade of the month of December, the ice gained consistency, the observed thicknesses exceeding 15 cm, in some cases. Besides, on the same temporal coordinates the transition to an evolved form of winter phenomenon was made as well—the ice bridge, which installed relatively quickly on the rivers Iza and Mara, due to the relatively low flow levels and the persistence of the very low temperatures on the hearth of Maramures Basin. In a first phase, the Vaser River has developed a form of transition to the compact ice bridge, respectively, *the bridge of ice with stitches of water*, due to the high speed of water flow induced by a corresponding slope and the numerous rapids, taking into account the mountain pass sector it crosses.

The river Tisa kept, throughout the period of study, the form of ice from bank, due to the constant flow, the ice with thicknesses around the value 5 cm, in the month of December and 10 cm in the second part of the interval.

The flow regime was disrupted, leading to a flow influenced even to 80% on the rivers Vaser and Mara and 30% on the river Iza. If in the first five pentads the coefficient of winter had oscillatory evolution, starting with the sixth pentad he remained at very low levels on the rivers Vaser and Mara and continued to fluctuate on the river Iza. This evolution continued until after the time of the period of rapid warming and liquid rainfall start, after which the coefficient climbed, again, to unit, as in the period before the production of winter phenomena, in November, 2016.

Following the sudden warming and with the arrival of the rains, the situation worsened, confirming the forecasts, and forming ice jam/river blockades, combined with the flash floods produced by the rapid warming and the fallen liquid precipitation: upstream from the Feresti hydrometric station, on a length of approximately 200 m; at the confluence of Tisa with the Viseu River in the village of Valea Viseului, on a length of 500 m, with a thickness of 3 m of the floes layer, both on the Viseu River as on the Tisa River, followed by a quick level rise, because of the constant discharges of the two rivers, on the Iza River, downstream and upstream from the confluence with Valea Băii, between the settlements of Barsana and Stramtura, on a length of about 1 km; on the Iza River, upstream of the Vadu Izei and Nănești villages, with a ice jam length of 7.5 km; on the Ruscova River, in the Ruscova locality, with a ice jam length of 250 m and a thickness of 2.45 m; on the Vaser River with ice blocks agglomerations, starting from Vișeu de Sus hydrometric station up to the settlement of Faina, thus on a length of 32 km.

During the manifestation of the winter phenomena, the water levels on the major rivers had *variations of small amplitudes* (generally less than 50 cm), specific to winter season, with two different exceptions from the points of view. The first of exceptions was the Vaser River, which has registered important variations of the water level (over 150 cm) starting from the third decade of the month of December 2016. The second exception refers to the time interval related to February 2017, when the fast warming of the air occurred, accompanied by liquid precipitation and the watercourses unleash following the emerging flash floods. The amplitude of water levels, between the moment of maximum, reached immediately before the breaking of the ice dams and the moment of pre-warming, was extremely high on the rivers: Iza at the Vadul Izei hydrometric station—271 cm; Tisa at the Valea Vişeului hydrometric station—260 cm; Mara at the Vadul Izei hydrometric station—185 cm; and much lower on the Vaser River at the Vișeul de Sus hydrometric station—60 cm. The maximum water level always exceeded the warning level (WL), or even the danger level (DL) at the Tisza River, Valea Vişeului hydrometric station, with the exception of the Vaser River, where not even warning level was reached (WL).

The mechanical effects are very commonly associated with floods, whereas, generally, behind the ice jams significant amounts of water accumulate and, by the sudden breaking of the ice dams, this quantity of water represents a devastating water front with solid objects, which move with very high speed and destroy everything in their path. The effects on the anthropogenic environment were one of the most visible, given the human nature and the location of the hearth of localities, the objectives with anthropogenic origin or the properties in the watercourses action area. In terms of value, they exceeded 1255,588 lei (275,347 EUR). Thanks to the prompt, concerted, and synchronized work of all involved authorities and institutions, the damage was much limited and, most importantly, there were no human victims. Both fish and batrachian populations, as well as the amphibious mammal identified in the study area, are vulnerable to winter phenomena (flood interruption, ice blocks transport, etc.) and floods (high turbidity effect, reduction of dissolved oxygen quantity, habitats destruction, etc.). Their habitats are affected, in large measure, by the phenomena, excepting some of the categories (Batrachia), for which the water surfaces remaining after floods at the end of spring season are favorable for hatching and breeding. Also, the space of the phenomena manifestation is limited to the periphery of the protected areas—just the Vaser valley with the related phenomena go deep into the area of Maramureş Mountains Natural Park, as well as to the riverside habitats exposed to the ice blocks dynamics and generated swift flash floods.

Acknowledgements We want to thank the Administration of the "Somes-Tisa" Water Basin, Cluj (S.T.W.B.A.), and Maramureș W.M.S. for the provided data and for the pictures captured during the extreme winter events in the analyzed basin. Thank you also all those who have offered or will offer suggestions for the improvement of the present paper.

References

Ardelean G, Béreş I (2000) The vertebrate fauna of Maramureş. The Universitaria Collection, Edit. Dacia, Cluj-Napoca, 378 p

Ashton GD (1986) River and lake ice engineering. Water Resources Publications, Littleton, Colorado, USA, 485 p

Bas N, Jenkins D, Rothery P (1984) Ecology of otters in Northern Scotland. V. The distribution of otter (Lutra lutra) faeces in relation to bankside vegetation in the river Dee in summer 1981. J Appl Ecol 21:507–513

Bashta AT, Potish L (2007) Mammals of the Transcarpathian region (Ukraine), Lviv, 260 p

Bates R.E., Billelo M.A. (1966) Defining the cold regions of the Northern Hemisphere. Cold Regions Research and Engineering Laboratory, Technical Report, Nr. 178

Beltaos S (1990) Fracture and breakup of river ice cover. Can J Civ Eng 17(2):173–183

Beltaos S (1993) Numerical computation of river ice jams. Can J Civ Eng 20(1):88–99

Beltaos S (2007) Hydro-climatic impacts on the ice cover of the lower Peace River. Hydrol Process 19

Beltaos S (2008) Progress in the study and management of river ice jams. Cold Reg Sci Technol 51:2–19

Beltaos S (2016) Extreme sediment pulses during ice breakup, Saint John River, Canada. Cold Reg Sci Technol 128(2016):38–46

Beltaos S, Burrell BC (2015) Hydroclimatic aspects of ice jam flooding near Perth-Andover, New Brunswick, Canadian. J Civ Eng 42(9) Special Issue: SI Pages: 686–695

Beltaos S, Burrell BC (2016a) Transport of suspended sediment during the breakup of the ice cover, Saint John River, Canada. Cold Reg Sci Technol 129:1–13

Beltaos S, Burrell BC (2016b) Characteristics of suspended sediment and metal transport during ice breakup, Saint John River, Canada. Cold Reg Sci Technol 123:164–176

Beltaos S, Prowse TD (2001) Climate impacts on extreme ice-jam events in Canadian rivers. Hydrol Sci J 46(1):157–181. https://doi.org/10.1080/02626660109492807

Boar N (2005) The Romanian-Ukrainian cross-border region of Maramureş, Edit. Presa Universitară Clujeană, 294 p

Boivin M, Buffin-Belanger T, Piegay H (2017) Interannual kinetics (2010–2013) of large wood in a river corridor exposed to a 50-year flood event and fluvial ice dynamics 3rd International Conference on Wood in World Rivers, Padova, Italy & Geomorphology, vol 279, Special Issue: SI Pages: 59–73

Chiş VT, Kosinszki S (2011) Geographical introductory characterization of the Upper Tisa River Basin (Romania-Ukraine), "The Upper Tisa River Basin". Transylvanian Rev Systematical Ecol Res 11:1–4
Ciaglic V (1965) The evolution of the freezing phenomenon on river Bistricioara in the winter of 1963–1964 (in Romanian). Hidrotehnica 10(2):92–101
Cocuţ M (2008) The characteristics of water flow in the Basin of Maramureş and in the limitrophe mountain zone. Doctoral thesis—manuscript. Babeş-BolyaiUniversity, College of Geography, Cluj-Napoca, 115 p
Cojoc G, Romanescu G, Tîrnovan A (2015) Exceptional floods on a developed river. Case study for the Bistrita River from the Eastern Carpathians (Romania). Nat Hazards 77(3):1421–1451
Colceriu R (2002) Ice jams, risk factors in the upper hydrographic basin of the Mures river (in Romanian), Edit. "Dimitrie Cantemir" Tg. Mureş, 154 p
Colceriu R (2003) The study of freezing phenomena in the upper course of Mures river (between the source and Tg. Mureş) (in Romanian). Doctoral thesis, The Geography Institute of the Bucharest Academy, manuscript
Colceriu R (2010) The strategy and the outlook against floods in the Mures hydrographic basin space (in Romanian). From "The water resources of Romania—vulnerability at anthropic pressures". In: Gâştescu P, Breţcan P (eds) The first Limnogeography national symposium works, 11–13 June, Valahia Univeristy, Târgovişte, Edit. Transversal, pp 175–183
Constantinescu C (1964) The factors that treats the existence and the duration of winter phenomena on Danube on the downstream sector from Tr. Severin (in Romanian). Meteorologia, Hidrologia şi Gospodărirea Apelor, 1
Daly SF (2009) Investigation of changes in conveyance of the St. Clair River over time using a state-space Model. US Army Engineering Research and Development Center, Cold Regions Research and Engineering Laboratory, ERDC/CRREL, Hanover, NH
Delibes M, Macdonald SM, Mason CF (1991) Seasonal marking, habitat and organochlorine contamination in otters (Lutra lutra): a comparison between catchments in Andalucía and Wales. Mammalia 55:567–578
Derecki JA, Quinn FH (1986) Record St. Clair River Ice Jam of 1984. J Hydraul Eng ASCE 112(12):1182–1194
Găman C (2014) Considerations on recent freezing phenomena on Bistriţa and Bistricioara rivers. PESD 8(2):225–242. De Gruyter Open, https://doi.org/10.2478/pesd-2014-0037
Gholamreza-Kashi S (2016) A forecasting methodology for predicting frazil ice flooding along urban streams using hydro-meteorological data. Can J Civ Eng 43(8):716–723
Giurma I, Stefanache D (2010) Winter phenomena on the Bistrita river between hazard and vulnerability (in Romanian). Paper N.I.H.W.M. National Institute of Hydrology and Water Management—Jubilee Conference, "Gheorghe Asachi" Technical University, Iaşi
Harka A, Bănărescu PM, Telcean I (1999) Fish fauna of the Upper Tisa, în Upper Tisa Valley, Tiscia monograph series, Szolnok-Szeged-Târgu Mures, pp 439–454
Huang Y, Sun J, Li W (2016) Experimental observations of the flexural failure process of snow covered ice. Cold Reg Sci Technol 129:14–30
Ichim I, Rădoane M (1986) The effects of dams in the landform dynamic. Geomorfological approach (in Romanian), Edit. Academiei, Bucureşti
Ichim I, Bătucă D, Rădoane M, Duma D (1989) The morphologic and the dynamic of river beds (in Romanian), Edit. Tehnică, Bucureşti
Jasek M (2003) Ice jam release surges, ice runs, and breaking fronts: Field measurements, physical descriptions, and research needs. Can J Civ Eng 30(1):113–127
Jurajda P, Reichard M (2006) Immediate impact of an extensive summer flood on the adult fish assemblage of a channelized Lowland River. J Freshw Ecol 21(3):493–501
Jurajda P, Hohausova E, Gclnar M (1998) Seasonal dynamics of fish abundance below a migration barrier in the lower regulated River Morava. Folia Zool 47:215–223
Koegel M, Das A, Marszelewski W et al (2017) Feasibility study for forecasting ice jams along the river Oder. Wasserwirtschaft 107(5):20–28

Kolerski T (2014) Modeling of ice phenomena in the mouth of the Vistula River. Acta Geophys 62(4):893–914

Kolerski T, Shen HT (2010) St. Clair River ice jam dynamics and possible effect on bed changes. In: Proceedings of the 20th I.A.H.R. International Symposium on Ice, 14–18 June 2010, Lahti, Finland

Kolerski T, Shen HT (2015) Possible effects of the 1984 St. Clair River ice jam on bed changes. Can J Civ Eng 42(9):696–703

Korytny LM, Kichigina NV (2006) Geographical analysis of river floods and their causes in southern East Siberia. Hydrol Sci J 51:3

Kowalczyk T, Hicks F (2003) Observations of dynamic ice jam release waves on the Athabasca River near Fort McMurray. Can J Civ Eng 34:473–484

Kraatz S, Khanbilvardi R, Romanov P (2017) A comparison of MODIS/VIIRS cloud masks over ice-bearing river: on achieving consistent cloud masking and improved river ice mapping. Remote Sens 9(3):229

Kruuk H (1995) Wild Otters. Predation and populations. Oxford University Press, Oxford

Kupferberg SJ, Palen WJ, Lind AJ et al (2012) Effects of flow regimes altered by dams on survival, population declines, and range-wide losses of California river-breeding frogs. Conserv Biol 26:513–524

Lagadec A, Boucher E, Germain D (2015) Tree ring analysis of hydro-climatic thresholds that trigger ice jams on the Mistassini River. Que Hydrol Process 29(23):4880–4890

Leopold LB, Maddock T (1953) The hydraulic geometry of stream channels and some physiographic implications, U.S. Geological Survey Professional Paper, 275 p

Leopold LB, Wolman MG, Miller JP (1964) Fluvial processes in geomorphology. W. Freeman and Co., San Francisco, p 522

Lindenschmidt KE, Das A, Rokaya P et al (2016) Ice-jam flood risk assessment and mapping. Hydrol Process 30(21):3754–3769

Liu L, Shen HT (2004) Dynamics of ice jam release surges. In: Proceedings of the 17th IAHR International Symposium on Ice, St. Petersburg, pp 244–250

Lu S, Shen HT, Crissman RD (1999) Numerical study of ice dynamics in upper Niagara River. J Cold Reg Eng ASCE 13(2):78–102

Lucie C, Nowroozpour A, Ettema R (2017) Ice jams in straight and sinuous channels: insights from small flumes. J Cold Reg Eng 31(3):04017006

Lucza Z (2017) This winter through the eyes of the hydrologist. FELSŐ-TISZA HIRADÓ, LVI. évfolyam 01.szám, pp 20–22 (in Hungarian) (https://www.fetivizig.hu/WEB/FETIKOVIZIG/VIZIGINFO_Start.nsf/Node.xsp?documentId=A7D2D748C29962FEC12580FA00315B18&action=editDocument)

Lukianets O, Obodovskyi I (2015) Spatial, temporal and forecast evaluation of rivers' streamflow of the Drainage Basin of the upper tisa under the conditions of climate change. Environ Res Eng Manage 71(1):36–46

Matthews VJ (1998) Patterns in freshwater fish ecology. Chapmann Hall, New York

Melquist WE, Hornocker MG (1983) Ecology of river otters in West Central Idaho. Wildl Monogr 83:1–60

Michener VK, Haeuber RA (1998) Flooding: natural and managed disturbances. Bioscience 48:677–680

Miţă P (1977) The freezing and the thermic regime of water courses in Romania (in Romanian). Doctoral thesis—manuscript, University of Bucharest

Morse B, Hicks F (2005) Advances in river ice hydrology 1999–2003. Hydrol Process 19(1):247–263. https://doi.org/10.1002/hyp

Mustăţea A (1996) Exceptional floods on Romanian territory. Genesis and effects (in Romanian), Edit. INHGA, Bucharest, 376 p

Nafziger J, She Y, Hicks F (2016) Celerities of waves and ice runs from ice jam releases. Cold Reg Sci Technol 123:71–80

Ocock J, Kingsford RT, Penman TD, Rowley JJL (2014) Frogs during the flood: Differential behaviours of two amphibian species in a dryland floodplain wetland. Austral Ecol 39:929–940
Oprea E, Irimia D (2015) Monitoring the conservation status of species and habitats in Romania on the basis of Article 17 of the Habitats Directive, a Project financed through SOP Environment, 2007–2013
Paparyha P, Pipash Ludmyla, Shmilo V, Veklyuk Anatoly (2011) Hydrochemical Status of Streams and Rivers of the Upper Tysa River Basin in the Ukrainian Carpathians. "The Upper Tisa River Basin". Transylvanian Rev Systematical Ecol Res 11:47–52
Păvăleanu I (2003) The ice jam phenomenon on Bistrita river, upstream of Izvoru Muntelui (in Romanian), Tehnical University "Gheorghe Asachi". Faculty of Hydrotehnics—manuscript, Iaşi
Pawlowski B (2016) Internal structure and sources of selected ice jams on the lower Vistula River. Hydrol Process 30(24):4543–4555
Poff XL, Allan JD, Bain MB, Karr JR, Prestegaard KL, Richter BD, Sparks RE, Stromberg JC (1997) The natural flow regime. Bioscience 47:769–784
Pop PG (2000) The Carpathians and sub-Carpathians of Romania, Edit. Presa Universitară, Cluj-Napoca, 264 p
Posea G, Moldovan C, Posea A (1980) County of Maramureş. Romanian Academy's Publishing, Bucureşti, p 179
Prenda J, López-Nieves P, Bravo R (2001) Conservation of otter (Lutra lutra) in a Mediterranean area: the importance of habitat quality and temporal variation in water availability. Aquatic Conserv Mar Freshw Ecosyst 11:343–355. https://doi.org/10.1002/aqc.454
Prowse TD, Beltaos S (2002) Climatic control of river-ice hydrology: a review. Hydrol Process 16:805–822
Prowse TD, Bonsal BR (2004) Historical trends in river-ice break-up: a review. Nord Hydrol 35(4–5):281–293
Prowse TD, Conly FM (1998) Effects of climatic variability and flow regulation on ice-jam flooding of a northern delta. Hydrol Process 12:1589–1610
Rădoane M (2004) Relief dinamics în the area of Izvoru Muntelui Lake (in Romanian), Edit. Universităţii Suceava, Suceava
Rădoane M, Ciaglic V, Rădoane N (2008) Researches on the ice jam formation in the upstream of Izvoru Muntelui reservoir. Analele Universității Ștefan cel Mare, Suceava, Secțiunea Geografie XVII:45–58
Rădoane M, Ciaglic V, Rădoane N (2010) Hydropower impact on the ice jam formation on the upper Bistriţa River, Romania. Cold Reg Sci Technol J 60(3):193–204
Rasmussen JL (1996) Floodplain management. Fisheries 21:6–10
Rhodes DD (1977) The b-f-m diagrame: graphical representation and interpretation of at-a-station hydraulic geometry. Am J Sci 277
Robb DM, Gaskin SJ, Marongiu J-C (2016) SPH-DEM model for free-surface flows containing solids applied to river ice jams. J Hydraul Res 54(1):27–40
Romanescu G (2003) Floods, between natural and accidental (in Romanian). The "Riscuri şi catastrofe" magazine, No. II, Edit. Casa Cărţii de Ştiință, Cluj-Napoca, pp 130–138
Romanescu G (2005) The upstream floods risk of Izvorul Muntelui Lake and the immediate effect of geomorfological characters of the river bed (in Romanian). The "Riscuri şi catastrofe" magazine, Edit. Casa Cărţii de Ştiință, Cluj-Napoca, pp 117–124
Romanescu G (2015) Water management. Hydrotechnical improvement of hydrographic basins and the wet zones (in Romanian), Edit. Terra Nostra, Iași, 324 p
Romanescu G and Bounegru O (2012) Ice dams and backwaters as hydrological risk phenomena—case study: the Bistrita River, upstream of the Izvorul Muntelui Lake (Romania). Flood Recovery Innovation and Response III, WIT Transactions on Ecology and The Environment, vol 159, WIT Press, pp 167–178, https://doi.org/10.2495/friar120141
Romanescu G, Stoleriu C (2017) Exceptional floods in the Prut basin, Romania, in the context of heavy rains in the summer of 2010. Nat Hazards Earth Syst Sci 17:381–396

Romanescu G, Cojoc GM, Sandu IG, Tîrnovan A, Dăscăliţa D, Sandu I (2015) Pollution sources and water quality in the Bistrita Catchment (Eastern Carpathians). Chem Mag 66(6):855–863

Sabău D, Bătinaş R, Roşu I, Şerban G (2017) Fresh Water Resources in the Natura 2000 Pricop-Huta Certeze and Tisa Superioară Protected Areas. "Air and Water—Components of the environment" Conference Proceedings. In: Şerban G, Croitoru A, Tudose T, Bătinaş R, Horvath CS, Holobâcă I(eds) Babeş-Bolyai University, Faculty of Geography, Cluj-Napoca, România, Edit. Casa Cărţii de Ştiinţă, pp 166–175

Semenescu M (1960) The freezing phenomenon in the sector of Porţile de Fier (in Romanian). Meteorologia Hidrologia şi Gospodărirea Apelor 4

Şerban G, Pandi G, Sima A (2012) The need for reservoir improvement in Vişeu river basin, with minimal impact on protected areas. In: Order to prevent flooding. Studia University "Babeş-Bolyai", Geographia, LVII, nr. 1, Cluj-Napoca, pp 71–80

Şerban G, Sabău A, Rafan S, Corpade C, Niţoaia A, Ponciş R (2016) Risks Induced by maximum flow with 1% probability and their effect on several species and habitats in Pricop-Huta-Certeze and Upper Tisa Natura 2000 Protected Areas. "Air and Water—Components of the environment" Conference Proceedings. In: Şerban G, Bătinaş R, Croitoru A, Holobâcă I, Horvath C, Tudose T (eds) Babeş-Bolyai University, Faculty of Geography, Cluj-Napoca, România, Edit. Casa Cărţii de Ştiinţă, pp 58–69

She Y, Hicks F (2006) Modeling ice jam release wave with consideration for ice effects. Cold Reg Sci Technol 20:137–147

Shen HT, Liu L (2003) Shokotsu River ice jam formation. Cold Reg Sci Technol J 37:35–49

Shen HT (2016) River ice processes. In: Wang LK, Yang CT, Wang MHS (eds) Handbook of environmental engineering, vol 16. Advances in Water Resources Management, pp 483–530

Shen HT, Gao L, Kolerski T, Liu L (2008) Dynamics of ice jam formation and release. J Coast Res S52:25–32

Shen HT, Su J, Liu L (2000) SPH simulation of River Ice dynamics. J Comput Phys 165(2):752–770

Sorocovschi V, Şerban G (2012) Elements of climatology and hydrology. Part II—Hydrology. ID Education form. Edit. Casa Cărţii de Ştiinţă, Cluj-Napoca, 242 p

Sorocovschi V, Şerban G, Bătinaş R (2002) Hydric risks in the lower basin of Aries river (in Romanian). "Riscuri şi catastrofe" magazine, Edit. Casa Cărţii de Ştiinţă, vol I, Cluj-Napoca, pp 143–148

Ştefănache D (2007a) Studies of the evolution of some dangerous hydrological phenomena (in Romanian). Doctoral thesis, Tehnical University "Gheorghe Asachi", Iaşi—manuscript

Ştefănache D (2007b) Le phénomène d"hiver sur les rivières de la Roumanie -L"évolution des barrages de glace du basin hydrographique supérieur de la rivière de Bistriţa, 14th Workshop on the Hydraulics of Ice Covered Rivers, Québec

Surdeanu V, Berindean N, Olariu P (2005) The natural and antrophic factors that leads to ice jams in the upper basin of Bistrita river (in Romanian). "Riscuri şi catastrofe" magazine, vol IV, no 2, Cluj-Napoca, pp 125–134

Telcean I, Cupşa Diana (2011) The occurrence of carpathian brook lamprey *Eudontomyzon danfordi* regan 1911 (petromyzontes, petromyzontidae) in the upper tisa tributaries from northern romania. Pisces hungarici 5:123–128

Thériault I, Saucet J-P, Taha W (2010) Validation of the mike-ice model simulating river flows in presence of ice and forecast of changes to the ice regime of the Romaine river due to hydroelectric project. In: Proceedings of the 20th IAHR International Symposium on Ice, Lahti, Finland

Tockner K, Klaus I, Baumgartner C, Ward JV (2006) Amphibian diversity and nestedness in a dynamic floodplain river (Tagliamento, NE-Italy). Hydrobiologia 565:121–133

Ujvári I (1972) The geography of Romanian waters, Edit. Ştiinţifică, Bucureşti, 578 p

USACE (1984) April 1984 Ice Jam Report; St. Clair River. Department of the Army, Corps of Engineers, Detroit District, Great Lake, Hydraulics and Hydrology Branch, Detroit, MI

Wang J, Hua J, Sui J et al (2016) The impact of bridge pier on ice jam evolution—an experimental study. J Hydrol Hydromechanics 64(1):75–82

Wang J, Shi F-Y, Chen P-P et al (2015a) Impact of bridge pier on the stability of ice jam. J Hydrodyn 27(6):865–871

Wang J, Shi F-Y, Chen P-P et al (2015b) Impacts of bridge piers on the initiation of ice cover—an experimental study. J Hydrol Hydromechanics 63(4):327–333

Wang J, Shi F-Y, Chen P-P, Wu P, Sui J (2014) Simulations of ice jam thickness distribution in the transverse direction. J Hydrodyn Ser B 26(5):762–769

White KD, Eames HJ (1999) CRREL Ice Jam Database, USA CRREL Report 99-2, CRREL, Hanover, NH. www.dtic.mil/get-tr-doc/pdf?AD=ADA362147

White KD, Tuthill AM, Vuyovich CM, Weyrick PB (2007) Observed climate variability impacts and river ice in the United States. CGU HS Committee on River Ice Processes and the Environment, 14th Workshop on the Hydraulics of Ice Covered Rivers, June 20–22, Quebec City, Quebec, Canada, 1–11 pp

Yatsyk AV, Byshovets LB et al (1991) Small rivers of Ukraine: Manual. In: Yatsyk AV (ed) K: Urozhai Publisher, 296 p (in Ukrainian)

Zare S, Moore S, Rennie CD et al (2016) Boundary shear stress in an ice-covered river during breakup. J Hydraul Eng 142(4):04015065

Zeleňáková M, Zvijáková L (2017) Using risk analysis for flood protection assessment. Springer International Publishing, Springer, 140 p, https://doi.org/10.1007/978-3-319-52150-3

* * * (1992) The atlas of water cadastre of Romania, Ministry of Environment and Aquaproject S.A., Bucureşti, 683 p

* * * (2015) Frogs in wetlands, http://www.environment.nsw.gov.au/wetlands/WetlandFrogs.htm

* * * (1996) Instructions of organization and the program of the hydrometric network's activity on rivers (in Romanian), N.I.M.H., Bucharest

* * * (2009) Sub-Basin Level Flood Action Plan Tisza River Basin, International Commission for the Protection of the Danube River—Flood Protection Expert Group, Hungary, Romania, Slovakia, Serbia, Ukraine

* * * (2013) Best Practice for Building and Working Safely on Ice Covers in Alberta. Government of Alberta, Occupational Health and Safety Contact Centre, http://work.alberta.ca/occupational-health-safety/274.html

* * * (2013) Guide for the activity of hydrometric stations on rivers, N.I.H.G.A., Bucharest

* * * Cold Regions Research and Engineering Laboratory (CRREL) of the U.S. Army Corps of Engineers, http://www.erdc.usace.army.mil/Locations/CRREL/

* * * Photo source 1: https://map.viamichelin.com/map/carte?map=viamichelin&z=4&lat=50.45043&lon=30.52449&width=550&height=382&format=png&version=latest&layer=background&debug_pattern=.*;

* * * Photo source 2: https://upload.wikimedia.org/wikipedia/commons/9/95/Romania_Ukraine_Locator.png

* * * Records of S.T.B.W.A. - M.W.M.S. (Maramureș Water Management System)

* * * Records of S.T.B.W.A. ("Someș-Tisa" Basin Water Administration, Cluj)

* * * (2016) Microsoft Office Home and Student

* * * (2008) The Climate of Romania. National Administration of Meteorology, Edit. of Romanian Academy, Bucharest

Chapter 8
Relation Between Air Temperature and Inland Surface Water Temperature During Climate Change (1961–2014): Case Study of the Polish Lowland

Włodzimierz Marszelewski and Bożena Pius

Abstract The chapter describes the relations between changes in mean annual and mean monthly (1961–2014) air and water temperatures in rivers and lakes, as well as their spatial variability in the Polish Lowland. The determination of the directions and rate of changes in air and water temperatures involved the application of linear regression. Its significance was verified by means of F test at a level of 0.05. The rate of increase in air temperature in the western part of the Polish Lowland was determined to average 0.32 °C·10 $years^{-1}$ and decrease towards its eastern part to 0.22 °C·10 $years^{-1}$. In the case of rivers, the mean increase in water temperature varied from 0.26 °C·10 $years^{-1}$ in the western part to 0.20 °C·10 $years^{-1}$ in the eastern part, and for lake waters, from 0.30 to 0.15 °C·10 $years^{-1}$. Increase in water temperature in particular months throughout the analysed period was much more variable. Its highest increase occurred in May and ranged in lakes from 3.1 to 3.9 °C, and in rivers from 1.9 to 2.6 °C. The lowest increases in water temperature (usually from 0.1 to 1.0 °C) were observed in autumn and winter months, whereas water temperature in some lakes showed a negative tendency in winter months. In the case of Lake Hańcza, in November it decreased during the analysed period by 1.3 °C. During longer periods such as a year or half-year, an evident increase was found to occur in air and inland water temperature. During shorter month-long periods, the increase is not always so evident, and sometimes even a decrease in temperature is observed.

Keywords Thermal regime · Changes · Rivers · Lakes

W. Marszelewski (✉) · B. Pius
Faculty of Earth Sciences, Department of Hydrology and Water Management, Nicolaus Copernicus University in Toruń, Ul. Lwowska 1, 87-100, Toruń, Poland
e-mail: marszel@umk.pl

B. Pius
e-mail: bpius@umk.pl

M. Zelenakova (ed.), *Water Management and the Environment: Case Studies*, Water Science and Technology Library 86,
https://doi.org/10.1007/978-3-319-79014-5_8

8.1 Introduction

Research on the relations between air and inland water temperature has a long history. It commenced already in the nineteenth century along with the popularisation of instrumental methods of hydrometeorological measurements. Due to short measurement series and a low number of sites of observations, however, the development of this kind of research occurred at the turn of the nineteenth and twentieth centuries. In the first half of the twentieth century, the investigations particularly concerned the natural relations between air and water temperature with consideration of their physical properties. This is exemplified by documenting different relations between mean monthly air and water temperatures in lakes during heat accumulation and during stagnation and heat loss (McCombie 1959). In the second half of the twentieth century, and particularly in the early twenty-first century, a rapid increase in interest in the issue occurred. This was and is related to climate warming and its increasingly visible effects in all types of inland waters.

Studies concerning changes in inland water temperature can be generally divided into two main groups. The first one includes publications discussing the rate of water temperature fluctuations and relations with air temperature (Gu and Li 2002; Dąbrowski et al. 2004; Pekarova et al. 2008; Mohseni and Stefan 1999; Jurgelėnaitė et al. 2012; North et al. 2013). The second group includes publications concerning effects of water temperature fluctuations on lake and river ecosystems with particular consideration of trophy and ichthyofauna (Caissie 2006; Hari et al. 2006; Davies 2010; Hudon et al. 2010; Leuven et al. 2011; Dugdale et al. 2013), as well as the effect of natural and anthropogenic factors on the modification of inland water temperature (among others Sridhar et al. 2004; Steel and Lange 2007; Webb et al. 2008; Xin and Kinouchi 2013). Particularly, interesting papers are those prepared based on results of observations for long measurement series (approximately 100 years), among others for the Danube River (Webb and Nobilis 1995), the Hudson and Potomac Rivers (Kaushal et al. 2010), as well as papers from the territory of Lithuania and Latvia, neighbouring with the Polish Lowland, and similar in terms of genesis (Jurgelėnaitė et al. 2012; Latkovska and Apsīte 2016). Currently, the development of measurement methods permits analysing increasingly longer observation series for different meteorological and hydrological elements. This allows for the application of modelling for simulating daily temperature fluctuations, and determining trends, among others of water temperature in lakes in periods even longer than 100 years (Magee et al. 2016). Such studies also permit the determination of the effect of other factors on the temperature of inland waters (among others wind speed and water transparency).

A review of the current knowledge on various effects of climate warming on European inland waters was presented by, among others, Dokulil (2014). The latest study results on water temperature increase in lakes around the globe and in Central Europe are included in papers by O'Reilly et al. (2015) and Woolway et al. (2016, 2017).

The majority of the aforementioned papers discuss the issues of temperature increase based on mean annual or mean seasonal values (e.g. from the warm half-year), as well as the spatial variability of the phenomenon. The primary objective of this chapter is the determination of the relations between changes in air and inland water temperature based on mean monthly values from the period 1961 to 2014, and their variability. The issue is discussed in the context of mean annual and seasonal values. Elements modifying water temperature fluctuations are also presented.

8.2 Study Methods

Calculations and analyses were performed based on data collected in the period 1961–2014 by the Institute of Meteorology and Water Management—National Research Institute. Data concerning air temperature were obtained from the following meteorological stations: Gorzów Wielkopolski, Chojnice, Toruń, and Białystok. Data on water temperature in rivers were obtained from the Oder River (in Gozdowice), Vistula River (in Toruń), and Biebrza River (in Burzyn), and data on temperature of lake waters from the following lakes: Sławskie, Charzykowskie, and Hańcza. The meteorological and hydrological observations were performed every day in the years 1961–2014 at 6 AM GMT: first at an altitude of 200 cm above ground level and then at a depth of 40 cm.

The study applied standard methods of determination of the direction and rate of air and water temperature fluctuations. Linear regression was used. Its significance was verified by means of F test at a level of 0.05. For the purpose of confirming or excluding the calculated statistics (trends) from linear regression, nonparametric Mann–Kendall test was applied (Kundzewicz and Robson 2000). The test requires no preliminary assumptions, and the estimation of significance is based on the zero hypothesis of independent observations with identical distribution. Detection of significant trends in time series by means of two tests reduces the probability of making an error of the first type, i.e. detection of changes in the case of their lack. The method was adopted due to the limitations in applying linear regression which requires the assumptions of distribution normality and values independent in a time series. Mann–Kendall's sum S was divided by the root of variance:

$$S = \sum_{i=t}^{n-1} \sum_{j=i+1}^{n} \operatorname{sgn}(x_j - x_i)$$

where: $\operatorname{sgn}(x) = 1$ for $x > 0$, 0 for $x = 0$, -1 for $x < 0$, where x denotes individual data series and n denotes the total number of years in a time series.

The analysis of the course of mean annual air and water temperature values was performed in reference to a year, seasons, and months. Dependencies between air temperature and water temperature in rivers and lakes were described by means of a Pearson coefficient of correlation, with significance at a level of 0.05 tested by t-student distribution with $n{-}2$ degrees of freedom.

8.3 Study Area

The study area covers the northern part of the Polish Lowland. The majority of the analysed objects are located in the young glacial area. More than half of the catchment of the Biebrza River is located in the area of the Central European glaciation, i.e. older in comparison with North Polish glaciation (Vistulian). The young glacial area is characterised by belt distribution of many geomorphological formations, deposits, and modern morphogenetic processes. The genesis of particular belts is related to the course of subsequent phases of the last glaciation. Many different relief elements occur here, including particularly: terminal moraine hills with a height of up to 328 m a.s.l., flat or undulating plateaus of the ground moraine, young network of river valleys, a dense network of subglacial channels, outwash plains, and numerous closed-drainage depressions. In addition to river valleys with usually meridional orientation, also wide and extensive Urstromtäler occurs with latitudinal orientation (Fig. 8.1). In valleys and Urstromtäler, permanent rivers occur. They are predominantly fed by groundwater and surface waters, as well as atmospheric precipitation. The majority of rivers flow through numerous lakes.

Among the rivers of the Polish Lowland, two longest ones stand out, namely the Vistula River (total length 1047 km) and the Oder River (854 km). The rivers flow out of springs in the mountains and then through uplands, and their lower sections run through the young glacial area. They are the most important rivers for the entire analysed area. They transport water to the Baltic Sea from almost all river systems of the Polish Lowland. In the lower sections of both rivers, gauging stations are located,

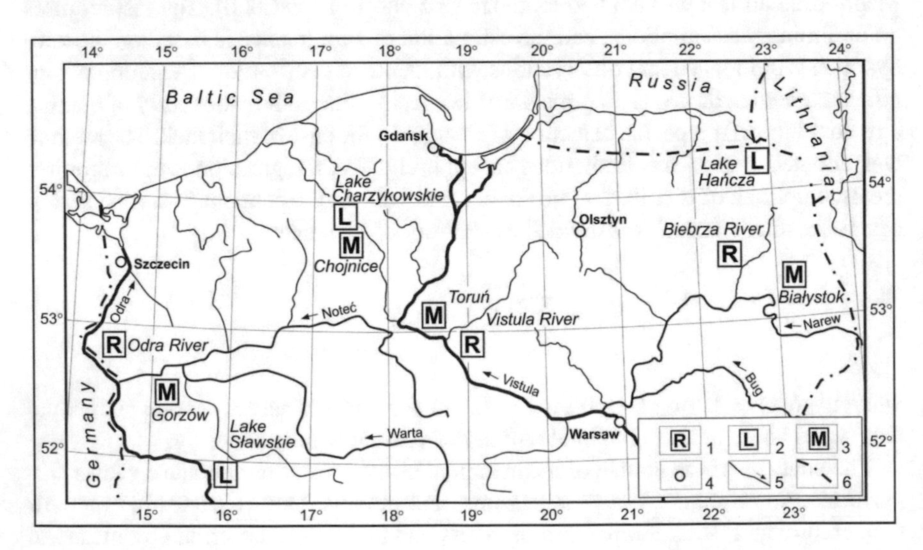

Fig. 8.1 Schematic map of the river network of the Polish Lowland with location of gauging sites. Description: 1—river gauging sites; 2—lake gauging sites; 3—meteorological stations; 4—cities; 5—main rivers; 6—country border. *Source* own elaboration

Table 8.1 Selected data on the analysed rivers

River/gauging station	Coordinates		$A \times 10^3$ km^2	Length of river		Qmean m^{3}s^{-1}
	φ	λ		L	LGS	
Oder/Gozdowice	52.764263	14.317719	109.7	855.0	645.3	523.0
Vistula/Toruń	53.008031	18.608269	181.0	1047.0	734.7	987.6
Biebrza/Burzyn	53.274332	22.459070	6.9	170.6	162.1	34.0

Description: *A*—catchment area; *L*—total river length; *LGS*—length of the river to the analysed gauging station; *Q*mean—mean discharge
Source own study based on data of the Institute of Meteorology and Water Management—National Research Institute

Table 8.2 Basic data on the analysed lakes

Lake	Coordinates		H m a.s.l.	S km^2	$V \times 10^6$m^3	Dmax	Dmean
	φ	λ					
Sławskie	51.891269	16.010868	56.9	8.22	42.66	12.3	5.2
Charzykowskie	53.778758	17.510357	120.0	13.64	134.53	30.5	9.8
Hańcza	54.264340	22.812267	227.3	2.91	120.36	108.5	38.7

Description: *H*—absolute height; *S*—surface area; *V*—volume; *D*max—maximum depth; *D*mean—mean depth
Source according to Inland Fisheries Institute in Olsztyn data

providing data on water temperature used in this chapter. The third gauging station is located on the Biebrza River flowing through a completely different valley. Unique in Europe for its marshes and peatlands, as well as its highly diversified fauna and flora, the Biebrza Valley was designated as a wetland site of global significance and has been under the protection of the Ramsar Convention since 1995 (Frąk et al. 2008). Some data on the analysed rivers are presented in Table 8.1.

The most characteristic objects of the young glacial area include post-glacial lakes. They developed cca 13 thousand years ago, after the end of the last glaciation on the northern hemisphere. They are dominated by channel and moraine lakes. The former have an elongated shape and are usually deep and narrow. Moraine lakes are usually oval, and substantially shallower. This chapter analyses channel lakes, considerably differing in terms of all morphometric parameters (Table 8.2). The maximum depth of the shallowest one—Lake Sławskie—amounts to 12.3 m and the deepest one—Lake Hańcza—108.5 m. Lake Hańcza is also the deepest lake in the Central European Lowland.

The study area is located in the moderate climate zone with evident transitional features between marine and continental moderate climate. Mean annual (1961–2014) air temperature decreases towards the east from approximately 8.7 °C in Gorzów Wielkopolski to 7.0 °C in Białystok (Table 8.3). Its higher variability (2.3 °C) is observed during the cold half-year (November–April) and considerably lower in the

Table 8.3 Location of meteorological stations and air temperature in the period 1961–2014

Meteorological station	Coordinates		Absolute height m a.s.l.	Mean temperature (1961–2014)		
	φ	λ		Year	May–Oct	Nov–Apr
Gorzów Wielkopolski	52.741231	15.277253	39	8.7	14.9	2.5
Chojnice	53.715288	17.532387	172	7.3	13.6	1.0
Toruń	53.042283	18.595451	66	8.1	14.6	1.6
Białystok	53.107221	23.162380	143	7.0	13.8	0.2

Source own study based on data of the Institute of Meteorology and Water Management—National Research Institute

warm half-year (1.1 °C). This results from more intensive influence of colder continental air masses during winter on the eastern part of the analysed area in comparison with the western part.

8.4 Relations Between Air and Surface Water Temperature

8.4.1 Western Part of the Polish Lowland

The relations between air temperature and surface water temperature in the western part of the Polish Lowland were discussed based on study results from the meteorological station in Gorzów Wielkopolski and from the gauging stations on the Oder River in Gozdowice and Lake Sławskie in Radzyń. The gauging stations on the river and lake are located at a distance of 55 and 70 km from the meteorological station, respectively.

Mean annual (1961–2014) air temperature in Gorzów Wielkopolski amounted to 8.7 °C, whereas in particular years, its high variability was observed: from 6.4 °C in 1996 to 10.7 °C in 2007. Nonetheless, an evident positive trend in air temperature occurred. During the analysed period, it averaged 0.32 °C·10 years^{-1} (Fig. 8.2). The trend is statistically significant at a level of 0.05.

Mean annual (1961–2014) water temperature in the Oder River and Lake Sławskie was higher than air temperature and amounted to 10.4 and 10.8 °C, respectively. Higher water temperature is a normal phenomenon resulting from completely different physical properties of water in comparison with air. Moreover, river valleys and lake basins are fed by groundwater which cool or warm surface waters depending on the season of the year (groundwater temperature is relatively stable throughout the year and amounts to approximately 8.5 °C). The phenomenon often occurs in young glacial areas (Jurgelėnaitė et al. 2012; Marszelewski and Pius 2016; Pius and Marszelewski 2016). It is one of the important factors due to which annual amplitudes

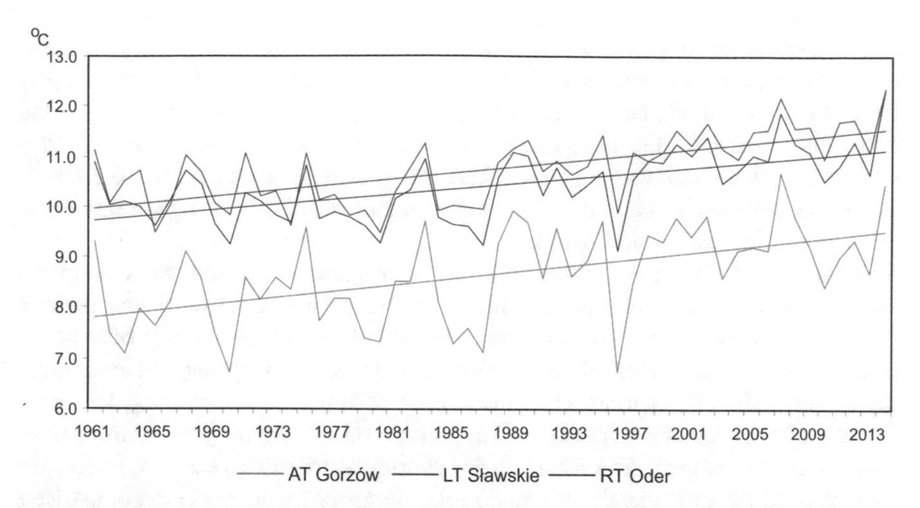

Fig. 8.2 Course of the mean annual air and surface water temperature in the period 1961–2014 in the western part of the Polish Lowland. Description: *AT*—air temperature in Gorzów Wielkopolski; *LT*—water temperature in Lake Sławskie; *RT*—water temperature in the Oder River. *Source* author's calculations based on data of the Institute of Meteorology and Water Management—National Research Institute

of surface water temperature were lower than those for air and amounted to 3.2 °C in the Oder River (from 9.2 to 12.4 °C) and 2.8 °C in Lake Sławskie (from 9.5 to 12.3 °C). In the period 1961–2014, a positive and statistically significant ($\alpha = 0.05$) trend of surface water temperature was also determined (compare Fig. 8.2). In the case of the Oder River, an increase in water temperature averaged 0.26 °C·10 years^{-1} and in the case of Lake Sławskie 0.30 °C·10 years^{-1}.

As shown by the above data, the rate of increase in mean annual air temperature (0.32 °C·10 years^{-1}) was somewhat faster in comparison with mean annual water temperature in the lake (0.30 °C·10 years^{-1}) and in the river (0.26 °C·10 years^{-1}). A question arises whether such dependencies are characteristic of all periods in a year (e.g. months or seasons), or only for the entire year. As shown by the analysis of relations occurring between air temperature and surface water temperature in shorter periods, such relations are much more complicated and show different patterns. It is evident both in shorter periods in a year (months) and during longer periods (half-years, seasons). In particular months of the year, the rate of increase in air temperature was not always the highest. This is particularly observed in November, when it is almost twice lower than the rate of increase in water temperature in Lake Sławskie (Table 8.4). This kind of situation results from an increase in cloudiness and reduction of insolation in that month, as well as in June and other autumn months (Żmudzka 2009). In the case of surface waters, the capacity of heat accumulation, particularly in lakes during summer and autumn, permits longer maintenance of higher than average water temperature. In the case of the Oder River (and other rivers), turbulent water movement and rapid heat exchange with the atmosphere result in lower monthly rate of increase in temperature amounting to only 0.07 °C·10 years^{-1}. For a similar reason, a lower rate of increase in air temperature in comparison with water temperature occurred in June, as well as in May and August (compare Table 8.4). In the remaining months, the rate of increase in air temperature was faster than that of water temperature, whereas it was the highest in April, when it equalled 0.53 °C·10 years^{-1}. The value is the highest among all the remaining ones characterising the rate of air temperature fluctuations in Poland.

Substantial differences between the rate of increase in air and water temperature also occurred in longer periods in a year, and especially in particular periods (Table 8.5). The effect of physical parameters of lake waters on the rate of increase in water temperature in so-called transitional seasons, i.e. in spring and autumn, is particularly evident. In spring, a vast mass of water subject to slow mixing decreases the rate of temperature increase, and in autumn on the contrary it contributes to longer retaining of heat. As a result, during shorter periods in a year, lake waters are characterised by an evidently different rate of temperature increase in comparison with river waters; although at the scale of the entire year, the rate is approximate.

8.4.2 Central Part of the Polish Lowland

The dependencies between air and water temperature in the central part of the Polish Lowland were investigated based on the example of the Vistula River in Toruń and Lake Charzykowskie in Charzykowy. Results of research on water temperature in the Vistula River were compared to those for air temperature from the meteorological station located 6 km from the river. In the case of Lake Charzykowskie, the meteorological station is located at a distance of approximately 5 km, in Chojnice.

Mean annual (1961–2014) air temperature in Toruń was lowered by 0.6 °C than in the western part of the Polish Lowland and amounted to 8.1 °C. In particular years in the analysed period, its considerable variability was recorded: from 5.9 °C in 1970 to 10.1 °C in 2007. An evident positive trend of air temperature also occurred, averaging 0.29 °C·10 years^{-1} (Fig. 8.3). In the case of the meteorological station in Chojnice, temperature values were somewhat lower and were as follows: mean annual 7.3 °C, variability of mean annual air temperature from 5.2 °C in 1970 to 9.3 °C in 2007, and value of temperature trend 0.31 °C·10 years^{-1} (Fig. 8.4).

Mean annual (1961–2014) water temperature in the Vistula River and in Lake Charzykowskie was higher than air temperature and amounted to 10.2 and 9.5 °C, respectively. Annual water temperature amplitudes were lower than those for air and equalled 2.6 °C (from 9.0 to 11.6 °C) in the Vistula River and 3.2 °C (from 8.3 to 11.5 °C) in Lake Charzykowskie. In the period 1961–2014, positive and statistically significant ($\alpha = 0.05$) trends of surface water temperature occurred (compare Figs. 8.3 and 8.4). In this period, an increase in water temperature in the Vistula River averaged 0.27 °C·10 years^{-1} and in the case of Lake Charzykowskie 0.24 °C·10 years^{-1}. Therefore, it turns out that in spite of somewhat lower values of air and water tem-

Table 8.4 Mean monthly and annual rate of air and surface water temperature fluctuations (in °C·10 years^{-1}) in the period 1961–2014 in the western part of the Polish Lowland

Element	Nov	Dec	Jan	Feb	Mar	Apr	May	Jun	Jul	Aug	Sep	Oct	Year
AT	0.09	0.27	0.44	0.37	0.38	***0.53***	***0.42***	0.13	***0.54***	***0.40***	0.17	0.12	***0.32***
WTL	0.17	***0.18***	***0.25***	***0.19***	***0.30***	***0.30***	***0.58***	***0.26***	***0.42***	***0.47***	***0.23***	***0.21***	***0.30***
WTR	0.07	0.10	***0.21***	0.20	***0.41***	***0.43***	***0.49***	***0.20***	***0.36***	***0.39***	0.16	0.10	***0.26***

Description: *AT*—air temperature in Gorzów Wielkopolski; *WTL*—water temperature in Lake Sławskie; *WTR*—water temperature in the Oder River. Statistically significant changes were marked with italics and bold font
Source author's calculations based on data of the Institute of Meteorology and Water Management—National Research Institute

Table 8.5 Mean seasonal rate of air and surface water temperature fluctuations (in °C·10 years^{-1}) in the period 1961–2014

Element	Cool half-year	Warm half-year	Winter	Spring	Summer	Autumn
AT	0.35	0.30	0.36	0.44	0.36	0.12
WTL	0.23	0.37	0.21	0.39	0.38	0.21
WTR	0.24	0.28	0.17	0.44	0.32	0.11

Description like in Table 8.4
Source author's calculations based on data of the Institute of Meteorology and Water Management—National Research Institute

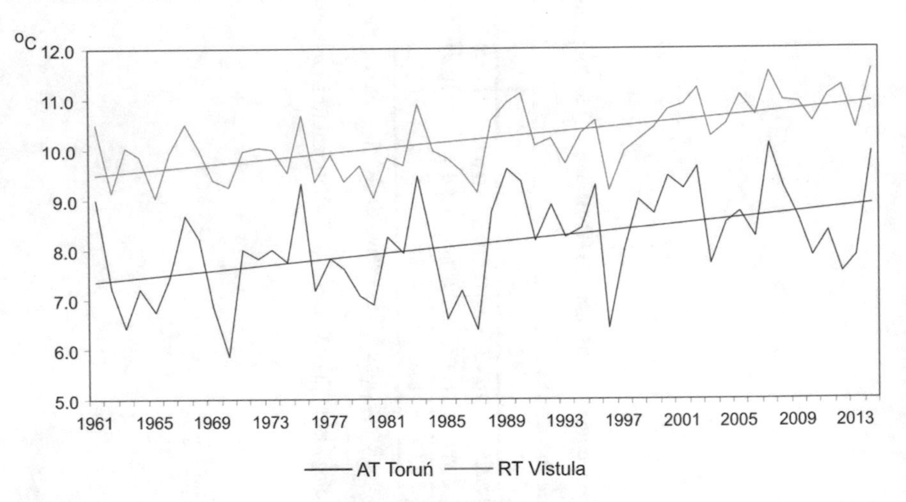

Fig. 8.3 Course of the mean annual air temperature in Toruń (*AT*) and water temperature in the Vistula River (*RT*) in the period 1961–2014. *Source* author's calculations based on data of the Institute of Meteorology and Water Management—National Research Institute

perature in the central part of the Polish Lowland, the rate of its increase was almost identical as in its western part.

Similarly as in the western part of the Polish Lowland, the rate of increase in air temperature was not identical in particular months (Table 8.6). The greatest differences between the rate of increase in air temperature and river water temperature occurred in as many as six months: November, May, June, August, September, and October. Air temperature in June deserves particular attention. It showed a slight, although not statistically significant, negative tendency (-0.01 °C·10 years^{-1}). Therefore, the dependencies between air and river water temperature in particular months in the central part of the Polish Lowland proved to be more complicated, although the general trend of temperature fluctuations was similar. This may result from substantially greater water mass in the Vistula River in Toruń (averaging 990 m^3·s^{-1}) in comparison with the Oder River in Gozdowice (averaging 545 m^3·s^{-1}), as well as from a slow but evident increase in climate continentalism. The disturbance of nat-

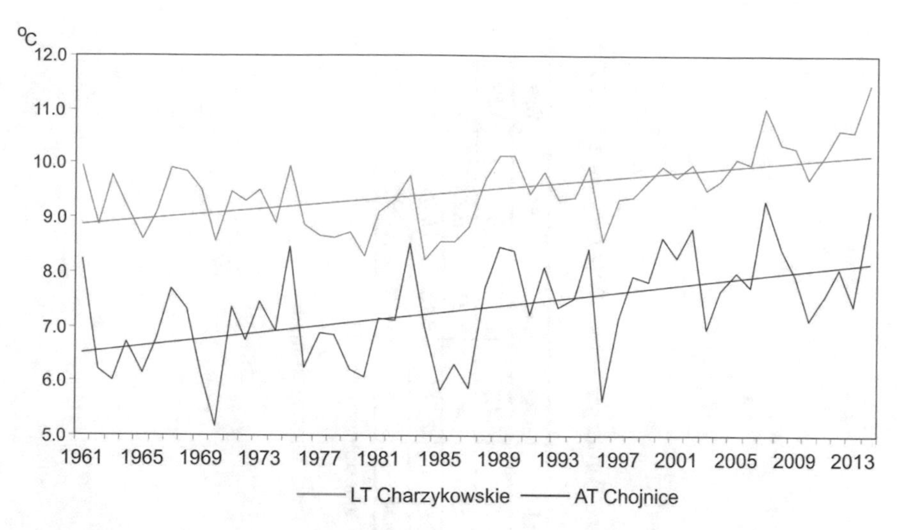

Fig. 8.4 Course of the mean annual air temperature in Chojnice (*AT*) and water temperature in Lake Charzykowskie (*LT*) in the period 1961–2014. *Source* author's calculations based on data of the Institute of Meteorology and Water Management—National Research Institute

ural relations between air and river water temperatures certainly also resulted from its pollution with sewage, lasting almost until the end of the twentieth century, and from the effect of hydrotechnical facilities.

Air and lake water temperature fluctuations in particular months had an equally complicated course (compare Table 8.6). The rate of increase in air temperature was lower only in two months (May and June)—similarly as in the western part of the study area. The negative tendency of mean water temperature fluctuations in lake Charzykowskie in November draws attention (–0.08 °C·10 years^{-1}). It is an effect of only a slight increase in water temperature in October and air temperature in November, as mentioned before.

Significant differences between the rate of increase in air and water temperature also occurred in longer periods of the year, including its individual seasons (Table 8.7). The lowest rate of temperature increase was observed in autumn, particularly in Lake Charzykowskie, where it amounted to only 0.04 °C·10 years^{-1}. In this case, similarly as in Lake Sławskie, the effect of physical properties of lake waters on the rate of increase in water temperature in so-called transitional seasons is manifested. Notice also that the rate of air temperature increase in the cool half-year is substantially faster than in the warm half-year (compare Table 8.7), unlike in the case of surface waters. This also results from the physical properties of waters. Their temperature can increase only from 0 °C, and an increase in air temperature also occurs in the range of negative values.

Table 8.6 Mean monthly and annual rate of air and surface water temperature fluctuations (in °C·10 years^{-1}) in the period 1961–2014 in the central part of the Polish Lowland

Element	Nov	Dec	Jan	Feb	Mar	Apr	May	Jun	Jul	Aug	Sep	Oct	Year
ATT	0.10	0.25	0.42	0.37	***0.37***	***0.51***	***0.35***	−0.01	***0.47***	***0.41***	0.17	0.07	***0.29***
WTR	***0.22***	***0.21***	***0.18***	0.11	***0.28***	***0.46***	***0.46***	0.15	***0.42***	***0.47***	***0.21***	***0.20***	***0.27***
ATC	0.14	0.29	0.44	***0.44***	***0.39***	***0.49***	***0.34***	0.03	***0.52***	***0.40***	0.20	0.08	***0.31***
WTL	−0.08	***0.15***	***0.20***	***0.15***	***0.30***	***0.47***	***0.65***	0.15	***0.35***	***0.35***	0.16	0.05	***0.24***

Description: *ATT*—air temperature in Toruń; *WTR*—water temperature in the Vistula River; *ATC*—air temperature in Chojnice; *WTL*—water temperature in Lake Charzykowskie. Statistically significant changes are marked with italics and bold font

Source author's calculations based on data of the Institute of Meteorology and Water Management—National Research Institute

Table 8.7 Mean seasonal rate of air and surface water temperature fluctuations (in °C·10 years^{-1}) in the period 1961–2014

Element	Cool half-year	Warm half-year	Winter	Spring	Summer	Autumn
ATT	***0.34***	***0.25***	0.34	***0.41***	***0.29***	0.11
WTR	***0.22***	***0.32***	***0.16***	***0.36***	***0.35***	***0.21***
ATC	***0.36***	***0.26***	***0.39***	***0.41***	***0.32***	0.14
WTL	***0.19***	***0.28***	***0.16***	***0.48***	***0.28***	0.04

Description as in Table 8.6
Source author's calculations based on data of the Institute of Meteorology and Water Management—National Research Institute

8.4.3 Eastern Part of the Polish Lowland

Air and surface water temperature fluctuations in the eastern part of the Polish Lowland are discussed based on results of research from the meteorological station in Białystok, as well as the Biebrza River (site Burzyn) and Lake Hańcza. The meteorological station in Białystok is located at a distance of approximately 40 km from the Biebrza River in Burzyn and approximately 100 km from Lake Hańcza.

Mean annual (1961–2014) air temperature in Białystok was the lowest among the analysed sites and amounted to 7.0 °C. The lowest mean annual temperature (5.1 °C) occurred in 1987 and the highest (8.8 °C) in 2007. Similarly as at other sites in the Polish Lowland, an evident, statistically significant, positive trend of air temperature was recorded, averaging 0.22 °C·10 years^{-1} (Fig. 8.5). It is worth emphasising that it is the lowest rate of temperature increase throughout the analysed area, lower by 0.1 °C·10 years^{-1} than in Gorzów Wielkopolski.

Mean annual (1961–2014) water temperatures in the Biebrza River and in Lake Hańcza were higher than air temperature and amounted to 9.1 and 8.5 °C, respectively.

Notice that in this case, unlike in the previous cases, higher mean annual water temperature occurred in the river, and not in the lake. This is related to their specific morphometric properties. The Biebrza River is sometimes described as a "wetland river". It is shallow, with slow water flow in a wide valley through numerous wetlands and flooded meadows. Due to this, water temperature during the summer season can reach much higher values than in a deep river with fast current. Due to the considerable depth and water volume, Lake Hańcza (compare Table 8.2) is characterised by a substantially longer period of water temperature increase in comparison with the remaining ones. Moreover, it is located in a region where the summer season lasts the shortest among all the regions of the Polish Lowland. For the above reasons, conditions for water temperature increase in Lake Hańcza in the warm half-year are unfavourable. As a result, water in the lake is characterised by temperature lower than in the river. It is a rare phenomenon.

Annual amplitudes of water temperature were lower than those for air and equalled 2.7 °C (from 7.9 to 10.6 °C) in the Biebrza River and 2.5 °C (from 7.2 to

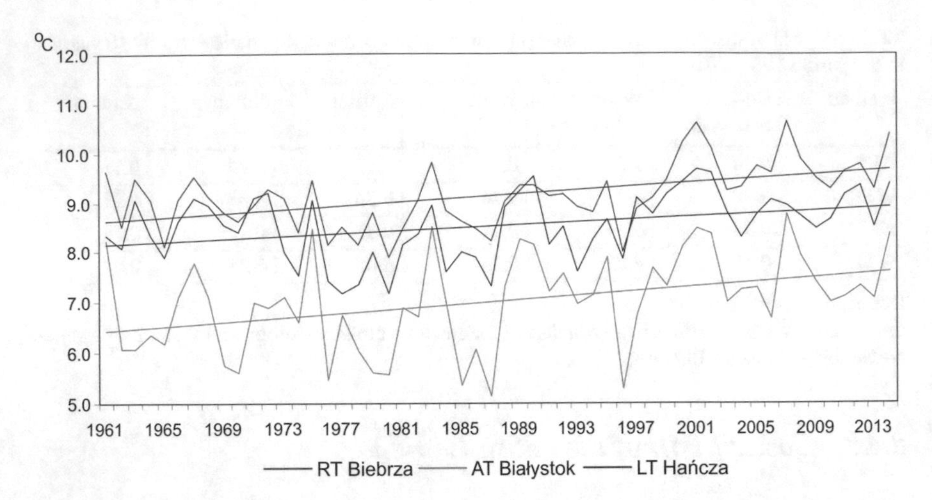

Fig. 8.5 Course of mean annual air temperature in Białystok (*AT*), water temperature in Lake Hańcza (*LT*), and water temperature in the Biebrza River (*RT*) in the period 1961–2014. *Source* author's calculations based on data of the Institute of Meteorology and Water Management—National Research Institute

9.7 °C) in Lake Hańcza. In the period 1961–2014, positive and statistically significant ($\alpha = 0.05$) trends of surface water temperature occurred (compare Fig. 8.5). In this period, an increase in water temperature in the Biebrza River averaged 0.20 °C·10 years^{-1} and in the case of Lake Hańcza 0.15 °C·10 years^{-1}. The rate of its increase, however, was the lowest among all analysed hydrological objects in the Polish Lowland.

The directions and rate of air temperature fluctuations in particular months were largely variable. In three months of the year (June, September, October), a slight negative tendency of mean air temperature fluctuations was recorded (Table 8.8), although such fluctuations were not statistically significant. Among the remaining months, the fastest increase in air temperature occurred in winter months, particularly in January, when it amounted to 0.58 °C·10 years^{-1}. Throughout the winter season (from December to February), mean air temperature increase reached 0.43 °C·10 years^{-1}, proving winter to be the season with the fastest temperature increase (Table 8.9). It is worth emphasising that the fast increase in air temperature during winter did not contribute to an increase in surface water temperature, because it occurred in the range of negative air temperatures. This is evidenced by a negative, although not statistically significant, trend of water temperature (−0.08 °C·10 years^{-1}) in Lake Hańcza during the winter season (compare Table 8.9).

The above hydrometeorological situations determined strong variability of the rate of air and water temperature fluctuations also in the winter half-year, namely: 0.35 °C·10 years^{-1} for air temperature, 0.19 °C·10 years^{-1} for river water temperature, and −0.01 °C·10 years^{-1} for Lake Hańcza. An opposite situation was observed during the warm half-year (compare Table 8.9).

Table 8.8 Mean monthly and annual rate of air and surface water temperature fluctuations (°C·10 years^{-1}) in the period 1961–2014 in the eastern part of the Polish Lowland

Element	Nov	Dec	Jan	Feb	Mar	Apr	May	Jun	Jul	Aug	Sep	Oct	Year
AT	0.09	0.28	0.58	0.42	0.40	***0.30***	0.13	−0.06	***0.37***	***0.23***	−0.02	−0.02	***0.22***
WTR	0.16	***0.16***	0.09	0.07	***0.20***	***0.47***	***0.35***	0.05	***0.37***	***0.29***	0.09	0.08	***0.20***
WTL	***−0.23***	***−0.23***	−0.01	−0.03	−0.05	***0.31***	***0.72***	***0.38***	***0.38***	***0.36***	0.02	−0.02	***0.15***

Description: *AT*—air temperature in Białystok; *WTR*—water temperature in the Biebrza River; *WTL*—water temperature in Lake Hańcza. Statistically significant changes are marked with italics and bold font

Source author's calculations based on data of the Institute of Meteorology and Water Management—National Research Institute

Table 8.9 Mean seasonal rate of air and surface water temperature fluctuations (in °C·10 years^{-1}) in the period 1961–2014

Element	Cool half-year	Warm half-year	Winter	Spring	Summer	Autumn
AT	***0.35***	0.10	***0.43***	***0.28***	***0.18***	0.02
WTR	***0.19***	***0.21***	***0.10***	***0.34***	***0.23***	0.12
WTL	−0.01	***0.31***	−0.08	***0.36***	***0.38***	−0.07

Description as in Table 8.8
Source author's calculations based on data of the Institute of Meteorology and Water Management—National Research Institute

8.5 Correlations Between Air and Surface Water Temperature Fluctuations

The previous parts of the chapter present the tendencies of air and surface water temperature fluctuations. Attention is paid to both similarities and differences in the scope of the rate of temperature fluctuations in three analysed environments: air, river water, and lake water. It was evidenced that local conditions considerably modify the value, and even directions of water temperature fluctuations. As a consequence, the existence of many differences between temperature fluctuations of air and surface water was documented. In spite of this, a strong correlation is observed between air temperature and surface water temperature.

The strongest correlations were recorded in the scope of annual temperature values, and particularly in the western part of the analysed area. Coefficients of correlation between air and surface water temperature vary from 0.86 to 0.94. Somewhat lower coefficients of correlation occur for the central part of the area, where almost all of them are higher than 0.81. The weakest correlations occur in the eastern part of the area, particularly between air temperature in Białystok and water temperature in Lake Hańcza (0.67) (Table 8.10).

Table 8.10 Coefficients of correlation between mean annual temperatures (grey colour designates the analysed groups of objects in the western, central, and eastern part of the Polish Lowland)

	Oder R.	*Gorzów*	*L. Sławskie*	*Toruń*	*Vistula R.*	*Chojnice*	*L. Charzykowskie*	*Biebrza R.*	*Białystok*	*L. Hańcza*
Oder River	1									
Gorzów Wielkopolski	0.93	1								
Lake Sławskie	0.94	0.86	1							
Toruń	0.87	0.97	0.79	1						
Vistula River	0.93	0.88	0.93	0.82	1					
Chojnice	0.93	0.99	0.86	0.97	0.89	1				
Lake Charzykowskie	0.90	0.81	0.91	0.72	0.88	0.81	1			
Biebrza River	0.89	0.79	0.88	0.75	0.90	0.82	0.85	1		
Białystok	0.87	0.94	0.78	0.94	0.85	0.95	0.74	0.82	1	
Lake Hańcza	0.75	0.63	0.75	0.58	0.75	0.63	0.73	0.80	0.67	1

Source author's calculations

Table 8.11 Coefficients of correlation of mean temperature in the cool half-year (grey colour designates the analysed groups of objects in the western, central, and eastern part of the Polish Lowland)

	Oder R.	*Gorzów*	*L. Sławskie*	*Toruń*	*Vistula R.*	*Chojnice*	*L. Charzykowskie*	*Biebrza R.*	*Białystok*	*L. Hańcza*
Oder River	1									
Gorzów Wielkopolski	0.81	1								
Lake Sławskie	0.92	0.80	1							
Toruń	0.74	0.98	0.75	1						
Vistula River	0.86	0.63	0.85	0.57	1					
Chojnice	0.79	0.98	0.79	0.98	0.62	1				
Lake Charzykowskie	0.84	0.67	0.86	0.60	0.83	0.66	1			
Biebrza River	0.79	0.54	0.82	0.51	0.84	0.55	0.81	1		
Białystok	0.76	0.95	0.76	0.97	0.60	0.97	0.62	0.52	1	
Lake Hańcza	0.62	0.59	0.60	0.57	0.57	0.57	0.59	0.50	0.52	1

Source author's calculations

Weaker correlations were observed in the case of temperature in shorter periods, e.g. in the cool half-year (Table 8.11). The values of coefficients of correlation between air and surface water temperature in the western part of the study area range from 0.80 to 0.92, but in the eastern part only from 0.50 to 0.52. High values of correlation coefficients describe the analysed phenomena in the central part of the Polish Lowland; although in the case of Lake Charzykowskie and air temperature, the value of correlation coefficient is lower and amounts to 0.66. The causes of weaker correlations between air and water temperatures in the cool half-year are presented above.

8.6 Final Remarks

Air temperature fluctuations are the most evident element accompanying climate fluctuations. Over the last several decades, they have usually been associated with an increase in its value. Such a tendency occurs both at the global and regional scale in reference to the basic unit of time, i.e. year. In shorter periods, the rate of air temperature fluctuations, however, is very variable. This also results in the variability of the rate of water temperature fluctuations. The dependencies are evident in the scope of river waters which respond to changes in air temperature more rapidly than lake waters.

This chapter concerns three rivers differing in both the discharge volume and the majority of morphometric and physiographic elements. It is very interesting that in spite of the evidenced differences in the rates of temperature fluctuations in particular rivers, the general course of such changes is similar (Fig. 8.6).

Inconsiderable changes, and even negative directions of changes, were observed in the period from November to February. Over the following three months, a growing increase in the rate of water temperature fluctuations occurred. Throughout the

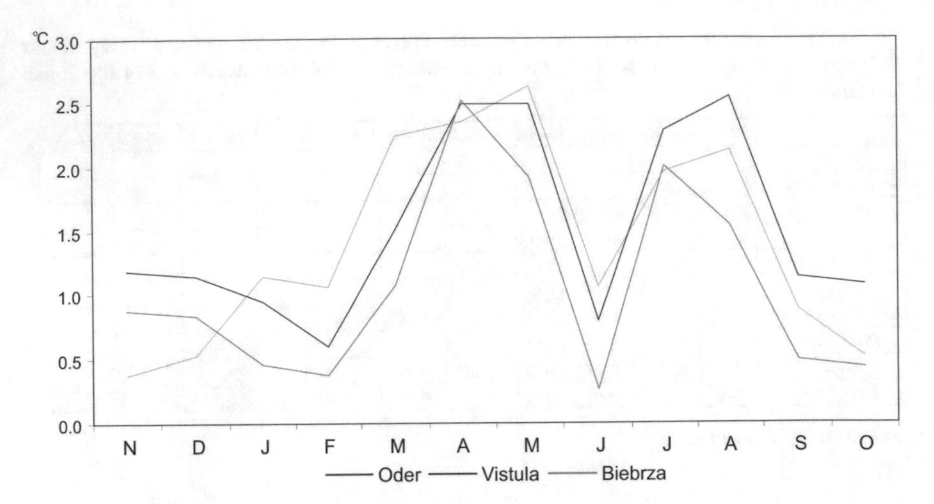

Fig. 8.6 Mean monthly (1961–2014) values of water temperature increase in rivers. *Source* author's calculations

analysed period, the highest increase in river water temperature was observed in April and May and varied from 1.9 to 2.6 °C. The evident discontinuation of the increase occurred in June as a result of a negative tendency of air temperature fluctuations. Over the following two months, another temperature increase occurred (by 1.5–2.5 °C in the years 1961–2014), and in September and October, a temperature decrease to a level characteristic of autumn and winter months (compare Fig. 8.6). In the case of rivers, in spite of the complicated course of water temperature fluctuations, a positive trend was recorded in each month; although in winter months, as well as in June, September, and October, temperature increase throughout the study period was lower than 0.5 °C.

Water temperature increase in lakes in particular months of the year showed higher regularity (Fig. 8.7). During autumn and winter months, it was low and amounted to approximately 1 °C in the case of Lakes Sławskie and Charzykowskie. In Lake Hańcza, however, a slight decrease in water temperature was recorded (to 1.2 °C), particularly in November and December. In the period from March to May, temperature increased in all lakes, even to 3–4 °C. In the period 1961–2014, the highest water temperature increase in lakes was observed in May (from 3.1 to 3.9 °C). Therefore, water temperature increase in lakes in May was by more than 1 °C higher in comparison with rivers. In June, like in rivers, an evident decline of water temperature increase occurred.

The relations between air and surface water temperature fluctuations are strong, although they are subject to modifications depending on the natural conditions and anthropo-pressure. Study results presented in this chapter confirm the earlier hypothesis that water temperature is a good indicator of climate changes, and particularly air temperature fluctuations.

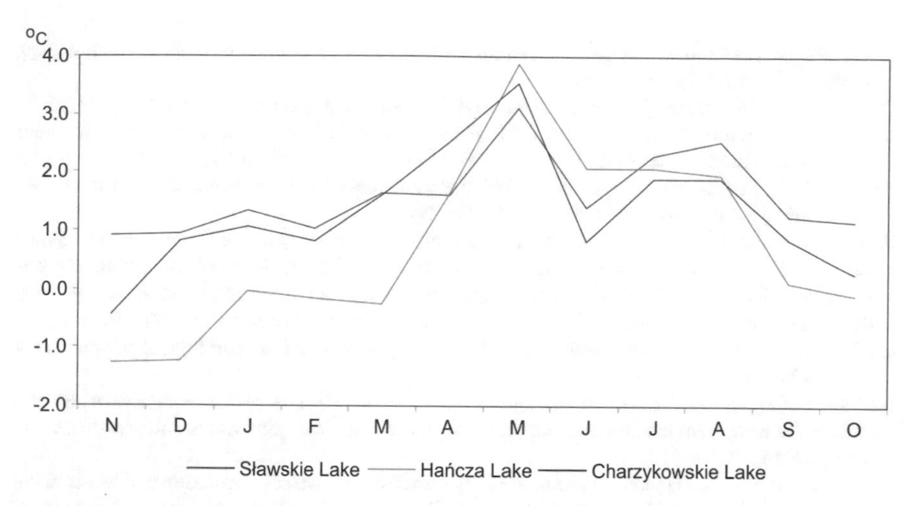

Fig. 8.7 Mean monthly (1961–2014) values of water temperature increase in lakes. *Source* author's calculations

The observed increase in the temperature of surface water results in changes in its physical properties, including among others a decrease in its density, viscosity, oxygen concentration. This is of high importance for aquatic ecosystems, because it contributes to an increase in the rate of sedimentation as a result of reduced ability of plankton to persist in the water. An increase in heat resources in inland waters also modifies the course of the development cycle of aquatic organisms (Adrian et al. 2006). Such changes show multiple directions and are difficult to determine based on research of only one element. In areas with transitional and cool moderate climate, an increase in water temperature in the winter season caused a reduction of persistence of ice phenomena and a reduction in the thickness of ice.

The research on the relations between air temperature and inland water temperature should be continued. Its results are useful for forecasting thermal conditions of rivers and lakes even until the end of the twenty-first century (Morrison et al. 2002; Ferrari et al. 2007; van Vliet et al. 2013).

Acknowledgements This work was supported by the National Centre for Research and Development of Warsaw (Poland) through the ERALECC project (ERA.Net RUS Plus; ID 226).

References

Adrian R, Wilhelm S, Gerten D (2006) Life-history traits of lake plankton species may govern their phenological response to climate warming. Glob Change Biol 12:652–661

Caissie D (2006) The thermal regime of rivers: a review. Freshw Biol 51(8):1389–1406

Dąbrowski M, Marszelewski W, Skowron R (2004) The trends and dependencies between air and water temperatures in lake in northern Poland from 1961–2000. Hydrol Earth Syst Sci 8(1):79–87

Davies PM (2010) Climate change implications for river restoration in global biodiversity hotspots. Restor Ecol 18(3):261–268
Dokulil MT (2014) Impact of climate warming on European inland waters. Inland Water 4(1):27–40
Dugdale S, Bergeron NE, St-Hilaire A (2013) Temporal variability of thermal refuges and water temperature patterns in an Atlantic salmon river. Remote Sens Environ 136:358–373
Ferrari MR, Miller JR, Russell GL (2007) Modeling changes in summer temperature of the Fraser River during the next century. J Hydrol 342:336–346
Frąk M, Kardel I, Stelmaszczyk M (2008) Phytoseston biodiversity of the Biebrza River against the background of selected water quality parameters—pilot study. In: Gołdyn R, Klimaszyk P, Kuczyńska-Kippen N, Piotrowicz R (eds) The functioning and protection of water ecosystems. Department of Water Protection, Faculty of Biology, Adam Mickiewicz University, Poznań
Gu RR, Li Y (2002) River temperature sensitivity to hydraulic and meteorological parameters. J Environ Manage 66:43–56
Hari RE, Livingstone DM, Siber R, Burkhardt-Holm P, Güttinger H (2006) Consequences of climatic change for water temperature and brown trout populations in Alpine rivers and streams. Glob Change Biol 12:10–26
Hudon C, Armellin A, Gagnon P, Patoine A (2010) Variations in water temperature and levels in the St. Lawrence River (Québec, Canada) and potential implications for three common fish species. Hydrobiologia 647(1):145–161
Jurgelėnaitė A, Kriaučiūnienė J, Šarauskienė D (2012) Spatial and temporal variation in the water temperature of Lithuanian rivers. Baltica 25(1):65–76
Kaushal SS, Likens GE, Jaworski NA, Pace ML, Sides AM, Seekell D, Belt KT, Secor DH, Wingate RL (2010) Rising stream and river temperatures in the United States. Front Ecol Environ 8(9):461–466
Kundzewicz Z, Robson A (2000) (eds) World climate programme—water, detecting trend and other changes in hydrological data, WCDMP, 45, 1–158
Latkovska I, Apsīte E (2016) Long-term changes in the water temperature of rivers in Latvia. Proc Latv Acad Sci B 70(701):78–87
Leuven RSEW, Hendriks AJ, Huijbregts MAJ, Lenders HJR, Matthews J, Van der Velde G (2011) Differences in sensitivity of native and exotic fish species to changes in river temperature. Curr Zool 57(6):852–862
Magee MR, Wu CH, Robertson DM, Lathrop RC, Hamilton DP (2016) Trends and abrupt changes in 104 years of ice cover and water temperature in a dimictic lake in response to air temperature, wind speed, and water clarity drivers. Hydrol Earth Syst Sci 20:1681–1702
Marszelewski W, Pius B (2016) Long-term changes in temperature of river waters in the transitional zone of the temperate climate: a case study of Polish rivers. Hydrol Sci J 61(8):1430–1442
McCombie AM (1959) Some relations between air temperatures and the surface water temperatures of lakes. Limnol Oceanogr 4:252–258
Mohseni O, Stefan HG (1999) Stream temperature/air temperature relationship: a physical interpretation. J Hydrol 218:128–141
Morrison J, Quick MC, Foreman MGG (2002) Climate change in the Fraser River watershed: flow and temperature projections. J Hydrol 263:230–244
North RP, Livingstone DM, Hari RE, Köster O, Niederhauser P, Kipfer R (2013) The physical impact of the late 1980s climate regime shift on Swiss rivers and lakes. Inland Waters 3:341–350
O'Reilly CM, Sharma S, Gray DK, Hampton SE et al (2015) Rapid and highly variable warming of lake surface waters around the globe. Geophys Res Lett 42:10773–10781
Pekarova P, Halmova D, Miklanek P, Onderka M, Pekar J, Skoda P (2008) Is the water temperature of the Danube river at Bratislava, Slovakia, rising? J Hydrometeorol 5:1115–1122
Pius B, Marszelewski W (2016) Effect of climatic changes on the development of the thermal-ice regime based on the example of Lake Charzykowskie (Poland) Bulletin of Geography. Phys Geogr Ser 11:27–33
Sridhar V, Sansone AL, Lamarche J, Dubin T (2004) Predictionof stream temperature in forested watersheds. J Am Water Resour Assoc 40(1):197–213

Steel EA, Lange IA (2007) Using wavelet analysis to detect changes in water temperature regimes at multiple scales: effects of multi-purpose dams in the Willamette River basin. River Res Appl 23:351–359

van Vliet MTH, Franssen WHP, Yearsley JR, Ludwig F, Haddeland I, Lettenmaier DP, Kabat P (2013) Global river discharge and water temperature under climate change. Glob Environ Change 23:450–464

Webb BW, Nobilis F (1995) Long-term water temperature trends in Austrian rivers. Hydrol Sci J 40:83–96

Webb BW, Hannah DM, Moore RD, Brown LE, Nobilis F (2008) Recent advances in stream and river temperature research. Hydrol Process 22:902–918

Woolway RI, Cinque K, de Eyto E, DeGasperi CL, Dokulil MT, Korhonen J, Maberly SC, Marszelewski W, May l, Merchant CJ, Paterson AM, Riffler M, Rimmer A, Rusak JA, Schladow SG, Schmid M, Teubner K, Verburg P, Vigneswaran B, Watanabe S, Weyhenmeyer GA (2016) Lake surface temperatures [in "State of the Climate in 2015"], Bull Am Meteorol Soc 97(8):S17–S18

Woolway RI, Dokulil MT, Marszelewski W, Schmid M, Bouffard D, Merchant ChJ (2017) Clim Change 142(3–4):505–520

Xin Z, Kinouchi T (2013) Analysis of stream temperature and heat budget in an urban river under strong anthropogenic influences. J Hydrol 489:16–25

Żmudzka E (2009) Contemporary changes of climate of Poland. Acta Agrophysica 13 (2):555–568

Chapter 9
Overview of River-Induced Hazards in Romania: Impacts and Management

Liliana Zaharia and Gabriela Ioana-Toroimac

Abstract River-related hazards, mainly floods and inundations are the most widespread and frequent of all the natural damaging phenomena, causing annually significant victims and economic losses. This chapter focuses on Romania, a country with one of the highest flood risk in Europe. According to EM-DAT database, within the period 1900–2016, more than half (55%) of the total number of disasters (94 events) were induced by floods. During the mentioned period, they were responsible for 57% of the total damage costs rclated to the top 10 disasters in Romania. The chapter presents a synthetic overview on the river-born hazards as floods and flooding, low-waters and hydrological droughts, and other damaging phenomena (water freezing, ice jams, sediment, and channel dynamics reservoirs' silting) with relevant examples, as well as with references to their impacts and management. Reducing the negative consequences of these hazards has become an important concern in the national water resources and related risks management policy. In this sense, two strategies are relevant: National Strategy for Drought Mitigation, Prevention and Combating Land Degradation and Desertification, on short, medium, and long term (2008), National Strategy for Flood Risk Management for medium and long term (2010), and Flood Risk Management Plans (2016).

Keywords Floods · Flooding · Flood damages · Hydrological droughts
Ice jams · Romania

L. Zaharia (✉) · G. Ioana-Toroimac
Faculty of Geography, University of Bucharest, Sector 1, 010041 Bucharest, Romania
e-mail: zaharialili@hotmail.com

G. Ioana-Toroimac
e-mail: gabriela_toroimac@yahoo.com

M. Zelenakova (ed.), *Water Management and the Environment: Case Studies*, Water Science and Technology Library 86,
https://doi.org/10.1007/978-3-319-79014-5_9

9.1 Introduction

Hydrological hazards are the most common risk phenomenon, each year generating significant societal and environmental damages. In the decade 2005–2014, almost half (192) of the average annual number of natural triggered disasters (380) had hydrological origin (mainly floods and subsidiary mass movements of hydrological origin). These disasters caused annually an average of 87.28 million victims (killed and total affected people) representing 44% of total disaster victims, and they were responsible for 22% of the total annual average cost of damages (34.55 from the total of 159.75 billion US$) (Guha-Sapir et al. 2016). The most widespread and damaging hydrological natural hazard are related to river-specific processes: flow-variation (floods, flooding, low-waters and droughts), water freezing/unfreezing and related ice jams, sediment, and channel dynamics (erosion and deposition processes). The floods have the largest share in natural disaster occurrence (almost half), and they are responsible for the biggest damages: between 2005 and 2014, the floods have caused annually 5993 fatalities and 99.6% of all disaster victims. The share of damage attributable to floods is annually more than 99% of total disaster costs (Guha-Sapir et al. 2016). In order to reduce the negative consequences of floods and related processes, over the past decades, flood mitigation has become a major concern of water policies and sustainable development strategies at different spatial scales (e.g., from European Union—EU, to national, regional, and local scale) (Zaharia and Ioana-Toroimac 2016).

This chapter focuses on Romania, a country with one of the highest flood risk exposure to floods and related damage costs in Europe (Pińskwar et al. 2012; Kundzewicz et al. 2013). It aims to highlight the main river-induced hazards and their features, based on relevant examples for different types of river-related damaging processes. After a statistical overview on the natural disasters in Romania, revealing the leading place occupied by floods, the chapter presents particularities of the specific river-born hazards as floods and flooding, low-waters and hydrological droughts, and other damaging phenomena (water freezing, sediment/channel dynamics and reservoirs' silting), as well as significant examples of different events/hazards and references to their impacts. Finally, some considerations on hydrologic risk management are presented. The chapter offers a synthetic and actualized overview on the whole range of natural hazards related to rivers in Romania. It completes the information from previous general works on hazards published in Romania (e.g., Romanescu 2008; Grecu 2016; Sorocovschi 2016) or the information on strictly certain hydrological hazards, mainly floods (e.g., Mustăţea 2005; Romanescu 2006). Within volumes "Riscuri şi catastrofe" (*Risks and catastrophes*) edited by Sorocovschi (2002–2017), papers on different river-induced hazards in Romania at regional or local scale can be found (most of them on floods).

The chapter is based mainly on a synthesis of significant scientific researches and processing of statistical data from different sources (EM-DAT database, scientific bibliography, etc.).

9.2 Statistic Overview on Natural Hazards and Related Damages in Romania

Located in the central part of Europe, on the parallel of 45°N, Romania (area of 238,391 km^2, with almost 20 million inhabitants) overlaps almost entirely (on 97% of its territory) the Danube River Basin. It has very diverse and complex geographic features that give it susceptibility to natural hazards. Its location at the jointing of several tectonic microplates favors earthquakes. The presence of the Carpathian mountains (with a maximum altitude of 2544 m), of the Black Sea coastline and of the Danube River (Fig. 9.1a), as well as a relatively dense hydrographic network (0.5 km/km^2, according to Pătru et al. 2006) are specific local factors favoring the natural hazards occurence, under a transitional climate between temperate oceanic and continental, with regional differences induced by orography and external influences (oceanic in the western part, Mediterranean in the south-west and south, Baltic in the northern part, excessive continental in the east and Pontic—related to the Black Sea—in the south-east).

The statistics (according to EM-DAT database, 2016, for the period 1900–2016) on the number of disasters by types in Romania show that out of a total of 94 events, 55% were induced by floods, 21% by extreme temperatures, 14% by earthquakes, and 7% by storms (Fig. 9.1b). The floods were responsible for more than half (57%) of the total damage costs related to the top 10 disasters in Romania (from 1900 to 2016), the earthquakes for about one-third (34%), and the droughts for 9% (Fig. 9.1c). The floods are the leader in the hierarchy of total affected people distribution by type of top 10 natural hazards in Romania. They hold 80%, followed by earthquakes (20%) (the floods occupy 9 of the 10 places in this top). Concerning the number of fatalities, the first place is occupied by earthquakes (61%), followed by floods (33%) and extreme temperatures (6%).

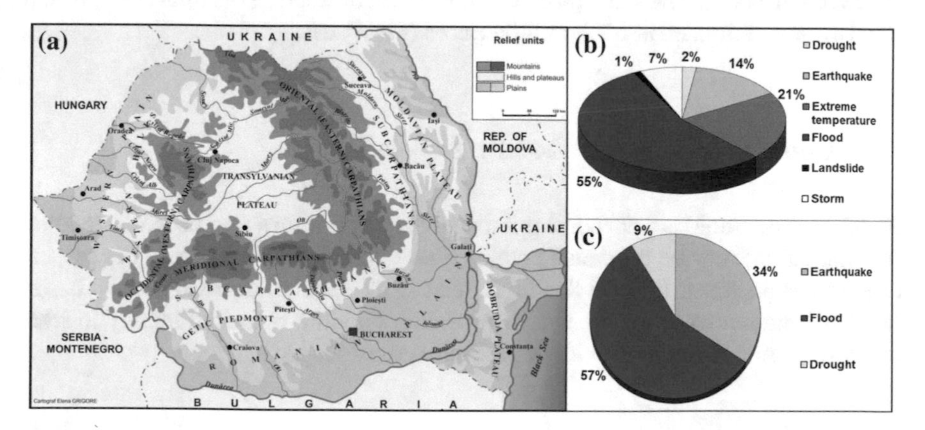

Fig. 9.1 **a** Map of major relief units in Romania; **b** distribution of number of disasters by types (1900–2016); **c** damage distribution by type of top 10 disasters (1900–2016). *Data source* for **b** and **c**: EM-DAT 2016

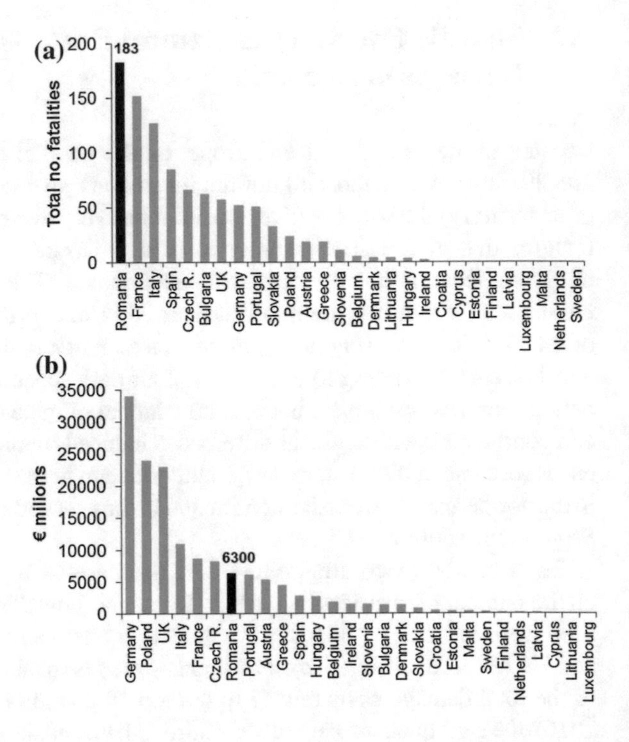

Fig. 9.2 Number of fatalities related to floods (**a**) and total costs over all flood events (**b**) in EU Member States (2001–2013). *Data source* EU 2014

At European spatial scale, Romania is in the top 10 of flood disasters statistics. According to EU (2014), between 2000 and 2013, Romania was the country with the highest number of fatalities caused by floods (183, representing 19% of the total number of deaths across the 28 EU Member States) (Fig. 9.2a). With 20 flood events recorded in the above mentioned period, Romania occupies the seventh place among EU Member States. The same place is occupied if considering the total costs overall flood events, estimated at 6300 million Euros (Fig. 9.2b).

9.3 Floods and Fluvial Flooding

Flood (or flash flood) is a phenomenon specific to the river flow regime while flooding (or inundation) may be generated both by rivers (fluvial flooding) and other causes: heavy rains or storms (pluvial flooding), sea advancement or encroachment (coastal flooding), groundwater rising, etc. In Romania, flooding is related mainly to river overflow during the periods with large discharges or ice jam.

Generally, the mean multiannual discharges of the Romanian rivers are less than 25 m^3/s, and only four rivers exceed 100 m^3/s: Siret (222 m^3/s), Olt (180 m^3/s), Mureş (179 m^3/s), and Someş (121 m^3/s) (Gâştescu et al. 1983). During the most important floods, maximum discharges of major internal rivers exceed 2000–3000 m^3/s (even 4000 m^3/s—Siret River). The Danube River, which develops on the Romanian territory its last 1075 km (i.e., 38% of its entire length of 2780 km) before reaching the Black Sea through the Danube delta, has a mean multiannual discharge rising from 5352 m^3/s at the entrance in the country to about 6450 m^3/s before entering the delta, but during the largest floods, the maximum discharge exceeded 15,000 m^3/s (Şerban et al. 2006; Sălăjan et al. 2011; Zaharia and Ioana-Toroimac 2013).

At the country scale, floods may occur every season, but they are more frequent in spring (30–50% of the total annual events) and summer (25–40%). In autumn and winter, it happens only 10–20%, respectively, 5–30% of the annual floods (Diaconu 1971a). The mean number of annual floods is 6–7 in the mountains and about 3 in the lowlands (Diaconu and Şerban 1994). The annual peak has mostly pluvial origins, with regional differences: between 50% in the north, and 100% in south-eastern regions. The annual floods originating from snow melting are less frequent (3–4% in the west and 0–2% in the rest of the country), while the mixed origins count for 25–30% in south and east, and for 10–15% in south-western areas (Diaconu 1971a; Diaconu and Şerban 1994).

In the small catchments, located mainly in mountainous regions, heavy rainfalls generate summer flash floods, lasting several hours. They have a high destructive potential, frequently amplified by dislocated material, turning sometimes into mud and debris flows. Some flash flood events had catastrophic consequences: the flash flood of 07/11–12/1999 from Râul Mare River's upper watershed (Southern Carpathians) led to 13 deaths and 24 injured (Bălteanu et al. 2004); the flash flood of 06/20/2006 from Ilişua River catchment (Someşul Mare watershed, in the western side of the Eastern Carpathians) caused 13 fatalities (Şerban et al. 2010); the flash flood of 09/05/2007 on the Tecucel River (south of Moldavian Plateau) affected Tecuci town, where three inhabitants died and over 200 houses were completely destroyed (Zaharia et al. 2009); the floods of 07/22–27/2008, on many rivers in the NE Carpathians (Suceava county) caused one death and major material damages (Bostan et al. 2009).

Within middle- and large-size watersheds (thousands of km^2), flood's duration is a few days or even few weeks. The Danube River, with a watershed of more than 800,000 km^2, generates slow floods, lasting up to two months (i.e., the flood of spring 2006), deriving usually from high amounts of spring precipitation overlapped to snow melt.

The largest floods and related inundations had dramatic consequences across many regions in Romania. One can mention as major inundations those occurred in: May 1912 (in Banat region), April and July 1932 (Crişuri and Mureş catchments), July 1940 (Ialomiţa watershed), July 1941 (Argeş—Vedea catchments), July 1969 (Upper Siret and Upper Prut watersheds), May 1970 (Someş—Tisa catchments), October 1972 (Jiu watershed), July 1975 (Olt, Argeş—Vedea, Ialomiţa—Buzău, Prut watersheds), August 1979 (Prut catchment), December 1995–January 1996 (Someş

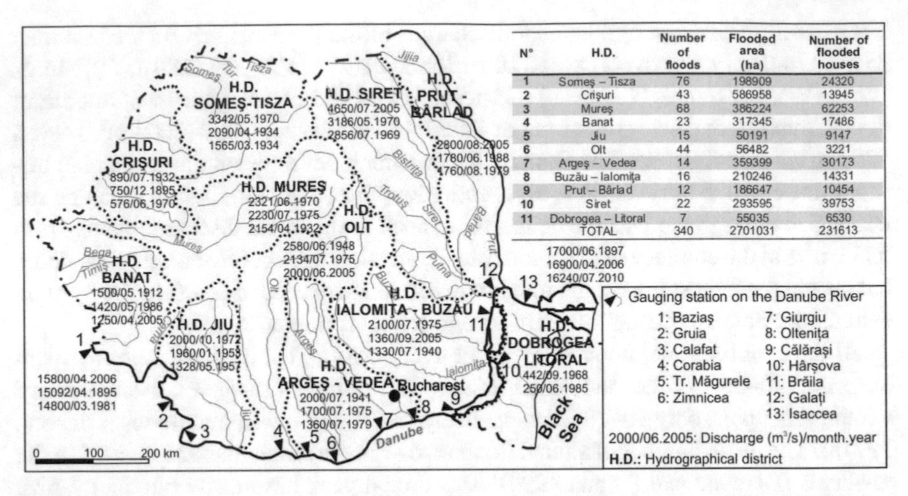

N°	H.D.	Number of floods	Flooded area (ha)	Number of flooded houses
1	Someş – Tisza	76	198909	24320
2	Crişuri	43	586958	13945
3	Mureş	68	386224	62253
4	Banat	23	317345	17486
5	Jiu	15	50191	9147
6	Olt	44	56482	3221
7	Argeş – Vedea	14	359399	30173
8	Buzău – Ialomiţa	16	210246	14331
9	Prut – Bârlad	12	186647	10454
10	Siret	22	293595	39753
11	Dobrogea – Litoral	7	55035	6530
	TOTAL	340	2701031	231613

Fig. 9.3 Largest three flood peaks recorded in Romania at the scale of hydrographical districts, and on the Danube River (the map). Data on the number of floods (exceeding the flooding stage), flooded areas and number of flooded houses (the table; reference period: 1970–2008). *Data sources* Mustăţea (2005), Mihailovici (2006), Oprişan (2006), Şerban et al. (2006), ANAR –INHGA (2012), Gâştescu and Ţuchiu (2012)

River watershed and Banat region), April, July, and October 2005 (almost the entire country), April–May 2006 (Danube River), July 2008 (Siret and Prut catchments), June–July 2010 (Prut, Siret, and Danube rivers), June–July 2014 (Jiu, lower Olt, and Argeş catchments), October 2016 (lower Siret catchment) (Mustăţea 2005; Şerbu et al. 2009; ANAR–INHGA 2012) (Fig. 9.3). Most of these floods are included in the Catalog of large floods in Europe in the twentieth century (Chorynski et al. 2012).

In recent years, an increase of frequency of large floods (with return period between 20 and 100 years) and their consequences has been identified in Romania. Thus, five of the nine floods included on the list of top 10 disasters by total people affected (recorded between 1900 and 2016), occurred in the period 2000–2006 (according to EM-DAT 2016). An increasing of flood damage costs was found after 2004 (Fig. 9.4).

In Romania, 83% of the administrative units (communes and towns) were affected by flooding between 1969 and 2008 (ANAR–INHGA 2012). In 2010, about 100,000 people have been exposed to floods (Pińskwar et al. 2012).

Many Romanian rivers reached their historic discharges in 2005, when the highest economic losses ever caused by floods and inundations in Romania were recorded: up to 1.3 billion Euros (MMGA 2006; Rădulescu et al. 2017) (Fig. 9.4). The floods in 2005 had large spatial and temporal extension affecting most of the country during seven successive episodes all year long (Mihailovici et al. 2006a; Zaharia et al. 2006). The most affected regions were south-west (Banat region), in April, Lower Siret Valley (in July), Argeş—Vedea and Ialomiţa catchment (in September). For example, in September 2005, the liquid discharge of Vedea River (in the central part of the Romanian Plain) exceeded 14 times the monthly average of the period

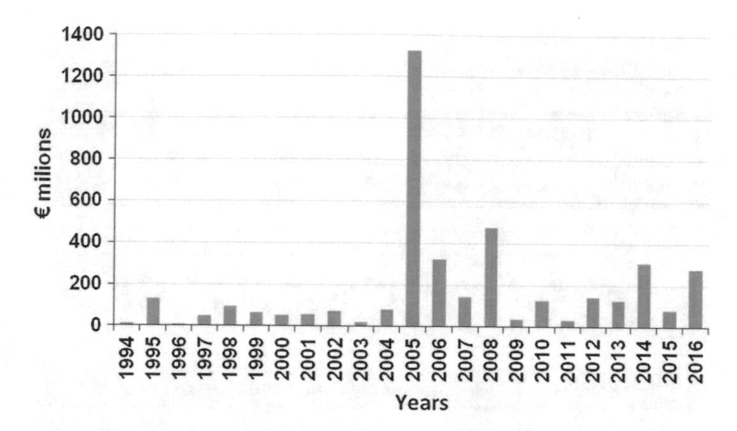

Fig. 9.4 Damage costs caused by flood events in Romania (1994–2016) *Data source* Rădulescu et al. (2017)

1961–2006, triggered by precipitation amounts over 4 times higher than the monthly average, at Alexandria weather station (Grecu et al. 2010).

In terms of human losses, the May 1970 floods killed the most people (215), while 76 deaths were registered in 2005 (EM-DAT 2016; MMGA 2006). Significant number of fatalities (108 persons) followed the flash flood of 07/29/1991, which caused Belci dam failure (in Eastern Carpathians), as well as the flood of July 1975 (60 death) (EM-DAT 2016; Mustăţea 2005). For discharges corresponding to the hundred-year return period, the total flooding area in Romania is estimated at 1028 million ha (5.8% of country's area), and the number of people exposed to flooding risk is about 929,000 (Oprişan 2006).

The flood of April–May 2006 can be considered historical for the Danube River. At the entry in Romania (at Baziaș gauging station), it was recorded the highest value since its establishment (in 1838), namely a maximum discharge of 15,800 m^3/s on 04/15/2006 (Mihailovici et al. 2006b). Along the Lower Danube, the maximum discharges in 2006 outclassed the hundred-year return period for many gauging stations (Baciu et al. 2006; Mihailovici et al. 2006b) (Fig. 9.5). The inundations affected 154 localities, and about 16,000 people were evacuated. Nearly 2000 houses were completely destroyed, and 443 were damaged (MMDD 2008). On the Lower Danube, historical discharges were registered during the flood of July 2010: 16,220 m^3/s at Galaţi and 16,240 m^3/s at Isaccea (Anghel et al. 2010). According to Şerban et al. (2006), the greatest discharges on Lower Danube were recorded in 1897, namely 17,000 m^3/s. In Fig. 9.6, the variation of the Danube River water level during the largest floods in the last half century (1970, 1981, 2006, and 2010) is presented.

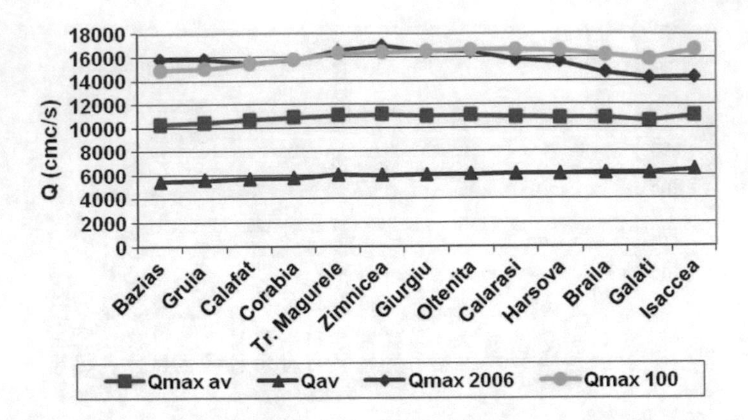

Fig. 9.5 Variability of the maximum flow in 2006 (*Q*max 2006) at gauging stations situated along the Romanian Danube River, compared with the average maximal discharge (*Q*max av), the average discharge (*Q*av), and the 100 years return period discharge (reference period: 1931–1994) *Data source* Baciu et al. 2006

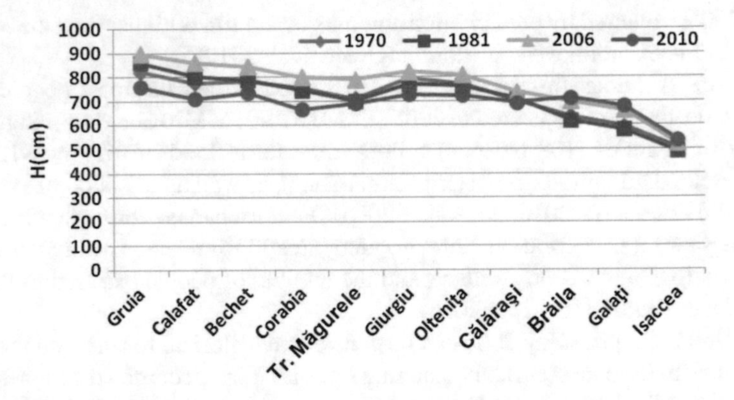

Fig. 9.6 Variability of the water level (H) at gauging stations situated along the Romanian Danube River during the largest floods in the last half century. *Data source* Anghel et al. (2010)

9.4 Low-Waters and Hydrological Droughts

Low-waters and hydrological droughts affect water resources as a consequence of precipitation deficit, but other factors, such as water freezing, permeable lithology, or water exploitation, are favoring such processes. In Romania, the low-waters and hydrological droughts are specific to summer–autumn as a consequence of precipitation deficit, and to winter, due to low amounts of precipitation and ice phenomena.

Depending on the frequency and duration of the hydrological droughts, the Romanian rivers are classified into permanent, semi-permanent (droughts only during very dry years), intermittent, or temporary (droughts every year) (Ujváry 1972). Permanent flow develops in mountainous rivers with catchments larger than 20 km^2, and semi-permanent flow occurs in the Subcarpathian hills and plateaus, where hydrological drought can last approximately 30 days in watersheds smaller than 1000 km^2 and 5–10 day in larger ones (Gâştescu et al. 1983). The hydrological drought reached 67 days in the Râmna River watershed in Curvature Subcarpathians (415 km^2) in 1952 (Zaharia 1999). Temporary flow is the result of heavy or long rainfalls and snowmelt, and it is specific to the western part of the Romanian Plain (Oltenia), the central and north-eastern part of the Moldavian Plateau (Bârlad), Piedmont areas and South Dobrudja (Diaconu and Şerban 1994). As relatively wide watersheds affected by hydrological droughts, one can distinguish: Vedea, Teleorman, Călmăţui, Sărata (in the Romanian Plain), Başeu, Bârlad, Vaslui (in the Moldavian Plateau), Teliţa, Taiţa, Casimcea (in the Dobrudja Plateau), and Amaradia (in the Getic Plateau). Spatial distribution of rivers according to drought types and data on droughts annual frequency can be found in the *Atlasul secării râurilor din România* (*Atlas of River's Droughts of Romania*) (IMH and IGFCOT 1974). Minimum annual mean monthly discharge with an exceeding probability of 95% (dilution flow) varies between 1 and 7 l/s.km^2 in mountainous regions and less than 0.1 l/s.km^2 in plateau and plain regions from south and east (Diaconu 1971b).

Between 1850 and 2008, four long periods of hydrologic deficit occurred in Romania (1858–1866, 1888–1908, 1942–1954, and 1982–1996 (Stanciu 2004); several years were excessively dry (1904, 1946, and 1990) (MADR 2008). After 2000, very dry years (2002, 2003, 2007, and 2012) alternated with years of high flows and historical floods (2005, 2008, and 2010). In the last decades, the increasing frequency of dry years and duration of hydrologic droughts have boosted the risk to associated hazards, especially in the western part of the Romanian Plain and in the Moldavian Plateau (MADR 2008).

In the case of Danube River, low-waters are specific to autumn and winter. During 1930–2010, the lowest flow recorded on the Romanian sector of Danube ranged between 990 m^3/s (at Gruia 1985) and 1790 m^3/s (before entering the delta, at Ceatal Izmail in 1947) (Gâştescu and Ţuchiu 2012).

The consequences of hydrological droughts, unlike those of floods, are hard to quantify, due to their complexity, higher spatial and temporal scales, as well as their relation to meteorological and pedological droughts. They affect social and economic activities, such as water supply for various demands and transportation on the Danube River. In Top 10 Natural Disasters in Romania, the complex drought of 2000 (meteorological, hydrologic, and pedologic) ranks the fourth place in terms of economic damages costs, estimated to 500 mil US$ (EM-DAT 2016).

9.5 Other River-Induced Hazards

Besides the extreme flows (high and low), rivers may cause negative impacts on society and environment by other damaging phenomena, such as water freezing/unfreezing, ice jams, sediment, and channel dynamics (erosion and deposition processes), as well as activation of slope geophysical processes (e.g., landslides, rockfalls, debris flow, and mudflow).

Freezing phenomena affect river flow, causing either floods and inundations, or low flow (up to river completely frost), leading to water supply deficiencies. The average total duration of river ice ranges from 20–40 days in the western part of Romania, to over 80–100 days in Northern Carpathians and Moldavian Plateau. In the winters 1953–1954 and 1963–1964, the river ice maintained for almost 150 days on rivers from eastern Transylvania and Moldavian Plateau (Gâştescu et al. 1983). Maximum duration of river ice may exceed 150 days, i.e., 151 days in winter 1986–1987, on Upper Bistriţa River, in Eastern Carpathians (Giurma and Ştefanache 2010).

On the Lower Danube, sheet ice maintains in average 15–20 days, but during the very cold winters, it can last up to 80–85 days (Miţă 1986), affecting the navigation and water supply in riparian localities.

The ice jams (called *zăpoare*, in Romanian) may cause severe flooding, so that they are important threats to settlements. The Bistriţa River (in the Eastern Carpathians) is representative for the high frequency of ice blocks. Their maximum duration was 84 days in the winter 2002/2003; ice blocks can reach 5–7 m height in critical areas (Giurma and Ştefanache 2010).

Sediment dynamics and reservoirs' silting may cause more or less significant economic and environmental damages. In Romania, the annual mean specific loads of suspended sediment range from less than 0.5 t/ha/year in mountainous regions and plains to over 20–25 t/ha/year in the Curvature Carpathians, with a national average of 2.06 t/ha/year (Diaconu 1971b; Mociorniţă and Birtu 1987). Reservoir dams stored about 200 million m^3 of silt during 15 years, with an annual rate of 13.4 million m^3, which represents more than 1/4 (27%) of the mean annual suspended sediment load in Romania (Rădoane and Rădoane 2005; Zaharia et al. 2011). The most affected are cascaded reservoirs on Olt and Argeş rivers, which filled almost half of their total volume with sediments. On the Argeş River, five of the 13 reservoirs are more than 70% silted, and one (Ogrezeni) is completely clogged (Teodor 1999; Rădoane and Rădoane 2005). Reservoirs' silting affects their water storage capacity and functionality, therefore generating economic losses.

Directly related to water and sediment flow is the **channel dynamics**, which is particularly active during floods. Erosion processes may affect the riverine area (buildings, infrastructures) causing economic damage. Sediment accumulation reduces the transport capacity of the river channel, favoring the flooding. Withal, sediment accumulation has negative consequences for fluvial navigation (e.g., on the Danube River). A recent study (Grecu et al. 2017) showed the role of floods and flooding on the river channel dynamics in Romania. As a result of large floods of 1970 and 1975, in Eastern Carpathians and Moldavian Subcarpathians, the river channels have incised by: 50 cm on Bistriţa River (at Frumosu), 80 cm on Trotuş River (at Ghimeş-Făget), 80 cm on Tazlăul Sărat River (at Lucăceşti), and 80–100 cm on Putna River (at Tulnici) (Rădoane et al. 1991). During the June–July 2010 flood, Siret River's channel (at Lespezi and Lungoci gauging stations) had aggradations ranging from 15 to 100 cm and degradations from 65 cm to 200 cm, in different flow phases (Obreja 2012). The flash flood occurred in Ilişua River basin on June 20, 2006 caused significant changes in the morphology of the river channel: in some places, it has widened by 2–3 m and up to 10–12 m; the sediment accumulations (mud, sand, clay, and gravel) have reached 0.2–1.5 m of thickness (Purdel 2011). Floods may also determine the evolution of braided islands; as an example, on a Subcarpathian reach of the Slănic River, after the flood of 2015 with a return period of approximately 7 years, braided islands reduced their number, grew in size, and became more elongated, therefore confirming their general model of evolution after disturbances by merging of several patches (Ioana-Toroimac 2018). In lack of high-magnitude and low-frequency floods, river channels are retracting especially when human pressures diminish the water and sediment resources; for example, the analysis of the lateral dynamics of the river channels in the area located at the contact between the Curvature Subcarpathians and the Romanian Plain have shown the narrowing of the active braided channels in the period 1980–2005: the mean width of the Prahova River channel diminished by 44%, the one of Milcov River by 43%, and the one of Cricovul Dulce River by 18% (Grecu et al. 2014).

9.6 Management of River-Related Risks in Romania

Reducing the negative consequences of river-induced hazards has become an important concern in the national water resources and related risks management policy. As a result of the significant damages caused by these hazards in the last decade (mostly floods, inundations, and droughts), Romania adopted some strategies and legislative regulations aiming to mitigate the risks induced by such phenomena. These strategies are consistent, generally, with the European Commission's Directives, of which, in the field of water management and associated risks, the most representative are *EU Water Framework Directive* (WFD) 2000/60/EC and the *EU Directive 2007/60/EC on the assessment and management of flood risks* (*Flood Directive*–FD). In Romania, the most important document regulating the flood risk management is the *National Strategy for Flood Risk Management* (*NSFRM*) *for medium and long term* (adopted in 2010). According to the requirements of the FD, in December 2015, *Flood Risk Management Plans* were elaborated for all the 11 hydrographic districts (River Basin Authorities) of the country (they were approved by the Government Decision 972/2016). These plans contribute to the achievement of the NSFRM objectives. Concerning the droughts, in 2008, the *National Strategy for Drought Mitigation, Prevention and Combating Land Degradation and Desertification, on short, medium and long term* was adopted. Unlike the previous documents, which emphasized the development of protection works (structural measures), the new strategies are more oriented toward non-structural measures and to develop society's capacity to adapt to extreme hydrometeorological hazards. In connection with this, noteworthy is *National Strategy for Climate Change* and the related *National Action Plan on Climate Change*, elaborated for the first time in 2005, and updated in December 2015. In 2008, the *Guide regarding the adaptation to climate change effects* was adopted; it foresees, among others, measures for flood risk management, including both hard (structural) and soft (non-structural) measures (Zaharia and Toroimac 2016).

9.7 Conclusions

In spite of the economic advantages they have, the rivers are responsible for significant societal and environmental damages. The most widespread and damaging river-induces hazards are the floods and flooding which cause annually major economic losses and human victims worldwide. This chapter focuses on Romania, European country with high exposure to river-related hazards, with the highest number of fatalities caused by floods (183) recorded between 2000 and 2013, representing 19% of the total number of death across the 28 EU Member States (EU 2014). At European spatial scale, Romania is in the top 10 of flood disasters statistics.

The chapter presents, in a synthetic way, the particularities of the specific river-born hazards in Romania, as floods and flooding (the most common river-induced hazards), low-waters and hydrological droughts, as well as other damaging phenomena (water freezing, ice jams, sediment, and channel dynamics reservoirs' silting), with relevant examples for the different hazards and references to their impacts.

In recent years (especially after 2004), an increase in the frequency of large floods and their consequences has been identified in Romania. Reducing the negative impact of these hazards has become an important concern in the national water resources and related risks management policy. In this sense, some strategies and action plans were adopted, of which the most relevant for flood risk mitigation are the *National Strategy for Flood Risk Management for medium and long term* (adopted 2010) and *Flood Risk Management Plans* (approved in 2016).

References

ANAR–INHGA (Administraţia Naţională Apele Române – Institutul Naţional de Hidrologie şi Gospodărirea Apelor) (2012) Sinteza studiilor de fundamentare a schemelor directoare de amenajare si management ale bazinelor hidrografice. http://www.mmediu.ro/protectia_mediului/evaluare_impact_planuri/2012-03-15_evaluare_impact_planuri_planamenajarebazinehidro2011.pdf. Accessed 30 Nov 2012

Anghel A, Frimescu L, Baciu O et al (2010) Caracterizarea viiturilor excepţionale din 2010. In: Proceedings of the annual conference of the National Institute of Hydrology and Water Management, Bucharest, 28–30 Sep 2010

Baciu O, Anghel E, Frimescu L et al (2006) Elaborarea prognozelor hidrologice pe Dunăre în perioada viiturii din intervalul aprilie – mai 2006. Hidrotehnica 51(5):21–30
Bălteanu D, Cheval S, Şerban M (2004) Evaluarea şi cartografierea hazardelor naturale şi tehnologice, la nivel local şi naţional. Studii de caz. In: Filip F, Simionescu B (eds) Fenomene majore de risc in Romania. București, Editura Academiei Române, pp 393–413
Bostan D, Mihăilă D, Tănasă I (2009) The abundant precipitations in the period 22nd–27th of July, 2008, from Suceava county and the surrounding areas. Causes and consequences. Riscuri şi catastrofe VII(6):61–70
Chorynski A, Pińskwar I, Kron W et al (2012) Catalogue of large floods in Europe in the 20th century. In: Kundzewich ZW (ed) Changes in flood risk in Europe. IAHS Special Publication 10, IAHS Press and CRC Press, Wallingford, p 27–54
Diaconu C (1971a) Râurile României. Monografie hidrologică, Institutul de Meteorologie şi Hidrologie, Bucureşti
Diaconu C (1971b) Probleme ale scurgerii de aluviuni a râurilor României. Studii de hidrologie XXXI:5–213
Diaconu C, Şerban P (1994) Sinteze şi regionalizări hidrologice. Editura Tehnică, Bucureşti
EM-DAT (2016) The emergency events database. Université catholique de Louvain (UCL) - CRED. www.emdat.be. Accessed 31 May 2016
EU (2014) Study on economic and social benefits of environmental protection and resource efficiency related to the European semester—DG Environment—February 2014. https://publications.europa.eu/en/publication-detail/-/publication/ef0c52c7-ed27-4b86-a4c3-e34d1bab4d1c. Accessed 13 July 2017
Gâştescu P, Diaconu D, Pişota I (1983) Apele. In: Badea L (ed) Geografia României. Geografia Fizică. I. Editura Academiei R.S.R, Bucureşti, pp 293–387
Gâştescu P, Ţuchiu E (2012) The Danube river in the lower sector in two hidrologycal hypostases—high and low waters. Riscuri şi catastrofe XI (10, 1):165–182
Giurma I, Ştefanache D (2010) Fenomene de iarnă pe râul Bistriţa între hazard şi vulnerabilitate. In: Proceedings of the annual conference of the National Institute of Hydrology and Water Management, Bucharest, 28–30 Sep 2010
Grecu F (2016) Hazarde şi riscuri naturale, 4th edn. Editura Universitară, Bucureşti
Grecu F, Zaharia L, Ghiţă C et al (2010) The dynamic factors of hydrogeomorphic vulnerability in the central sector of the Romanian plain. Metal Int XV 9(Special issue):139–148
Grecu F, Ioana-Toroimac G, Molin P et al (2014) River channel dynamics in the contact area between Romanian Plain and Curvature Subcarpathians. Revista de geomorfologie 16:5–12
Grecu F, Zaharia L, Ioana-Toroimac G et al (2017) Floods and flash-floods related to river channel dynamics. In: Rădoane M, Vespremeanu-Stroe A (eds) Landform dynamics and evolution in Romania. Springer, Cham, pp 821–844
Guha-Sapir D, Hoyois P, Below R (2016) Annual disaster statistical review 2015: the numbers and trends. http://www.cred.be/sites/default/files/ADSR_2015.pdf. Accessed 12 July 2017
IMH, IGFCOT (1974) Atlasul secării râurilor din România, Bucureşti
Ioana-Toroimac G. (2018) Decadal response of braided islands to disturbances: case studies in the Romanian Subcarpahians. Analele Universităţii din Bucureşti. Seria Geografie, accepted
Kundzewicz ZW, Pińskwar I, Brakenridge GR (2013) Large floods in Europe, 1985–2009. Hydrol Sci J 58(1):1–7. https://doi.org/10.1080/02626667.2012.745082
MADR (Ministerul Agriculturii şi Dezvoltarii Rurale) (2008) Strategia naţională privind reducerea efectelor secetei, prevenirea şi combaterea degradării terenurilor şi deşertificării, pe termen scurt, mediu şi lung. http://old.madr.ro/pages/strategie/strategie_antiseceta_update_09.05.2008.pdf. Accessed 15 July 2017
Mihailovici M (2006) Convieţuind cu viiturile. World Water Day Conference, Bucharest, 23 Mar 2006
Mihailovici M, Gabor O, Rândaşu S et al (2006a) 2005 floods in Romania. Hidrotehnica 51(6):23–35
Mihailovici M, Gabor O, Pătru Ş et al (2006b) Soluţii propuse pentru reamenajarea fluviului Dunărea pe sectorul românesc. Hidrotehnica 51(5):9–20

Miţă P (1986) Temperatura apei şi fenomenele de îngheţ pe cursurile de apă din România. Studii şi cercetări de hidrologie 54:3–182

MMDD (Ministerul Mediului şi Dezvoltării Durabile) (2008) Nota de fundamentare. http://www.mmediu.ro/legislatie/acte_normative/gospodarirea_apelor/HG_Lunca_Dunarii_MFP.pdf. Accessed 30 Nov 2012

MMGA (Ministerul Mediului şi Gospodăririi apelor) (2006) Raport privind efectele inundaţiilor şi fenomenelor meteorologice periculoase produse în anul 2005. http://www.mmediu.ro/vechi/departament_ape/gospodarirea_apelor/inundatii/raport_cmsu.pdf. Accessed 30 Nov 2012

Mociorniţă C, Birtu E (1987) Unele aspecte privind scurgerea de aluviuni în suspensie în România. Hidrotehnica 32(7):241–245

Mustăţea A (2005) Viituri excepţionale pe teritoriul României. INHGA, Bucureşti

Obreja F (2012) The sediment transport of the Siret River during the floods from 2010. Forum geografic. Studii şi cercetări de geografie şi protecţia mediului XI 1:90–99. https://doi.org/10.5775/fg.2067-4635.2012.064.i

Oprişan E (2006) Gestionarea situaţiilor de criză. Vulnerabilitatea la inundaţii. Ph.D. thesis, Universitatea Tehnică de Construcții București

Pătru I, Zaharia L, Oprea R (2006) Geografia fizică a României. Climă, ape, vegetaţie, soluri. Editura Universitară, Bucureşti

Pińskwar I, Kundzewicz ZW, Peduzzi P et al (2012) Changing floods in Europe. In Kundzewicz ZW (ed) Changing in flood risk in Europe. IAHS Special Publication 10, CRC Press Book, Wallingford, p 83–96

Purdel A (2011) Analiză asupra modificărilor geomorfologice majore produse de viitura excepţională de pe râul Ilişua din data de 20.06.2006. In: Conferinţa Ştiinţifică Anuală a Institutului Naţional de Hidrologie şi Gospodărire a Apelor, București, 1–3 Nov 2011

Rădoane M, Rădoane N (2005) Dams, sediment sources and reservoir silting in Romania. Geomorphology 71:112–125

Rădoane M, Ichim I, Pandi G (1991) Tendinţe actuale în dinamica patului albiilor de râu din Carpaţii de Curbură. Studii şi cercetări de geografie XXXVIII:21–31

Rădulescu D, Mătreaţă M, Chendeş V et al (2017) Considerations on flood risk in Romania. Conference air and water. Components of the environment, Cluj-Napoca, 17–19 March 2006

Romanescu G (2006) Inundaţiile ca factor de risc. Editura Terra Nostra, Iaşi

Romanescu G (2008) Evaluarea riscurilor hidrologice. Editura Terra Nostra, Iaşi

Sălăjan L, Francu A, Duţă A, (2011) Analisys of maximum flow on the Danube River in 2010. In: Sesiunea anuală de comunicări Institutul Naţional de Hidrologie şi Gospodărire a Apelor, Bucureşti, 1–3 Nov 2011

Sorocovshi V (ed) (2002–2017) Riscuri şi catastrofe. 1–20, Editura Casa Cărţii de Ştiinţă, Cluj-Napoca

Sorocovschi V (2016) Riscuri naturale. Aspecte teoretice şi aplicative, Editura Casa Cărţii de Ştiinţă, Cluj-Napoca

Stanciu P (2004) Caracteristicile viiturilor şi secetelor. Hidrotehnica 49(2–3):27–33

Şerban P, Gălie-Şerban A, Buţă C (2006) Analiza viiturii produse pe Dunăre în perioada aprilie – mai. Hidrotehnica 51(5):3–8

Şerban G, Selagea H, Máthé E (2010) Efecte produse de viitura din 20.06.2006 în bazinul râului Ilişua (bazinul Someşul Mare). Conference Air and Water. Components of the environment, 156–165

Şerbu M, Obreja F, Olariu P (2009) Viiturile din anul 2008 în bazinul superior al Siretului. Cauze, efecte, evaluare. Hidrotehnica 54(12):38–44

Teodor S (1999) Lacul de baraj şi noua morfodinamică. Studii de caz pentru râul Argeş, Editura Vergiliu, București

Ujváry I (1972) Geografia apelor României. Editura Ştiinţifică, Bucureşti

Zaharia L (1999) Resursele de apă din bazinul râului Putna. Editura Universităţii din Bucureşti, București, Studiu de hidrologie

Zaharia L, Ioana-Toroimac G (2013) Romanian Danube River management: impacts and perspectives. In: Arnaud-Fassetta G, Masson E, Reynard E (eds) European continental hydrosystems under changing water policy. Friedrich Pfeil Verlag, München, pp 159–170

Zaharia L, Ioana-Toroimac G (2016) Developing soft measures for flood risk mitigation and adaptation in Romania: public informing and awareness. Riscuri si catastrofe XV(18,1):7–22

Zaharia L, Beltrando G, Nedelcu G et al (2006) Les inondations de 2005 en Roumanie. In: Actes du XIXème Colloque International de Climatologie, Epernay, 6–9 Sep 2006

Zaharia L, Catană S, Popa D et al (2009) Synergetic factors of the catastrophic flood affecting Tecuci City (Romania) in September 2007. In: Proceedings of the final conference of the COST action C22 Urban Flood Management in cooperation with UNESCO-IHP, Paris, 26–27 November 2009

Zaharia L, Grecu F, Ioana-Toroimac G et al (2011) Sediment transport and river channel dynamics in Romania—variability and control factors. In: Manning AJ (ed) Sediment transport in aquatic environments. InTech, Rijeka, pp 293–316

Part III
Polluted Water in Urban Areas

Chapter 10
Contributors to Faecal Water Contamination in Urban Environments

Lisa Paruch and Adam M. Paruch

Abstract Faecal contamination of water has both anthropogenic and zoogenic origins that can shade various point and nonpoint/diffuse sources of pollution. Due to the dual origin and number of sources of faecal contamination, there are immense challenges in the implementation of effective measures to protect water bodies from pollution that poses threats to human and environmental health. The main health threats refer to infections, illnesses and deaths caused by enteric pathogenic microbes, in particular those responsible for waterborne zoonoses. To detect and identify the origins and sources of faecal pollution simultaneously, various methods and indicators have been compiled into a comprehensive measuring toolbox. Molecular diagnostics using genetic markers derived from *Bacteroidales* 16S rRNA gene sequences are quite prevalent in the current methodological implementation for the identification of faecal contamination sources in water. For instance, a culture- and library-independent microbial source tracking toolbox combining micro- and molecular biology tests run as a three-step procedure has been implemented in Norway. Outcomes from the Norwegian studies have identified two general trends in dominance of contributors to faecal water contamination in urban environments. Firstly, there is a tendency of higher contributions from anthropogenic sources during the cold season. Secondly, the identification of the dominance of zoogenic sources in faecal water contamination during warm periods of the year.

Keywords *Bacteroidales* · Faecal indicator bacteria · Gene markers
Microbial source tracking · Water pollution

L. Paruch · A. M. Paruch (✉)
Division of Environment and Natural Resources, NIBIO—Norwegian Institute of Bioeconomy Research, Pb 115, 1431 Aas, Norway
e-mail: adam.paruch@nibio.no

L. Paruch
e-mail: lisa.paruch@nibio.no

M. Zelenakova (ed.), *Water Management and the Environment: Case Studies*, Water Science and Technology Library 86,
https://doi.org/10.1007/978-3-319-79014-5_10

Table 10.1 General concentrations of the most common microorganisms in human faecal matter of healthy or infected individuals (Edberg et al. 2000; WHO 2006)

Organisms	Numbers per gram of faecal matter
Bacteria:	
– *Escherichia coli (E. coli)*	10^9
– *Salmonella* spp.	10^4–10^8
– *Campylobacter jejuni*	10^6–10^9
– *Shigella* spp.	10^7
– *Vibrio cholerae*	10^7
Viruses:	
– Enteroviruses	10^4–10^9
– Rotaviruses	10^7–10^{11}
Protozoa:	
– *Cryptosporidium parvum*	10^7–10^8
– *Giardia intestinalis*	10^5–10^8
– *Entamoeba histolytica*	10^5–10^8
Helminths:	
– *Ascaris lumbricoides*	1–10^5
– *Schistosoma mansoni*	1–10^3
– *Clonorchis sinensis*	10^2

10.1 Concise Facts on Faecal Contamination

In general, faecal contamination refers to any kind of pollution caused solely or partially by faecal matter, or pollution that contains any portion of this matter. The faecal matter characterises wastes from metabolic processes occurring in a gastrointestinal/digestive tract (gut) of humans and other animals, and are defecated as faeces (solid or semisolid wastes) through anus (in most mammals) or as excreta (faeces and urine) through cloaca (in birds, reptiles and amphibians).

The gut is a habitat of trillions of various organisms among which bacteria dominate with approximately 500 different species (Marotz and Zarrinpar 2016; Quigley 2013). The gut bacteria comprise the major number of the microbes in the whole body and constitute about 10 times more than all body cells (Quigley 2013). Taking a vital part in the metabolic process, these microbes are continuously defecated; hence, faecal matter contains an abundance of live microorganisms (Marotz and Zarrinpar 2016). The number and variety of faecal microbes depend greatly on animal species, but even within the same sort, there are substantial variations. In humans, concentrations of faecal microbes vary according to, e.g., gender, age, health conditions, diet, physical activities, lifestyle and region of living (Table 10.1).

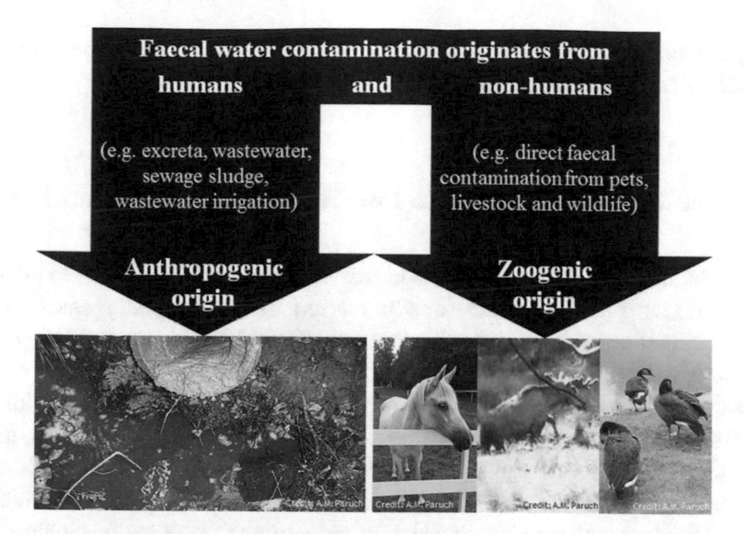

Fig. 10.1 Anthropogenic and zoogenic origins and multiple sources of faecal water contamination *Source* Authors

10.2 Origins and Sources of Faecal Water Contamination

Faecal contamination of water has both anthropogenic and zoogenic origins that can shade multiple sources of pollution (Fig. 10.1). The main human sources include direct disposal of excreta, leakages from sewers, overflows from sewage pumping stations, uncontrolled discharges from treatment plants, inadequate sewage sludge handling and insufficient performance of decentralised wastewater treatment systems. Non-human sources are characterised by animal faecal contamination directly from pets, livestock and wildlife, or indirectly from the improper utilisation of manure, slurry and other materials containing animal faeces. From all of these sources, a high number of faecal microbes (viruses, bacteria and parasites) can directly contaminate groundwater and surface water bodies (drinking-, irrigation- and recreation water) or indirectly from soil and vegetation through agricultural drainage, irrigation and organic fertilisation, particularly during and after heavy precipitation and subsequent run-offs (Paruch et al. 2015a).

Due to the dual origin of various point and nonpoint/diffuse sources of faecal water pollution, there are immense challenges in the implementation of efficient measures protecting water bodies from contamination that poses threats to human and environmental health. Normally, it is possible to localise point sources of faecal water pollution (typically, direct discharge of wastewater), while it is more problematic to locate inflows from diffuse faecal pollution sources (usually, storm water-, urban water- and agricultural run-offs contaminated with faecal matter from humans, pets, livestock and wild animals). It is therefore crucial to identify the primary origins of faecal contamination as to act upon their elimination, if possible directly at the source or if this is not applicable then at least to minimise the exposure to various contaminating agents, and hence to reduce potential health risks of waterborne infections and diseases to humans and animals.

10.3 Health Risks Related to Faecal Water Contamination

The main health threats associated with faecal water contamination refer to infections, illnesses and, in plenty of cases, deaths caused by enteric pathogens (infectious agents causing diseases), in particular those causing waterborne zoonoses (zoonotic infections and diseases transmitted between animals and humans through water). It has been recently reported that 5 million people, including 1.5 million children, die every year as a result of water-related diseases (IWFA 2017). The numbers of infected and ill individuals are vastly higher, although not completely reported as some cases have been ignored due to minor abdominal and diarrhoeal symptoms. Mortality and burden of disease resulting from faecal contamination of water represent almost 10% of the total burden of human disease worldwide (WHO 2017).

Water-related infections and diseases are normally characterised by four main categories (Moe 2004), in which water is the major, but not the only transmission route of pathogens. These categories are defined as follows:

- water-borne infections representing classic examples in which pathogenic organisms enter water sources through faecal contamination,
- water-washed infections that occur due to lack of adequate water and sanitation facilities, hence poor hygienic manners,
- water-based infections that are caused generally by pathogens spending part of their life in the aquatic environments and
- water-related insect vectors which are associated with infections transmitted by insects breeding in or near water.

Water-related pathogenic organisms and toxins produced by these organisms (which are of particular health concern as they are significant virulence causes of microbial pathogenicity), as well as the infections and disease symptoms they cause, may be characterised by more than one category. In addition, the pathogenic microbes and toxins have various transmission pathways (Pond 2005; WHO 2011), with the most common routes through:

- ingestion (e.g. *E. coli*, *Salmonella* spp., *Vibrio cholerae*, *Shigella* spp., *Campylobacter* spp., *Helicobacter* spp., *Cryptosporidium parvum*, *Giardia intestinalis*, enteroviruses, noroviruses, hepatoviruses and rotaviruses),
- direct contact (e.g. *Pseudomonas aeruginosa*, *Aeromonas* spp., *Mycobacteria* spp., *Acanthamoeba* spp., *Naegleria* spp., *Leptospira* spp. and *Schistosoma* spp.) and
- inhalation and aspiration (e.g. *Legionella* spp. and *Mycobacteria* spp., adenoviruses and enteroviruses).

Table 10.2 General classification of water-related zoonoses (Moe 2004)

Classes	Examples of zoonotic diseases
1. Waterborne through drinking water	Balantidiasis, campylobacteriosis, cryptosporidiosis, cysticercosis, *E. coli* O157:H7, giardiasis, microsporidiosis, salmonellosis, toxoplasmosis, tularaemia and yersiniosis
2. Waterborne through recreational water	Cryptosporidiosis, giardiasis and leptospirosis
3. Water-based infections	Dracunculiasis and schistosomiasis
4. Water-washed infections	Cryptosporidiosis, giardiasis, hepatitis viruses
5. Water-related insect vectors	West Nile virus, Rift Valley fever virus, yellow fever virus, sleeping sickness
6. Water/wastewater aerosols inhalation	Legionellosis
7. Aquatic food consumption	Paragonimiasis

Some pathogenic microbes (e.g. *Aeromonas* spp., *Pseudomonas* spp., *Vibrio vulnificus* and *Vibrio parahaemolyticus*) can also colonise through wound infections (Pond 2005). Furthermore, other pathogens (e.g. *Campylobacter* spp., *E. coli*, *Salmonella* spp., *Vibrio* spp. and *Shigella* spp.) and toxic species (e.g. *Gambierdiscus, Gonyaulax*, *Gymnodinium* and *Paragonimus*) might be transmitted through a raw edible consumption of infected and faecally contaminated aquatic animals and plants (Moe 2004).

Based on the criteria of water-related infections and diseases, and transmission routes of pathogens and toxins, a general classification of water-related zoonotic diseases has been suggested by World Health Organization (Moe 2004). Seven main classes exemplifying common zoonoses have been distinguished, as presented in Table 10.2.

The criteria defining water-related infections and diseases are not only characterised by zoonoses as the pathogen pathways, but also have person-to-person and vector-borne transmissions (Moe 2004). Although the pathogenic organisms might come from both humans and animals, and/or occur naturally (e.g. *Legionella* spp. causing legionellosis), the zoonotic pathogens still comprise 75% of emerging infectious diseases (Bolin et al. 2004). Furthermore, human and animal faecal contaminations constitute the largest load of pathogens associated with waterborne disease transmission. The faecal pathogen groups with their associated organisms most common in human and animal excreta, as well as symptoms and diseases caused by these pathogens are presented in Table 10.3.

Table 10.3 Examples of faecal pathogens causing acute disease outcomes (Kanarat 2004; Pond 2005; Suresh and Smith 2004; WHO 2006)

Pathogen groups and organisms	Symptoms and diseases
Bacteria:	
– *Escherichia coli (E. coli)*	Urinary tract infection, haemolytic-uraemic syndrome, colitis with diarrhoea
– *Salmonella* spp.	Fever, abdominal pain, diarrhoea or constipation
– *Campylobacter* spp.	Diarrhoea, abdominal pain, fever
– *Helicobacter pylori*	Gastritis, intestinal metaplasia and gastric cancer
– *Enterococcus faecalis*	Endocarditis and bacteraemia, urinary tract infection, meningitis
– *Clostridium perfringens*	Abdominal cramping and diarrhoea
– *Shigella* spp.	Shigellosis, abdominal pain, diarrhoea, fever
– *Vibrio cholerae*	Cholera, muscle cramps, vomiting, diarrhoea
Viruses:	
– Hepatitis A and E	Hepatitis, fever, abdominal pain, diarrhoea
– Adenoviruses	Respiratory disease, eye infection
– Rota-, noro-, enteroviruses	Gastroenteritis, fever, vomiting, abdominal pain, diarrhoea
Protozoa:	
– *Cryptosporidium parvum*	Cryptosporidiosis, fever, crampy abdominal pain, watery diarrhea
– *Giardia lamblia*	Giardiasis, fever, abdominal pain, diarrhoea
– *Entamoeba histolytica*	Amoebiasis, abdominal pain, bloody diarrhoea
Helminths:	
– *Ascaris lumbricoides*	Ascariasis, fever, abdominal swelling and pain, diarrhoea
– *Schistosoma mansoni*	Schistosomiasis, abdominal pain, diarrhoea, urinary tract infection
– *Clonorchis sinensis*	Clonorchiasis, abdominal pain, nausea, diarrhea
– *Diphyllobothrium latum*	Diphyllobothriasis, vomiting, abdominal discomfort, diarrhea
– *Fasciolopsis buski*	Fasciolopsiasis, abdominal pain, chronic diarrhoea, anemia

10.4 Detection of Faecal Water Contamination and Identification of Pollution Sources

To detect and identify the origins and sources of faecal pollution simultaneously, various methods and indicators have been combined into a comprehensive measuring toolbox. In general, there are two main methods that have been applied for tracking of faecal water contamination, chemical source tracking (CST) and microbial source tracking (MST). Various chemical substances and components as well as different microbial indicators and markers have been employed in these methods, which are therefore also known under other terms as, e.g., bacterial source tracking or faecal source identification (Field 2004).

When using the CST method, chemical detection may provide supplementary evidence on the faecal source (Staley et al. 2016; Harrault et al. 2014; Hartel et al. 2008). Caffeine, faecal sterols and stanols, bile acids, laundry brighteners, fragrances and pesticides can be used as chemical indicators and molecular tracers to aid in the identification of faecal inputs, but they do have limits to their use, as the chemical indicators respond differently to many environmental factors (Tran et al. 2015). Therefore, the CST methods should be applied in the combination with the MST methods.

Overall, there are two main categories within MST, i.e. culture-based and culture-independent methods. Both categories can further be subdivided into library-dependent and library-independent approaches (Hagedorn et al. 2011). Notably, under the first category, antibiotic resistance mapping (Olivas and Faulkner 2008) and other phenotypic methods, e.g. carbon-source utilisation profiling (Smith et al. 2010) and fatty acid methyl ester profiling (Duran et al. 2009) for source tracking, utilise the biological traits (phenotypes) to classify the sources. Genotypic library-dependent methods, like ribotyping, repetitive extragenic palindromic polymerase chain reaction, amplified fragment length polymorphism and pulsed-field gel electrophoresis, are DNA fingerprinting techniques based on the established amplicons' library (Field 2004). Sorting/clustering of microbe groups is accomplished by directly comparing the generated DNA polymorphisms (Carson et al. 2003). This is quite technically demanding, and the results are less reproducible. In comparison, the culture- and library-independent methods are remarkably more time efficient, are less labour intensive and are more accurate.

In the molecular culture- and library-independent methods, some faecal viruses have been selected as good candidates for detection purpose. For instance, human-specific adenoviruses and enteroviruses (Bambic et al. 2015), and bovine/ovine adenoviruses (Ahmed et al. 2013) are highly host specific. However, due to the small size of viruses and low viral load, a large amount of water is normally required for a concentrated sample. An enrichment step to facilitate the capture of viruses is also required. In terms of anaerobic bacterial genes, animal-specific *Bifidobacterium* spp. (e.g. *B. dentium* and *B. adolescentis*) became targets in markers development (Venegas et al. 2015). In addition, host-specific toxin genes in *E. coli* and Enterococci can be targeted for source determination, for example human-specific ST1b toxin (Moyo et al. 2007), pig-specific ST1b toxin (Khatib et al. 2003) and enterococcal surface protein (Scott et al. 2005). As the toxin target genes are rare and thus need the enrichment procedure, the final detection may only be semi-quantitative and it also inherits instability due to the horizontal transfer of genes (Böhm et al. 2015). Host-specific *Bacteroidales* genetic markers are by far the most tested/optimised and exhibited in most cases geographical stability across USA, Canada, Europe, New Zealand and Japan (Kobayashi et al. 2013; Mieszkin et al. 2013; Sowah et al. 2017).

Regardless of the variety of markers that have been applied in MST surveys, many of them are still under comparable testing and verification processes, while others are less applied in practice. Yet, the molecular diagnostics using genetic markers derived from *Bacteroidales* 16S rRNA gene sequences are quite prevalent in the current methodological implementation for the identification of faecal contamination sources in water.

10.5 Methodological Toolbox Discriminating Dominant Sources of Faecal Water Contamination

Numbers of host-specific *Bacteroidales* genetic markers have been developed to discriminate faecal pollution between human and different warm-blooded animal species (Dick et al. 2005; Layton et al. 2006; Reischer et al. 2007; Shanks et al. 2008; Tambalo et al. 2012). These can be further employed in various attempts focussing on providing detailed profiling of markers contributions in the faecal contamination and defining the dominant source(s) of this pollution. One of such attempts has been undertaken in Norway, where a culture- and library-independent MST toolbox has been utilised since 2013.

The Norwegian approach focusses on faecal contamination of aquatic ecosystems, mainly in urban and agricultural landscapes, as well as catchments of drinking water reservoirs, as these significantly influence human and environmental health. The developed methodological toolbox has been described in greater detail elsewhere (Paruch et al. 2015b). Briefly, it combines micro- and molecular biology and consists of three independent steps:

1. microbial analyses of faecal water contamination based on the detection of *E. coli*,
2. molecular DNA tests using real-time quantitative polymerase chain reaction (RT-qPCR) for the detection and quantification of host-specific *Bacteroidales* 16S rRNA genetic markers and
3. profiling of the genetic markers contribution in the detected faecal contamination.

In step one, *E. coli* bacteria have been used as the historical and most frequent faecal indicator employed. These bacteria have also often been applied in other MST studies on faecal water contamination (Åström et al. 2015; Shahryari et al. 2014; Tambalo et al. 2012). Although *E. coli* greatly satisfies most of the criteria of faecal indicator bacteria, i.e. has dominant faecal origin, is present in large numbers in faeces of human and warm-blooded animals and is rapidly detectable by simple methods (Paruch and Mæhlum 2012), it does not satisfactorily fit into the criteria of a source identifier. This is due to low host specificity, possible replication in the environment, as well as geographic and temporal variability (Farnleitner et al. 2010; Field and Samadpour 2007; USEPA 2005). Therefore, bacteria belonging to the phylum *Bacteroidetes*, especially *Bacteroidales* spp., have widely been applied in various MST studies with molecular diagnostics based on RT-qPCR (Dick et al. 2005; Lamendella et al. 2009; Layton et al. 2006; Reischer et al. 2007; Shanks et al. 2008; Tambalo et al. 2012).

In step two, the performance of *Bacteroidales* genetic markers, in terms of sensitivity and specificity, needs to be evaluated prior to their adaptation. Furthermore, analyses of melting curves are strongly recommended as it has been proved that they are essential in discriminating strains of intestinal and non-intestinal *Bacteroidales* bacteria (Paruch and Paruch 2017). This is of high importance as *Bacteroidales* are environmental bacteria, but still species of the genus *Bacteroides* comprise the largest portion of the gut microbes and normally constitute about 30% of total faecal bacteria (Layton et al. 2006). They can even make up to 52% of human faecal flora (Dick et al. 2005) with concentrations up to 10^{11} organisms per gram of faeces (McQuaig et al. 2012). In addition, these bacteria are strictly anaerobic, having little potential for growth in the environment (Dick et al. 2005; USEPA 2005) and are highly host-specific, thus enabling distinguishable host's identification (Layton et al. 2006).

Since there are no significant correlations between *E. coli* bacteria and the host-specific *Bacteroidales* genetic markers (Harwood et al. 2014), only the percentage profile of markers contribution in the measured faecal contamination can be further assessed in step three.

10.6 Exposure of Urban Catchments to Various Faecal Pollution Sources

When faecal contamination of water occurs in urban areas, it is quite common to assume that it was caused primarily by leaks from sewer systems, uncontrolled discharges of wastewater, overflows from sewage pumping stations and floods after extreme precipitation. In addition, the practices of collecting urban run-off (mainly storm water, but also wash water after maintenance of roads, railways, bridges and tunnels) jointly with sewage into the sewer system contribute to overflows and overloading of wastewater treatment plants. Furthermore, climate change predictions on frequent episodes of extreme precipitation will also expect more overflows as the urban drainage systems get easily overloaded. These scenarios have already been observed, in particular when related to abrupt rainfall (usually short but intensive) followed by run-off predominating over water infiltration, especially in tight urban areas with dense surfaces (concrete, steel and asphalt). It is therefore often taken for granted that faecal water contamination in urban areas is mainly derived from sewer systems; hence, it is entirely of anthropogenic origin. Such assumptions must, however, be revised since even the modern cities are also living areas for a variety of animal species, not just pets like dogs and cats, but also wildlife, and these can make a significant zoogenic contribution to faecal water contamination in urban catchments.

Not all that comes from the sewer system is anthropogenic. Sewerage pipelines offer suitable environments for common sewer rats, also known as Norwegian rats (*Rattus norvegicus*), which thrive with food residues and fat rests deposited in the sewer systems all year-round. This, in fact, is a historical problem of growing cities in which the rat concentration might exceed the city's population; e.g., there were reported more rats (up to one million individuals) than citizens of Oslo (Aftenposten 2013), three times more rats than people in Stockholm (DN 2016) and as many as four rats per 100 m of sewage pipeline in Copenhagen (Fettvett 2016). It has also been documented that excreta of rats represent a risk for public health as they contain both zoonotic and multiresistant bacteria (Guenther et al. 2013). In general, the large number and diversity of pathogens enter sewer systems through four main routes (Gerardi 2006), representing both anthropogenic and zoogenic origins:

1. domestic wastewater,
2. industrial wastewater, e.g. food production and processing,
3. inflow and infiltration of animal excrements and
4. excreta of inhabitants of sewer systems, mainly rats.

In addition, human and livestock wastes in urban areas can attract various animals not only to be fed occasionally, but also to live around (e.g. rats, pigeons and crows), and there is a high prevalence of human pathogens in such wildlife (Benskin et al. 2009; Scheffe 2007).

The intensive growth of the human population results in extending urban areas to huge metropolises, great agglomerations, megacities and supercities (megalopolises). These expanding areas reduce the natural habitat of wildlife, which in many cases adopt their lifestyle to the new situations and move close to, or into the cities, where food and settlements can be easily found. In addition, an increased trend in the development of "green cities" and an extensive evolution of "blue-green" solutions for urban run-off also open various options for habitats of different animals in the cities. For instance, the re-opening of watercourses that previously run in pipes is not only an important strategy for meeting challenges of changing climate, but it offers an attractive landscape element for the city population. These areas also attract wildlife and create new inner-city ecosystems. Consequently, a large variation in urban wildlife can be found and contribute to faecal water contamination. This wildlife is represented mostly by different species of birds (e.g. gulls, pigeons, crows, rooks, ravens), in particular waterfowl (swans, geese and ducks), and wild mammals (e.g. raccoons, foxes, rats, beavers and bats).

Apart from the habitats of wildlife in cities, there are also an increasing number of pets, particularly dogs and cats, in proportion to the city's population. Dogs, in particular, are the significant hosts for pathogenic microbes, e.g. *Giardia* spp. and *Salmonella* spp. (Schueler 2000). Furthermore, it becomes quite popular to

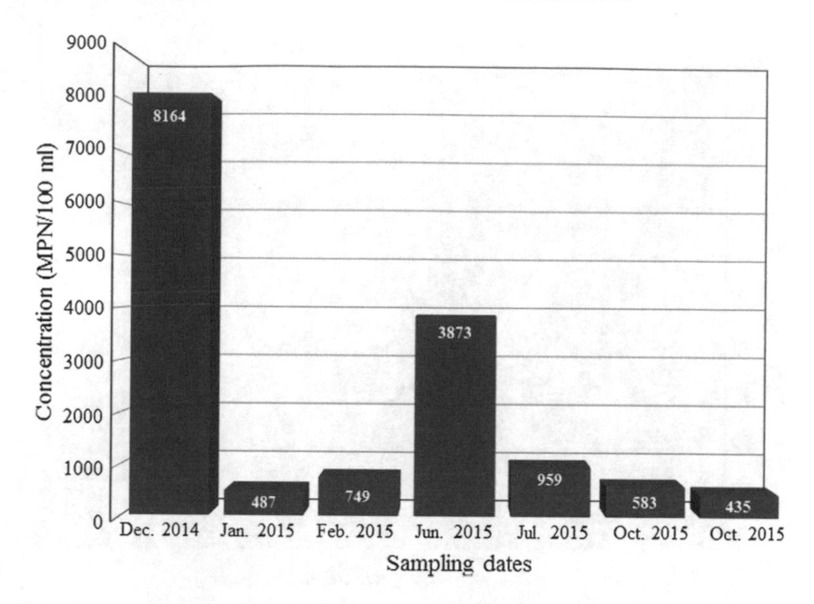

Fig. 10.2 Concentrations of *E. coli* bacteria expressed as the most probable number (MPN)/100 ml of water samples collected from Aker River, Oslo, Norway *Source* Authors

establish "educational" or "hobby" farms with livestock to be presented year-round for city people. Another trendy wave that comes to the cities is to organise riding clubs/centres. It is therefore quite common that cattle and horses grazing pastures become natural elements and landscapes of the urban catchments.

The zoogenic contribution to faecal water contamination in urban areas can be dominant in some periods, as reported by recent Norwegian MST studies (Paruch et al. 2017). One of the studies was conducted on water samples from Akerselva (Aker River) that flows through Oslo, the capital and the most populated city in Norway. Although the samples were collected at irregular intervals (from December 2014 to October 2015), all of them were faecally polluted (Fig. 10.2). The highest *E. coli* concentration was observed at the same occasion as the human marker revealed its dominance (96%) in faecal water contamination (Fig. 10.3). Anthropogenic origin of this contamination was detected in all water samples, but most samples had dominant zoogenic origin, with highest genetic marker contribution of 97% to faecal water contamination. Another study was performed on water samples from Blåveisbekken, a stream which in large part flows in a culvert in Ski town located approximately 20 km south-east of Oslo. All the samples collected during the course of nearly two years (from November 2014 to September 2016) revealed faecal contamination (Fig. 10.4). Almost all the highest *E. coli* concentrations represented anthropogenic origin, while zoogenic origin was dominant in only one-third of the samples (Fig. 10.5).

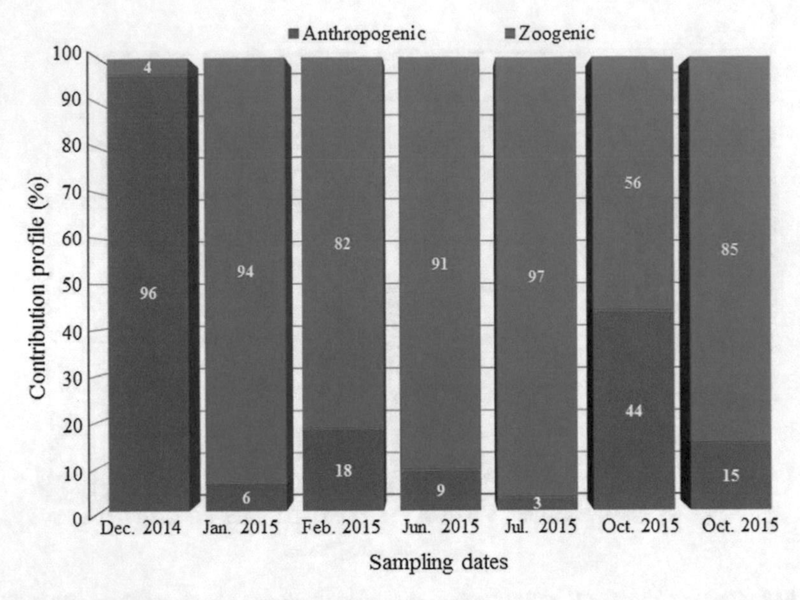

Fig. 10.3 Contribution profile of genetic markers in faecal contamination of water samples collected from Aker River, Oslo, Norway *Source* Authors

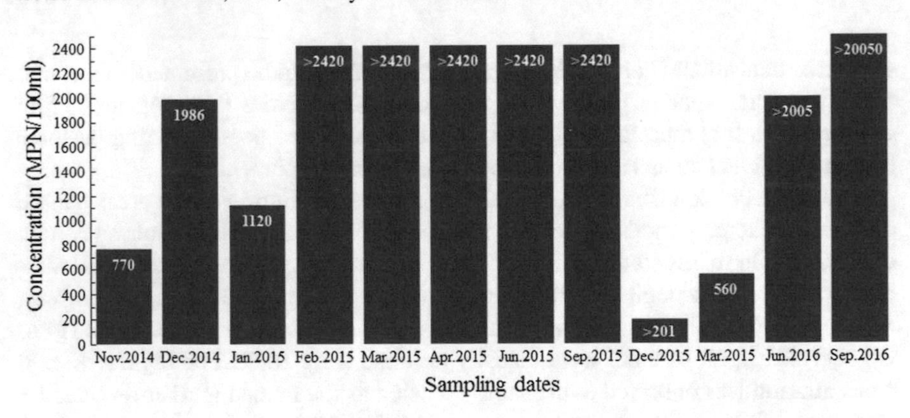

Fig. 10.4 Concentrations of E. coli bacteria expressed as the most probable number (MPN)/100 ml of water samples collected from Blåveisbekken stream, Ski, Norway *Source* Authors

Results from the Norwegian MST studies exhibited two general trends in dominance of contributors to faecal water contamination in urban environments. In the cold season, particularly in autumn, winter and spring, the observed tendency shows higher faecal contributions from anthropogenic sources. While during warm periods of the year, the tendency shifts to dominance of zoogenic sources in faecal water contamination.

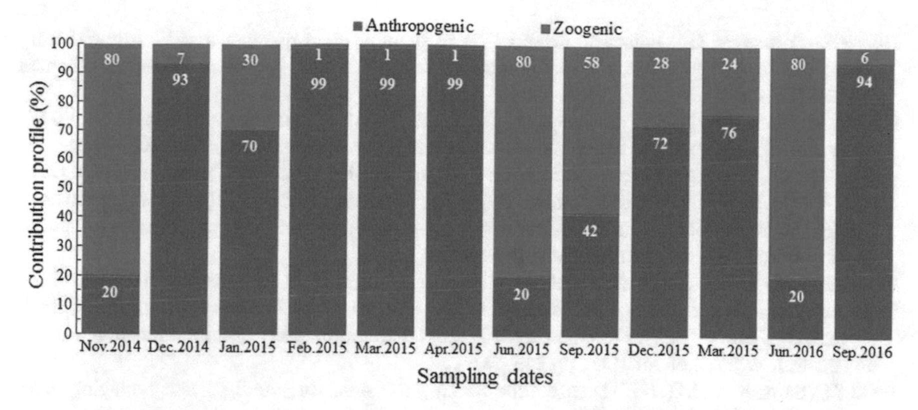

Fig. 10.5 Contribution profile of genetic markers in faecal contamination of water samples collected from Blåveisbekken stream, Ski, Norway *Source* Authors

References

Aftenposten (2013) Opp mot en million rotter i Oslo. https://www.aftenposten.no/osloby/i/212oy/Opp-mot-en-million-rotter-i-Oslo. Accessed 6 Aug 2017

Ahmed W, Sritharan T, Palmer A, Sidhu JP, Toze S (2013) Evaluation of bovine feces-associated microbial source tracking markers and their correlations with fecal indicators and zoonotic pathogens in a Brisbane, Australia, reservoir. Appl Environ Microbiol 79(8):2682–2691

Åström J, Pettersson TJ, Reischer GH, Norberg T, Hermansson M (2015) Incorporating expert judgments in utility evaluation of *Bacteroidales* qPCR assays for microbial source tracking in a drinking water source. Environ Sci Technol 49(3):1311–1318

Bambic DG, Kildare-Hann BJ, Rajal VB, Sturm BS, Minton CB, Schriewer A, Wuertz S (2015) Spatial and hydrologic variation of *Bacteroidales*, adenovirus and enterovirus in a semi-arid wastewater effluent-impacted watershed. Water Res 15(75):83–94

Benskin CMcWH, Wilson K, Jones K, Hartley IR (2009) Bacterial pathogens in wild birds: a review of the frequency and effects of infection. Biol Rev 84:349–373. https://doi.org/10.1111/j.1469-185X.2008.00076.x

Böhm ME, Huptas C, Krey VM, Scherer S (2015) Massive horizontal gene transfer, strictly vertical inheritance and ancient duplications differentially shape the evolution of *Bacillus cereus* enterotoxin operons *hbl*, *cytK* and *nhe*. BMC Evol Biol 10(15):246. https://doi.org/10.1186/s12862-015-0529-4

Bolin C, Brown C, Rose J (2004) Emerging zoonotic diseases and water. In: Cotruvo J, Dufour A, Rees G, Bartram J, Carr R, Cliver DO, Craun GF, Fayer R, Gannonp VPJ (eds) Waterborne zoonoses: identification, causes, and control. WHO, London, UK, pp 21–26

Carson CA, Shear BL, Ellersieck MR, Schnell JD (2003) Comparison of ribotyping and repetitive extragenic palindromic-PCR for identification of fecal *Escherichia coli* from humans and animals. Appl Environ Microbiol 69:1836–1839

Dick LK, Bernhard AE, Brodeur TJ, Santo Domingo JW, Simpson JM, Walters SP, Field KG (2005) Host distributions of uncultivated fecal *Bacteroidales* bacteria reveal genetic markers for fecal source identification. Appl Environ Microbiol 71(6):3184–3191

DN (2016) Dagens Nyheter, Råttinvasion i Stockholm—och de föredrar Östermalm. http://www.dn.se/sthlm/rattinvasion-i-stockholm-och-de-foredrar-ostermalm/. Accessed 21 Sept 2016

Duran M, Yurtsever D, Dunaev T (2009) Choice of indicator organism and library size considerations for phenotypic microbial source tracking by FAME profiling. Water Sci Technol 60(10):2659–2668

Edberg SC, Rice EW, Karlin RJ, Allen MJ (2000) Escherichia coli: the best biological drinking water indicator for public health protection. Symp Ser Soc Appl Microbiol 29:106S–116S

Farnleitner AH, Ryzinska-Paier G, Reischer GH, Burtscher MM, Knetsch S, Kirschner AKT, Dirnböck T, Kuschnig G, Mach LR, Sommer R (2010) *Escherichia coli* and enterococci are sensitive and reliable indicators for human, livestock and wildlife faecal pollution in alpine mountainous water resources. J Appl Microbiol 109:1599–1608

Fettvett (2016) Rotterace i avløpsnettet. http://fettvett.no/rotterace.html. Accessed 6 Aug 2017

Field KG (2004) Faecal source identification. In: Cotruvo J, Dufour A, Rees G, Bartram J, Carr R, Cliver DO, Craun GF, Fayer R, Gannonp VPJ (eds) Waterborne zoonoses: identification, causes, and control. WHO, London, UK, pp 349–366

Field KG, Samadpour M (2007) Fecal source tracking, the indicator paradigm, and managing water quality. Water Res 41:3517–3538

Gerardi MH (2006) Wastewater bacteria. John Wiley and Sons Inc, Hoboken, NJ, USA

Guenther S, Wuttke J, Bethe A, Vojtěch J, Schaufler K, Semmler T, Ulrich RG, Wieler LH, Ewers C (2013) Is fecal carriage of extended-spectrum-β-lactamase-producing *Escherichia coli* in urban rats a risk for public health? Antimicrob Agents Chemother 57(5):2424–2425. https://doi.org/10.1128/AAC.02321-12

Hagedorn C, Harwood VJ, Blanch A (2011) Microbial source tracking: methods, applications and case studies. Springer, New York

Harrault L, Jarde E, Jeanneau L, Petitjean P (2014) Development of the analysis of fecal stanols in the oyster *Crassostrea gigas* and identification of fecal contamination in shellfish harvesting areas. Lipids 49(6):597–607

Hartel PG, Rodgers K, Moody GL, Hemmings SN, Fisher JA, McDonald JL (2008) Combining targeted sampling and fluorometry to identify human fecal contamination in a freshwater creek. J Water Health 6(1):105–116

Harwood VJ, Staley C, Badgley BD, Borges K, Korajkic A (2014) Microbial source tracking markers for detection of fecal contamination in environmental waters: relationships between pathogens and human health outcomes. FEMS Microbiol Rev 38:1–40

IWFA (2017) Institute Water for Africa, Water and health. https://www.water-for-africa.org/en/health.html. Accessed 26 June 2017

Kanarat S (2004) What are the criteria for determining whether a disease is zoonotic and water related? In: Cotruvo J, Dufour A, Rees G, Bartram J, Carr R, Cliver DO, Craun GF, Fayer R, Gannonp VPJ (eds) Waterborne zoonoses: identification, causes, and control. WHO, London, UK, pp 136–150

Khatib LA, Tsai YL, Olson BH (2003) A biomarker for the identification of swine fecal pollution in water using the STII toxin gene from enterotoxigenic *Escherichia coli*. Appl Microbiol Biotechnol 63(2):231–238

Kobayashi A, Sano D, Okabe S (2013) Effects of temperature and predator on the persistence of host-specific Bacteroides-Prevotella genetic markers in water. Water Sci Technol 67(4):838–845

Lamendella R, Santo Domingo JW, Yannarell AC, Ghosh S, Di Giovanni G, Mackie RI, Oerther DB (2009) Evaluation of swine-specific PCR assays used for fecal source tracking and analysis of molecular diversity of swine-specific "*Bacteroidales*" populations. Appl Environ Microbiol 75:5787–5796

Layton A, McKay L, Williams D, Garrett V, Gentry R, Sayler G (2006) Development of Bacteroides 16S rRNA gene TaqMan-based real-time PCR assays for estimation of total, human, and bovine fecal pollution in water. Appl Environ Microbiol 72(6):4214–4224

Marotz CA, Zarrinpar A (2016) Treating obesity and metabolic syndrome with fecal microbiota transplantation. Yale J Biol Med 89(3):383–388

McQuaig S, Griffith J, Harwood VJ (2012) Association of fecal indicator bacteria with human viruses and microbial source tracking markers at coastal beaches impacted by nonpoint source pollution. Appl Environ Microbiol 78(18):6423–6432
Mieszkin S, Caprais MP, Le Mennec C, Le Goff M, Edge TA, Gourmelon M (2013) Identification of the origin of faecal contamination in estuarine oysters using *Bacteroidales* and F-specific RNA bacteriophage markers. J Appl Microbiol 115(3):897–907
Moe CL (2004) What are the criteria for determining whether a disease is zoonotic and water related? In: Cotruvo J, Dufour A, Rees G, Bartram J, Carr R, Cliver DO, Craun GF, Fayer R, Gannonp VPJ (eds) Waterborne zoonoses: identification, causes, and control. WHO, London, UK, pp 27–45
Moyo SJ, Maselle SY, Matee MI, Langeland N, Mylvaganam H (2007) Identification of diarrheagenic *Escherichia coli* isolated from infants and children in Dar es Salaam, Tanzania. BMC Infect Dis 9(7):92. https://doi.org/10.1186/1471-2334-7-92
Olivas Y, Faulkner BR (2008) Fecal source tracking by antibiotic resistance analysis on a watershed exhibiting low resistance. Environ Monit Assess 139:15–25
Paruch AM, Mæhlum T (2012) Specific features of *Escherichia coli* that distinguish it from coliform and thermotolerant coliform bacteria and define it as the most accurate indicator of faecal contamination in the environment. Ecol Indic 23:140–142
Paruch L, Paruch AM (2017) The importance of melting curve analysis in discriminating faecal and environmental *Bacteroidales* bacteria. Microbiol 86(4):536–538. https://doi.org/10.1134/S0026261717040117
Paruch AM, Mæhlum T, Robertson L (2015a) Changes in microbial quality of irrigation water under different weather conditions in Southeast Norway. Environ Process 2:115–124. https://doi.org/10.1007/s40710-014-0054-2
Paruch L, Paruch AM, Blankenberg A-GB, Bechmann M, Mæhlum T (2015b) Application of host-specific genetic markers for microbial source tracking of faecal water contamination in an agricultural catchment. Acta Agric Scand 65(S2):164–172
Paruch AM, Paruch L, Mæhlum T (2017) Kildesporing av fekal vannforurensing med molekylærbiologiske metoder—Eksempler på undersøkelser i Norge (Source tracking of fecal water contamination by molecular biology methods—Examples of surveys in Norway). NIBIO Rapport 3/66, Aas, Norway
Pond K (2005) Water recreation and disease: plausibility of associated infections, acute effects, sequelae and mortality. WHO/IWA, London
Quigley EM (2013) Gut bacteria in health and disease. Gastroenterol Hepatol (NY) 9(9):560–569
Reischer GH, Kasper DC, Steinborn R, Farnleitner AH, Mach RL (2007) A quantitative real-time PCR assay for the highly sensitive and specific detection of human faecal influence in spring water from a large alpine catchment area. Lett Appl Microbiol 44(4):351–356
Scheffe L (2007) Reducing risk of *E. coli* O157: H7 contamination. Nutrient Management Technical Note No 7. USDA, NRCS, Washington, DC, USA
Schueler TR (2000) Microbes and urban watersheds: concentrations, sources, and pathways. In: Schueler TR, Holland HK (eds) The practice of watershed protection. Center for Watershed Protection, Ellicott City, Md, pp 68–78
Scott TM, Jenkins TM, Lukasik J, Rose JB (2005) Potential use of a host associated molecular marker in *Enterococcus faecium* as an index of human pollution. Environ Sci Technol 39(1):283–287
Shahryari A, Nikaeen M, Khiadani Hajian M, Nabavi F, Hatamzadeh M, Hassanzadeh A (2014) Applicability of universal *Bacteroidales* genetic marker for microbial monitoring of drinking water sources in comparison to conventional indicators. Environ Monit Assess 186(11):7055–7062. https://doi.org/10.1007/s10661-014-3910-7
Shanks OC, Atikovic E, Blackwood AD, Lu J, Noble RT, Santo Domingo J, Seifring S, Sivaganesan M, Huagland RA (2008) Quantitative PCR for detection and enumeration of genetic markers of bovine fecal pollution. Appl Environ Microbiol 74(3):745–752
Smith A, Sterba-Boatwright B, Mott J (2010) Novel application of a statistical technique, random forests in a bacterial source tracking study. Water Res 44(14):4067–4076

Sowah RA, Habteselassie MY, Radcliffe DE, Bauske E, Risse M (2017) Isolating the impact of septic systems on fecal pollution in streams of suburban watersheds in Georgia, United States. Water Res 108:330–338

Staley ZR, Grabuski J, Sverko E, Edge TA (2016) Comparison of microbial and chemical source tracking markers to identify fecal contamination sources in Humber River (Toronto, Ontario, Canada) and associated storm water outfalls. Appl Environ Microbiol 82(21):6357–6366

Suresh K, Smith HV (2004) Tropical organisms in Asia/Africa/South America. In: Cotruvo J, Dufour A, Rees G, Bartram J, Carr R, Cliver DO, Craun GF, Fayer R, Gannonp VPJ (eds) Waterborne zoonoses: identification, causes, and control. WHO, London, UK, pp 93–112

Tambalo DD, Fremaux B, Boa T, Yost CK (2012) Persistence of host-associated *Bacteroidales* gene markers and their quantitative detection in an urban and agricultural mixed prairie watershed. Water Res 46(9):2891–2904

Tran NH, Gin KY, Ngo HH (2015) Fecal pollution source tracking toolbox for identification, evaluation and characterization of fecal contamination in receiving urban surface waters and groundwater. Sci Total Environ 15(538):38–57

USEPA (2005) United States Environmental Protection Agency, Microbial source tracking guide document. Office of Research and Development, EPA-600/R-05/064, Washington, DC

Venegas C, Diez H, Blanch AR, Jofre J, Campos C (2015) Microbial source markers assessment in the Bogota River basin (Colombia). J Water Health 13(3):801–810

WHO (2006) World Health Organization, Guidelines for the safe use of wastewater, excreta and greywater. In: Wastewater and excreta use in aquaculture, vol 3. WHO Press, Geneva, Switzerland

WHO (2011) World Health Organization, Guidelines for drinking-water quality, 4th edn. WHO Press, Geneva, Switzerland

WHO (2017) World Health Organization, Mortality and burden of disease from water and sanitation. http://www.who.int/gho/phe/water_sanitation/burden/en/index2.html. Accessed 26 June 2017

Chapter 11
Occurrence and Removal of Emerging Micropollutants from Urban Wastewater

Petr Hlavínek and Adéla Žižlavská

Abstract In the last few years, the issue of new dangerous micropollutants, penetrating from municipal wastewater into the environment, which can pose a threat to animals and plants, but also humans, has become very topical. These substances, the originator of which is our developed society, are completely out of the ordinary standards of wastewater treatment, and most of the commonly used purification technology is ineffective against them. The persistent and bioaccumulative nature of these substances brings major risk to wild live plants and animals because it affects their endocrine systems. It causes transcription of DNA and RNA. The immediate danger to humans is primarily the bioaccumulation potential of these substances and their binding to a solid soil matrix where they can be leached into groundwater and contaminated with drink water sources. The escape of these substances into the environment is primarily connected with the abundant use of chemical additives in various industries, with a separate chapter being agriculture, where the massive use of chlorine-based chemical pesticides for long years has completely infested a large amount of soil. Wastewater is one of the main gates of micropollutants entering the aquatic environment, especially those that are part of pharmaceutical products, such as hormonal contraceptives, which contain synthetic hormones that cause severe sex-mutagens changes in fish. However, the increased occurrence of micropollutants is not only related to the fact that these substances are abundantly overused, but also that the detection methods were able to capture these substances during the sub-20 years. From the time perspective, we are not yet able to determine their future environmental developments, and we do not know about interactions exist between them and other substances in the frame of sewage and treatment process, when that may induce new substances that can often be more dangerous than origins mother substances. Our aim should, therefore, be to make the processes and technologies for removing these substances from wastewater as efficiently as possible so as to prevent their free distribution to the environment. This section of the publication attempts to

P. Hlavínek (✉) · A. Žižlavská
Centre AdMaS, VUT Brno, Brno, Czech Republic
e-mail: hlavinek.p@fce.vutbr.cz

A. Žižlavská
e-mail: gotzingerova.a@fce.vutbr.cz

M. Zelenakova (ed.), *Water Management and the Environment: Case Studies*, Water Science and Technology Library 86,
https://doi.org/10.1007/978-3-319-79014-5_11

summarize the current knowledge of hazardous micropollutants in wastewater and the possibilities of their elimination.

Keywords Urban wastewater · Micropollutants · Xenobiotics · Wastewater treatment · Environment

11.1 Introduction

Due to the advanced development of technology and methods used in the field of water composition analysis, the occurrence of micropollutants, that is not inherent in the natural aquatic environment, has been observed over the past twenty years. This is a consequence of the development of the chemical and pharmaceutical industry and the massive use of chemicals in various industries that have become indispensable for our society. Together with this comfort that this progress has brought us, we have created a completely new kind of pollution. It is a very wide group including a large number of diverse substances and their metabolites, such as pharmaceuticals, pesticides, detergents, paints, varnishes, plastic packaging, food additives, cosmetics. (Fatta-Kassinos et al. 2010). The scientific community marked them as xenobiotics (from lat.xenos—foreign, bios- biological) from the point of view of its foreign origin to ordinary organic pollution. Their specificity lies is that their concentrations in water in the order of ng/l–g/l have far more negative effects on living organisms than normal organic pollutants such as toxicity (acute or chronic), bioaccumulation effects, carcinogenesis, mutagenicity, affect the reproductive potential and It's able to intervene in these effects with several subpopulations in succession (Jobling and Tyler 2003).

Xenobiotics are persistent in relation to normal biological and chemical degradation processes, and many of these substances are practically unchanged through the water treatment cycle and are freely dispersed into surface waters and thus into subterranean waters. Additionally, they come into waters not only in the original forms, but also in the form of metabolites that can be deconjugate[1] back into the parent form or form new conjugates with other substances present in water during transport from the originator to the recipient (even during purification of WWTP). Another drawback is that we are not currently at such a technological level as to be able to monitor or anticipate these marches.

[1]Deconjugation—the most famous example of this process is the conversion of gluconic acid and sulfuric acid (conjugates) that occur in the female body after using hormonal contraceptives. Hormonal contraceptives containing ethinylestradiol, for example, go into the intestine where it is distributed and absorbed through the liver. From the liver, some of them continue to spread into the body, and they form conjugates that once again enter the intestine, where the bacteria are deconjugated back to the hormone that is absorbed again and the whole cycle is repeated (enterohepatic circulation) according to the latest studies. The process of conjugation and deconjugation takes place not only in the bodies of animals, but virtually throughout the ecosystem, also in sewerage, WWTP.

Xenobiotics pose a very dangerous threat to aquatic biota in natural waters, and not only to aquatic biota but also to all other organisms, including humans, who come into contact with this medium. Currently, the elimination of these substances is also in the interest of the European Union and the UN.

11.2 Terminology

The terminology of the marking "Xenobiotics" is not completely settled for these types of substances or their individual subgroups there is a larger number of the name below, then I give some other frequent indications for this group of substances or its individual fractions. In the labeling of xenobiotics in professional publications, other names are often used; this group is referred to as "Persistent Toxic Substances" (PTS); this name is derived from their persistence capability in degradation-resistant environments. Often, also because of their low detection concentrations, it is possible in the literature to meet the name of micropollutants.

Sometimes we can see the abbreviation PBT—persistent, bio-accumulative and toxic substances. The subgroup of persistent toxic substances is "Persistent Organic Pollutants" (POPs)—this term is used primarily in international conventions. In the literature on the purity of rain and surface waters, we can also see a name of xenobiotic organic compounds included polycyclic aromatic hydrocarbons (PAHs), pesticides, phthalates, alkylphenol, ethoxylates, dioxins, furans, and polycyclic chlorinated biphenyls.

11.3 Legislation of Environmental Protection Against Pollution of Micro-pollutants

The basic documents (European Commission) dealing with the issue of environmental micropollutants within the EU are the Water Framework Directive, the Marine Strategy Framework Directive, the Nitrates Directive, the Plant Protection Products Directive, the Urban Waste Water Treatment Directive, the Framework Directive on Sustainable Use of Pesticides, Drinking Water, and Stockholm Convention on POPs.

The main part of the EU strategy for the protection of surface water against harmful pollutants is contained in the Water Framework Directive 2000/60/EC which contains Articles 4, 10, 11, and 16 and Annex V, VIII, IX, and X dealing with chemical pollution. On December 16, 2008, environmental quality standards (EQS) for the first 33 priority substances (already known eight substances) were established under Directive 2008/105/EC. Of these, 13 of them were identified as priority hazardous substances (these are highly persistent, toxic substances showing the age of bioaccumulation capacity).

On 31 January, a proposal for a directive supplementing WFD 2000/60/EC (European Commission, Directive 2000/60/EC) and EQSD 2008/105/EC was adopted, adding 15 priority substances to the list, 6 of which identified priority hazardous substances. This Directive has also tightened the EQS for some selected substances.

The proposal for a directive also sets out several new theses in the context of the protection of the natural aquatic environment:

- Biotics EQS setting for existing and newly discovered priority substances;
- Ensuring better and more comprehensible monitoring for PBT substance (PBT—persistent, bioaccumulative, toxic);
- Establishment of supervisory mechanisms to review the list of priority substances on a rolling basis based on WISE monitoring (see below).
- A list of priority substances will be reviewed and updated every 4 years.
- Substances identified as priority hazardous will be completely excluded from all production processes within the EU over the next 20 years (since December 2008) and that the EU will aspire to reduce, controlling and regulating emissions, discharges and losses of priority substances into the water.

In terms of time, the priority substances are able to affect negatively on the two levels, both on a long-term basis (primarily accumulation effects that cause chronic problems) and then on a short-term level (acute exacerbation).

For this reason, the EU created two types of assessments called "environmental quality standards" for annual average concentrations and maximum allowable concentrations. In line with the requirements of the Water Framework Directive and the EQSD, the Commission subsequently reviewed the list of priority substances and submitted to it in 2012 a proposal for a Directive amending the Water Framework Directive and the EQSD as regards priority substances (The Water Framework Directive: Tap into it! 2002).

On the basis of proposal (COM (2011)876) (MESAUC, European Commission) entered into force amending directive 2013/39/EU content a revised list of priority substances (PPs), and precautions leading to legislative improvement of environmental protection. The main changes are as follows:

- Additions of the 15 priority substances, of which 6 of them was marked such as priority hazardous substances;
- Approved of the new EQS, strict parameters for four existing priority substances, and EQS for three others were slightly revised;
- Two priority substances were marked as priority hazardous substances;
- Assessment of the biota standards;
- Performance of an effective monitoring and reporting system for substances that act as PBT substances;
- Establishment of a watch-list mechanism, targeting on the monitoring of substances in EU.

The European Union, on the WFD public library Web site in CIRCABC, lists a number of interesting documents on the impact of dangerous micropollutants in the environment, or it is possible to find the so-called list of proposed micropollutants that will be added to the list of priority substances. In 2017, the predominant majority of the substances on the list are pharmaceuticals.

List of monitoring substances for 2017:

- 17alpha-ethinylestradiol
- 17beta-estradiol
- Aclonifen
- Bifenox
- Cybutryne
- Cypermethrin
- Dichlorvos
- Diclofenac
- Dicofol
- Dioxins
- HBCDD
- Heptachlor
- PFOS
- Quinoxyfen
- Terbutryn.

In the frame of EU, other documents legislating on water protection from micropollutant pollution are: The Marine Strategy Directive was adopted on June 17, 2008 by the Marine Strategy Framework, and its educative objective is to achieve good ecological status by 2020 in order to preserve its biodiversity, the Directive, Nitrates Directive, Plant Protection Products Directive, and Urban Waste Water Treatment Directive.

Another very interesting document is Council Directive 91/271/EEC on Urban Wastewater Treatment which was adopted on May 21, 1991. Its aim is to protect the environment from discharges and discharges of certain municipal wastewater from discharges and discharges from certain industrial sectors (see Annex III of the Directive) and concerns disposal, cleaning and discharges:

- Wastewater from households;
- A mixture of wastewater;
- Sewage from certain industries (see Annex III to the Directive).

Annual reports on the achievement of the set objectives in the field of wastewater treatment and new monitoring proposals are also published annually in the framework of this Directive. The 2017 report right from the start warns that the most alarming for the status of European waters is quoting "The unmanaged and uncleaned wastewater produced by the EU's 500 million people is one of the main sources of pollution that affect the quality of freshwater and marine waters and pose a risk to human health and biodiversity."

Within the framework of the Structured Implementation and Information Framework (SIIF) project, a pilot project was launched to disseminate the date from the wastewater treatment state within the EU. The platform is available online and is currently a pilot-only platform, and all data cannot yet be traced. In the future, we are expected to be able to find information about individual WWTP's throughout Europe.

Framework Directive about the sustainable use of pesticides, in particular, the management and use of pesticides within the European Union. Pesticides represent a relatively broad group of pollutant micropollutants. The European Commission, through the implementation of roadway, restricts pesticide bloating and constantly checks for possible potential negative effects on honey. An overview of prohibited and authorized pesticides can be found on the EU's pesticides database.

Stockholm Convention on POPs is in the framework of international cooperation, dangerous persistent substances are also controlled by the United Nations Environment Program (UNEP), which in 2001 created the so-called Stockholm Agreement, which undertook to eliminate the most important perishable organic substances (POPs). Additionally, the agreement is also included in the B Restriction and C Unintentional Production appendices. At present, 29 substances are on the lists of the Stockholm Treaty and 174 countries, including the Czech Republic and the EU, have ratified it. On March 22, 2007 the European Environment Agency in cooperation with the European Commission (DG Environment, Joint Research Center and Eurostat) launched a Water Information System for Europe (WISE) system to record and retain data from 54,000 surface water monitoring stations and 51,000 groundwater monitoring stations on their chemical, physical, and biological water status.

Within the European environment, micronutrients and their impact on the environment are dealt with FATE (European Commission) project carried out at the Institute for Environment and Sustainability (IES) of the Joint Research Centre (JRC). The aim of the FATE project is to provide support for the implementation and validation of directives, strategies, instruments, and international conventions in the frame EU area that deals with pollutants.

FATE project committed to providing crucial information about monitoring and modeling of contaminants in the environment, to the institutions competent such as policymakers and the scientific community. All information is freely available directly on the project Web site (European Commission).

The fear of the uncontrolled proliferation of large quantities of substances produced and delivered to Europe has led to the adoption of the REACH regulation.

The primary objective of REACH is making a pressure on industry branches in order to gradually replace substances of very high concern in all manufacturing processes within the EU. (REACH, European Commission). It was through the evaluation of the research and its subsequent implementation at international, national, and local level that it was possible to prohibit the use of many of these harmful substances (e.g., DDT, various groups of phthalates) or at least to regulate.

11.4 History

The first pioneer in identifying of hazardous substances in the environment was Rachel Carson, an American zoologist and marine biologist. In 1962, she published a legendary book called Silent Spring, which highlighted the negative effects of the massive use of DDT and insecticide sprays. The book "Silent Spring" has awakened the public in both America and overseas. As a result of the media pressure raised by the book, the US government was preparing a detailed study of the impact of DDT and selected insecticides on the environment, which led to the banning of DDT in the USA in 1972 (Lear 1997).

In 1988, American zoology professor Theo Colborn revealed in the research of the state of the Great Lakes region of North America that persistent synthetic chemicals occurring in these waters affect the genetic evolution of animals. Considering this fact, in 1991, it convened 21th International Scientific Conference, inviting scientists from 15 different disciplines. The conference took place in the American Wingspan and was called "Changes in Sexual Development Changes: Human-Nature Connectivity." This was the first time that the term endocrine disruption was mentioned and explained.

On the basis of a wide public interest, Theo Colborn published a book entitled "Our Stolen Future" in 1996 (dealing with the impact of xenobiotics on nature and, as a consequence, on human health). In 1997, the G8 issued a statement on the environmental impact on children's health (Lemonick 2007). This statement was devoted to chemicals that disturb and threaten the health of children, I quote:

"There is scientific evidence that many environmental pollutants can have adverse effects on health through their ability to change hormone function in the body. These effects, which include cancer, reproductive disorders, behavioural changes and immune dysfunction, have been observed in both laboratory animals exposed to particular chemicals and wild populations in several widely-contaminated ecosystems such as the Great Lakes region. Some of these substances are also able to cause long-term neurological damage. Infants and children may be at risk of the potential impact of these impurities. We recommend ongoing efforts to establish an international list of research activities, develop an international science assessment, identify and prioritize research needs and missing data, and establish a mechanism for coordination and cooperation in meeting research needs. These activities should complement the initiatives that are being pursued in international fora [such as the Intergovernmental Chemical Safety Forum (FISC)] and through the work of agencies such as the United Nations Environment Agency. We are committed to developing strategies for pollution prevention, identifying the main sources and effects of endocrine-disrupting chemicals and continuing to inform the public of the knowledge they have gained."

A Convention on the elimination of the most dangerous Persistent Organic Pollutants (POPs) was issued in Stockholm, Sweden, under United Nations Environment Program (UNEP) on May 23, 2001. Today, the Convention has ratified 174 countries including the Czech Republic and the European Union. In the Czech Republic, the Convention entered into force on May 17, 2004.

At the time of its inception, a list of 12 substances in the list of substances in May 2009 in the Fourth Conference of the Parties to the Stockholm Convention in Geneva included nine other substances and their metabolites on the list. In 2011, the twenty-second endosulfan item was added to the list.

11.5 Analysis, Classification, and Occurrence of Micropollutants

Xenobiotics are substances that occur at very low concentrations, and LC-MS and GC-MS are commonly used for their analysis. To determine the overall sample potential, in vitro bioassays or combinations thereof with LC-MS and GC-MS are used.

Xenobiotics are a heterogeneous substances group of a synthetic or natural nature that are used across all industrialized industries, from the metallurgical industry until of dermatological preparations production. For this reason, we can classify them from many points of view.

One of the most monitored groups is a pharmaceutical that can be divided according to Table 11.1.

To illustrate how wide a group with different properties are xenobiotics, we have the Table 11.2. Major pathways of these substances into the surface water are WWTP, runoff from fields, runoff from roads, water management of landfills, sludge, old environmental burdens.

Within the European Union, the Institute for Health and Consumer Protection has developed a Technical Handbook for Risk Analysis (2003 edition) which sets out three possible approaches for evaluation:

- Quantitative PEC/PNEC estimates of predicted environmental concentration (PEC) subcenters if they are below the so-called no-effect PNEC (predicted no-effect environmental concentration), which is the concentrations of the substances that do not yet have an adverse effect on the aquatic biota;
- Qualitative environmental risk analysis of substances;
- PBT estimate estimates the ability of substances to withstand natural degradation processes, bioaccumulation potential, and toxicity.

PEC values are derived from already measured data or from model calculations. PNECs are determined on the basis of laboratory tests in several cases on derived model calculations.

However, according to a team of Danish scientists from the Danish Technical University (A. Baun, E. Eriksson, A. Ledin, P.S. Mikkelsen), these methods are completely unsuitable for rainwater and surface water intended for re-use. Therefore, they developed two methodologies, in 2009 RICH process (Ranking and Identification of Chemical Hazards) and in 2007 CHIAT process (Chemical Hazard Identification and Assessment Tool) (Eriksson et al. 2007a, b).

Table 11.1 Pharmaceuticals detected in wastewater *Sources* Caliman and Gavrilescu (2009), Adeel et al. (2017), Heberer (2002), Ingerslev et al. (2003), Vrcek (2017), Snyder et al. (2001, 2003), Suarez et al. (2008)

Steroidal pharmaceuticals	Estrogens	17—estradiol (E2)—component of menopause medicines Estriol—(E3)—name-pour cream, sclerosis multi-sclerosis 17a-ethinylestradiol (EE)—anti-	Estradiol is the most potent natural estrogen. EE activates estrogen receptor (estrogen activation)—endocrine disruption
	Progesterons	Norethisterone (or 19-year-17α-ethynyltestosterone) combined oral contraceptives, menstrual treatment, menopause, or postponing menopause, to prevent uterine bleeding, prevent premature labor. Progesterone is used to support pregnancy in assisted reproduction (ART)	These are substances that, in combination with estrogen, are estrogenic—act as an endocrine disrupter
	Antiestrogens	Hormonal treatment of breast cancer	Prevents estrogen from binding to its receptor
	Androgens and glukokortikoids	Testosterone—anabolic steroids, growth stimulation, osteoporosis treatment Beclomethasone—a sculpture of various drugs, for example, the treatment of asthma, skin essence Hydrocortisone—treatment of allergic reactions, psoriasis, excizia	Endocrine disruption
	Fytoestrogens	Alleviating the symptoms of menopause and menstrual pain	Exhibit estrogenic activity
	Veterinary growth hormones	Termination of cows' pregnancies and support for the growth of livestock and slaughter animals	Endocrine disruption

(continued)

Table 11.1 (continued)

Non-steroidal pharmaceuticals	Agents for the treatment of blood and blood-forming organs	Acetylsalicylic acid—reducing blood clotting Pentoxyphyllin improves the blood flow of peptic vessels	Pentoxifylline is also a receptor antagonist. Acetylsalicylic acid increases the activity of hepatic enzymes
	Agents for the treatment of heart and circulatory system	Lowering blood cholesterol—statins	Endocrine disruption
	Antibiotics	Treatment of infectious conditions	Multidrug resistant of the bacteria
	Analgesics	A group of pain-relieving medicines (ibuprofen, carbamazepine, clofibric acid)	Increased inhibition of the organism, bioaccumulative effects
	Antidepressants	Monoamine reuptake inhibitors (fluoxetine, citalopram)	They can affect influencing signaling by dopamine and noradrenaline, mutagenesis
	Agents for the treatment of allergies and asthma	Glucocorticoids Antihistamines Sympathomimetic	Endocrine disruption
	X-ray media	Magnetic resonance, radiology, during surgery	Highly persistent, carcinogenic
	Cytostatics	Treatment of cancer	They show carcinogenicity, mutagenicity or embryotoxic parameters
	Antiepileptics	Treatment of epilepsy	Strongly persistent
	Beta-blockers	Beta-blockers to reduce the effect of stress hormones. Treatment of the cardiovascular system for the treatment of anxiety disorders and migraine	Strongly persistent

Table 11.2 Selected types of xenobiotics

Substance/Group of substances	Uses	Effects
Alkyphenols (AP)	Industrial detergents, emulsifiers, textile and leather coatings, as additives for pesticides and other agro-products, water based dyes, shampoos and personal hygiene products. Nonylphenol derivatives (NP) are also used as antioxidants in some plastics	Ability to mimic the effects of natural estrogenic hormones. The human health risks associated with AP have not been sufficiently investigated. Other negative features: DNA damage in human lymphocytes, bioaccumulation ability, toxic to aquatic organisms According to the Integrated Pollution Register of the Czech Republic, the main sources of NP and NPE (2013) are the wastewater treatment plants. (Randák 2004; REACH)
Phenols	Production of plastics, destruction of bacteria and algae in water slurries, disinfection, production of some drugs (against sore throat, skin diseases), the main chemical among the product for the production of phenolic resins and chemical fibers, slimicides in industrial aquatic systems, chlorophenols—disinfectant and antiseptic agents, pesticide additive	Negative phenomena—excessive phenol can cause damage to the brain, digestive tract, eyes, heart, kidney, liver, lungs, peripheral nerves, skin, and unborn children. (ATSDR; IRZ; Arnika; Gong and Han 2006)
Phthalates	PVC plasticizers, cosmetics, insecticides, adhesives and paints, di (2-ethylhexyl) phthalate (DEHP) is most commonly used as a plasticizer in PVC in flooring, building materials, interior household piping	According to animal studies, DEHP is known to be a poisonous substance endangering reproductive capacity (damaging male and female reproductive tract) causing birth defects (e.g., skeletal defects, ocular defects, defects of the nerve group forming the nucleus of the older embryo), cardiovascular problems and infertility. DEHP also damages the kidneys and the liver where it accumulates. (ATSDR; IRZ; Arnika)

(continued)

Table 11.2 (continued)

Substance/Group of substances	Uses	Effects
Hexachlorobutadiene	Abroad, it is used for the production of chloroprene rubber as a solvent (for us this production has been completed), for the production of lubricants in flywheels, such as heat and hydraulic fluid (replaced by PCB), is produced in the production of perchlorethylene and carbon tetrachloride by perchloration	There are no studies to show effects on humans. Most of its effects on human health are thus based on animal studies—irritation of the upper respiratory tract after inhalation of high concentrations over a short period of time, reduction in fetal weights in mothers that inhaled high concentrations of the substance. Nephrotoxin (damages kidneys and liver). (Maxa et al. 2002; EPA; IRZ; Arnika)
Chloroalkanes	PCB substitutes, plasticizers, lubricants, flame retardants, as additive in the manufacture of dyestuffs, sealant, adhesives, etc.	High bioaccumulation capacity, dermatitis and respiratory disease. A mixture of chlorinated paraffins C12 with 60% chlorine content is considered by the IARC as a possible carcinogen for humans, very toxic to aquatic organisms, may cause long-term adverse effects in the aquatic environment. Toxicity to aquatic organisms has been demonstrated at a concentration range of 0.12–1.45 μg/(EPA; IRZ; Arnika)
Triclosan	Cosmetic preparations as well as preservatives, biocides, detergents, soaps, deodorants, creams, toothpastes, or as additives in plastics and textiles	It binds to a specific bacterial receptor, inhibits triclosan microbial fatty acid biosynthesis as well as antibiotics so it can induce cross-resistance to them. In addition, triclosan is a different substance in E coli and P aeruginosa by subtracting the mechanism of ejection from their cells, thus triclosan may contribute to cross-resistance to clinically important antibiotics. (Ni et al. 2005; Arnika; Nghiem and Coleman 2008)
Oktachlorstyren	As a by-product of high-temperature chemical processes of chlorine chemistry or by the combustion of chlorinated wastes in the Czech Republic is included in the group of chlorinated pesticides	The effect of octachlorostyrene on human health has not yet been sufficiently explored. (EPA)

(continued)

Table 11.2 (continued)

Substance/Group of substances	Uses	Effects
Pentachlorobenzen	Production of pentachloronitrobenzene (quintozene), flame retardant, intermediate or unintended by-product in the production of chlorinated aliphatic hydrocarbons (tetrachloroethylene), an intermediate of natural degradation of lindane and hexachlorobenzene	Long-term exposure may affect the liver and kidneys and cause lesions. Animal studies have shown that pentachlorobenzene may have a negative effect on reproduction and is therefore classified as reprotoxic. (WHO 1991; EPA; Arnika)
Pesticides	Pesticides are a wide range of products used to protect products from various industries (predominantly agriculture) from pests or directly to kill them	The ability of endocrine disruption to other large amounts of negative effects such as bee hives, surface water contamination, Parkinson's disease, brain damage, and others. Many of the pesticides are currently on the list of the Stockholm Agreement. (EPA; IZR; Arnika)
Polybromate Diphenylethers	Pesticide	The immune response of the organism, the reproduction cycle, and the effect on the development of the next generation negatively affect the hormonal balance of the organism. (Pulkrabová et al. 2009; REACH)
Polychloride Dibenzothiofenes	By-products in chemical processes and during combustion	It is believed that due to its similarity to dioxins, it affects primarily the human hormone and immune system. (Kopponen et al. 1994)
Polychlorated Bifenyles	Used in transformer and condenser oils, colors, plasticizers, dyeing papers, inks, lipsticks, like dioxins like unintended by-products in a number of industrial productions (metallurgy, waste incineration, production of various chlorine compounds or car combustion engines with the combustion of leaded petrol etc.)	PCBs cause liver disease, circulatory disturbances, fatigue, prolong the pregnancy and cause reproductive problems. Fish living for a long time in water contaminated with trace concentrations of PCBs have concentrated these substances up to a thousand times. The distribution of PCBs in fish bodies is not uniform. (Holoubek et al. 2000)
Styrens	It is used as a monomer for the production of rubber, plastics, insulation, fiberglass, piping, automotive components, packaging for carpets, carpets, polystyrene production	It is shown that certain styrene substances released from plastic food packaging can be classified as esters. It has binding affinities to human estrogenic receptors. (EPA)

(continued)

Table 11.2 (continued)

Substance/Group of substances	Uses	Effects
Polycyclic aromatics hydrocarbons	Combustion of fossil fuels (incomplete combustion processes), waste incineration, road transport, cracking of crude oil, aluminum production, metallurgical processes, coke, asphalt, cement production, refineries, crematoria, fires.	Animal studies have shown an effect on fertility reduction and developmental defects. (IRZ; EPA; Arnika; Stockholm Agreement; SCF 2002; Harrigan et al. 2004; Nebert et al. 2000)
Mercury	Battery manufacture, glass additive, various alloys and soldering paints, anti-corrosion coatings. Organic compounds tetraethyllead additive to benzine (by introducing unleaded fuels, their consumption has dropped significantly)	Lead can affect hematopoietic and nervous system, kidney, immune mechanisms, digestive and reproductive system. (IRZ; EPA; Arnika)
Tributyltin	Additive in ship coatings, in paper mills and then in paper products including disposable diapers and liners, in cooling systems and towers, additive in wood preservatives, molluscs), rodenticides	Tributyltin blocks the conversion of testosterone into estrogens. CIT-pesticide risks the female, due to overproduction of testosterone and exhibits some male characteristics (developing a penis). In extreme cases, it leads to infertility. (REACH)
Epichlorohydrin	Production of epoxy resins	Easily penetrates cell membranes, proven to be a carcinogen. (IARC; REACH)
Zearalenone	Molds attack cereals such as corn, barley, oats,	Endocrine disruptors. (Hughes 1988)
Butylhydroxianizol	Preservatives in grass, food packaging, feed, cosmetics, rubber, and petroleum products. BHA is also commonly used in medicine such as isotretinoin and lovastatin and simvastatin.	Reacts with estrogenic receptor rainbow trout and humans. (REACH; IARC)
Parabens	Preservatives, in shampoos, commercial moisturizers and shaving gels and personal lubricants and topical/parenteral drugs, spray tanning solution, makeup, dentifrices	Imitate estrogens. (American cancer society; Arnika)

(continued)

Table 11.2 (continued)

Substance/Group of substances	Uses	Effects
Bisphenol A	Manufacture of epoxy resins, additives in plastics, food and beverage packaging (baby bottles prohibited), home electronics, glasses, lenses, and many others	Estrogenic effects in rats and hormonal effects that increase the risk of breast cancer in humans are documented. Described is the ability to act antiandrogenically, which in men causes femininity. (IARC)
Dioxins	In the combustion of waste containing chlorinated substances or as by-products in chemical production where chlorine is used (in the production of pesticides, bleaching of paper with chlorine, etc.)	Cause hormonal disorders, endanger the reproduction of animals including humans, cause damage to the immune system and some cause cancer, highly chemically stable. (IARC)
Mosuses	Perfumes, colognes and toilet waters, cosmetics, soaps, detergents	Endocrine disruption. (Snyder et al. 2003)
UV-protection additives	Protective creams before UV light	Processed interactions with an estrogen, androgen and progesterone receptor. (Snyder et al. 2003; Caliman and Gavrilescu 2009)

These methods also calculate the value of the hazard to be taken into account when calculating the overall impact of xenobiotics on the aquatic ecosystem.

Generally, it is necessary to create and evaluate potential chemical hazards in rainwater, sustainable control of rainwater quality and drainage before it penetrates into groundwater or the surface recipient because the identification of diluted xenobiotics in containers is often quite impossible. In many cases, input concentrations are too low, and delays or degradation processes may increase disproportionately. Therefore, it is necessary to consider the results of the accuracy of the quantitative analysis for the evaluation of individual substances.

Current EU assessments have numerous gaps in knowledge of potential effects or transformation processes that xenobiotics can cause in rainwater. This is also related to the identification of compounds present in the organic matter of wastewater in terms of degradation mechanisms.

A common problem with the classification and risk analysis of xenobiotic hazards in wastewater is also the choice of chemicals to be included because analyzing these substances is relatively costly.

Following a comprehensive classification and risk assessment of micropollutants, there is another problem with testing and analysis of these substances. Due to the complex interaction relationships between these substances, the evaluation of the concentration of samples based on the identification of individual substances does not seem to be always effective. Possible solution, therefore, is the testing of wastewater by means of in vitro bioassays (possibly a combination of these tests with classical chemical analyses), which assesses the overall potential of the sample. The ability of the substances contained in the effluent to affect the living organisms in the recipient, e.g. in the estrogenic or androgenic disposition.

For a comprehensive assessment of the sources from which xenobiotics penetrate into the aquatic ecosystem, it is best to use the scheme according to Fig. 11.1.

11.6 Factors Affecting Degradation of Xenobiotics

As already stated in the introduction, xenobiotics represent a relatively wide spectrum of compounds that do not always have the same physicochemical properties, as a result of which we can say that the efficiency of different purification processes to the level of their degradation will be different. Therefore, the design and treatment of rainwater and sewage water should always be based on a detailed water analysis, taking into account the possible interaction of each substance. According to the most recent surveys, it has been found that the rate of removal of xenobiotic substances is based on a number of crucial factors: season, temperature, intensity of solar radiation, hydraulic delays (Bester et al. 2010; Caliman and Gavrilescu 2009). In particular, hydraulic delays have a very strong impact on the degradation process of these substances.

In the field of water management, it is necessary to prevent undesirable leaks into surface water by the application of efficient cleaning technologies. As stated,

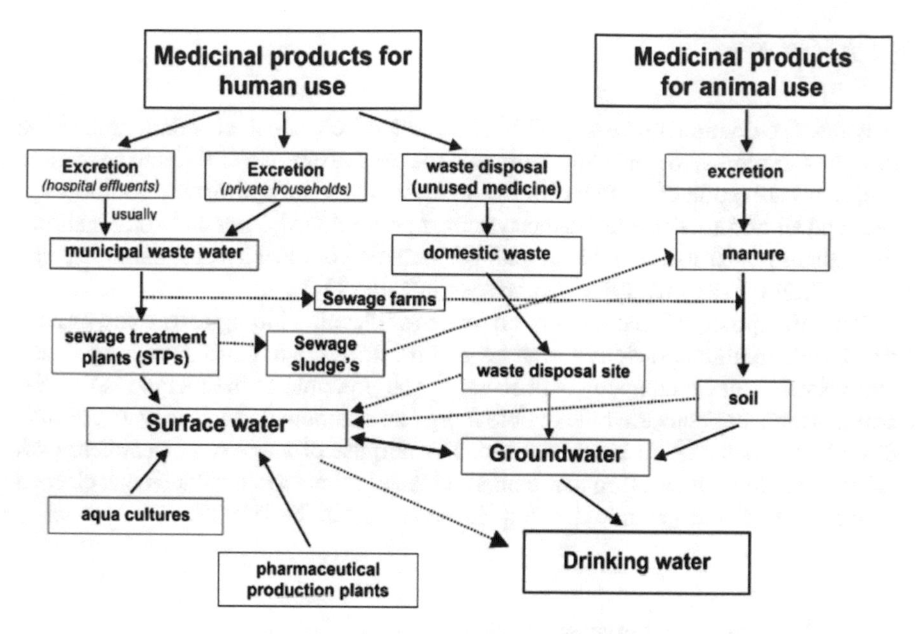

Fig. 11.1 Scheme showing possible sources and pathways for the occurrence of pharmaceutical residues in the aquatic environment. *Source* (Heberer 2002)

biological wastewater treatment methods provide different degradation efficiencies depending on the particular type of substance (derived from the log $_{Kow}$ value). For example, the monochloropropyl derivative TCPP releases into rainwater mainly from construction, demolition, and reconstruction. Based on the hydrophobic properties of the activated sludge biodegradation method, the existing parameters have virtually no influence on the reduction of their concentrations in the effluent from the WWTP. In some cases, it is advisable to turn attention toward promising innovative purification technologies such as advanced oxidation processes or membrane technologies.

11.7 Activated Sludge Methods (Conventional Method)

Conventional sewage treatment at WWTPs shows a relatively low efficiency of degradation of xenobiotic substances in sewage (Auriol et al. 2006). However, it has been noted that certain biological processes in combination with adsorption on solid sludge particles can lead to 45–99% of their degradation from wastewater. Very nitrification has been shown to be effective in this regard, which has led to the degradation of steroidal estrogens. However, in the conventional method of purification, it can be stated that different substances have different effects, for example, in the analgesic group, acetylsalicylic acid is well degraded; and however, diclofenac is completely persistent against the conventional method of purification.

11.8 AOP's

Advanced oxidation process (AOPs) is a set of chemical cleaning procedures designed to remove organic and inorganic materials from water through oxidation. Pollutants are oxidized by four different agents: ozone and hydrogen peroxide, oxygen, and air and are delivered to the system at programmed doses and combinations. Procedures may also be combined with UV radiation or other specific catalysts, e.g., use of Fenton's reagent. Types can be seen in Table 11.3.

The AOP process is particularly effective for cleaning biologically toxic or non-degradable materials such as aromatics and pesticides, and petroleum components and volatile organic compounds in sewage. The principle of the method is that the contaminants are converted into stable inorganic compounds such as water, carbon dioxide, and salts; i.e., it is mineralized. The purpose of wastewater treatment with AOP is to reduce chemical contaminants and their toxicity to a level such that cleaned water can be discharged into the recipient or at least to the WWTP.

11.9 Sorption Processes

Currently, sorption methods are used to improve biological cleaning processes, which use activated charcoal as sorbent, either in the form of powdered activated carbon (PAC) or granulated (GAC) powders. In active coal, it has been shown to be able to sorb some steroid hormones or nonylphenol on the other hand, wrongly removing the X-ray contrast agent iopromide and ibuprofen, meprobamate, sulfamethoxazole, and diclofenac.

Another interesting feature is that activated carbon obtained from cork pulp and plastic residues showed improved sorption potential in ibuprofen removal than commercial sorbents (Mestre et al. 2009).

11.10 Membrane Technology

Membrane filtration is the separation of substances by means of a polymeric membrane which functions on the principle of a semi-permeable physical barrier with pores or molecular channels. The incoming concentrate is divided into permeate (containing substances that pass through the membrane) and the retentive (contains substances that retain the membrane). The most important types of membrane filtration are pressurized processes including microfiltration (MF), ultrafiltration (UF), nanofiltration (NF), and reverse osmosis (RO), (Zhou and Smith 2001).

The effectiveness of membrane removal depends on (Liu et al. 2009):

(1) Physicochemical properties of xenobiotics: molar mass, solubility in water, metal, electrostatic properties;

Table 11.3 Types of advanced oxidation methods (AOPs) investigated for the removal of xenobiotics from wastewater (Caliman and Gavrilescu 2009)

AOPs	Basic principle	Principle of the main chemical reaction
O_3/UV	Photolysis of ozone produces hydroxyl radicals (OH·). Higher pH promotes the formation of radicals. The presence of hydrogen peroxide can increase the efficiency of the process	$O_3 + h\nu + H_2O \rightarrow 2\,OH\cdot + O_2$
O_3/H_2O_2	Under alkaline conditions, free hydroxyl radicals are formed. The presence of peroxide leads to intensification of the reaction (up to 1000 times). Processes can be enhanced by UV light	$O_3 + H_2O_2 \rightarrow 2\,OH\cdot + 3\,O_2$
UV/H_2O_2	Direct photolysis of peroxide is the fission reaction of the bond between its two oxygen atoms	$H_2O_2 + h\nu \rightarrow 2\,OH\cdot$
Fenton reaction	A bivalent iron-containing substance is a catalyst for the formation of radicals. The efficiency of the process can be improved by UV radiation	$Fe^{2+} + H_2O_2 \rightarrow Fe^{3+} + OH^- + OH\cdot$
Photocatalysis TiO_2/UV	The process involves the absorption of energy-rich radiation with a semiconductor and the subsequent formation of electron-hole and hydroxyl radicals. Efficiency improvements can be achieved by the addition of oxidizing agents such as O_3, O_2 or H_2O_2	
Sonolysis	Water ultrasound leads to cyclical growth, growth, and subsequent collapse of the tubes, which is repeated in extremely short time intervals. In close proximity to the dying bubbles, a lot of energy is released and reactive molecules such as hydroxyl radicals are formed here. The presence of light or hydrogen peroxide can increase the efficiency of the process	

(2) Membrane properties that affect the rejection mechanism: difference in pore size, charge, and particle adsorption capability;
(3) Water properties: pH, ionic strength, concentration of natural organic substances in water.

The presence of bivalent cations (calcium, magnesium) in water reduces the retention of xenobiotics due to the limitation of the association of pollutants with natural organic matter).

The efficiency of the process increases in the order of: MF < UF < NF < RO.

Micropollutants are substances ranging from ng/l to μg/l, so they are captured by nanofiltration or reverse osmosis. In other processes, they are captured by hydrophobic interaction.

It has been shown that when removing triclosan using two NF membranes (NF-270 and NF-90) and one RO membrane (BW-30), only the reverse osmosis 10 h was used to filter the membrane tritosan permeate in the presence and thickness of the membrane (Coleman et al. 2005).

The effectiveness of retention of xenobiotics using commercially available NFs can be supposedly improved by chemical modifications (Kim et al. 2009). Nanofiltration and reverse osmosis are considered to be the most effective methods of removing endocrine disrupters in wastewater recycling. However, using these methods, it has been observed that some substances can escape from the membrane during filtering removal or at high pH changes. For this reason, it is appropriate to use diaphragm processes only for purification (Kvanli et al. 2008)

11.11 Bioremediation

Bioremediation is a process where we remove pollutants using metabolism of microorganisms. Microorganisms used to perform bioremediation functions are known as bioremediators.

A promising solution in the field of wastewater treatment is the use of so-called white root fungi, which is able to degrade a wide range of persistent substances, including xenobiotics such as DDT, PCB, PAU, and other enzymes. However, there is no study yet to show the effects of white root fungi in practice. However, this method is also sensitive to high heavy metal contents, low temperatures, or the competitiveness of other fungi. The species used for wastewater treatment are Trametes versicolor, Irpes lacteum, Phanerochaete chrysosporium and Ganoderma lucidum (Marco-Urrea et al. 2009)

Decontamination technologies operating on the principle of biochemical processes are (Vaněk et al. 2002):

- Phytoextraction—sowing or planting of selected plants on the contaminated area;
- Rhizofiltration—absorption, concentration, and precipitation of xenobiotics from polluted water by roots of living plants;

- Phytodegradation—uses symbiotic interactions or associations between plants and microorganisms in the rhizosphere environment for the decontamination of pollutants;
- Phytovolatilization—based on the ability of some microorganisms to enzymatically reduce mercury ions to metallic mercury, which, due to its physical properties, diffuses into the environment in the form of vapors.

Particularly effective in the degradation of some types of xenobiotics is the process of phytodegradation and rhizofiltration. Phytodegradation has been tested so far only on soil but has been found to be capable of removing TPH, PAH, chlorinated pesticides, PCBs, TCE, explosives, nitrates, organophosphate pesticides, and detergents. Rhizofiltration was tested only for capture of radionuclides, and some toxic metals were able to remove these substances from water even at low concentrations

Phytoremediation is the process by which we remove contaminants such as metals and pesticides and solvents and explosives and petroleum and its derivatives and various other impurities by plants (herbs, shrubs, and trees).

Phytoremediation is considered an economically efficient and non-interfering technology for the environment. Over the past 20 years, this technology has become increasingly popular. However, its main disadvantage is that the process is dependent on plant growth, tolerance to toxicity, and bioaccumulation ability of plants.

Phytoremediation is a method that has many modifications for purification of wastewater which is determined by the method called "Rhizofiltration" where the wastewater is filtered through the roots of plants. Pollutants remain absorbed or adsorbed on roots. This method can also be combined with the above-mentioned white root fungi (e.g., for root treatment plants) (Kučerová et al. 1999).

11.12 Conclusions

It should be emphasized that the existing traditional WWTP treatment technologies are unable to effectively eliminate most xenobiotics The biggest problem of the treated wastewater which contains the very wide range of these substances is the high variability of the degradation efficiency, due to the other factors such as temperature, water retention, sludge aging and nitrification-enhancing environments.

Although the Water Framework Directive (Gasperi et al. 2009) has forced the member states, that to devote more attention to priority substances and to control of their entry into surface waters, it is expected to focus in the future on further tightening of legislation for precise determination of the occurrence and importance of PPs in different diffusion and local sources and identification of the origin, cycle, and transmission of PPs in surface waters. These provisions will make it as easy as possible to detect and control sources of pollutants, which will make an improvement of the surface water monitoring.

Acknowledgements This chapter has been worked out under the project No. LO1408 "AdMaS UP—Advanced Materials, Structures and Technologies", supported by Ministry of Education, Youth and Sports under the "National Sustainability Programme I".

References

Adeel M, Song X, Wang Y et al (2017) Environmental impact of estrogens on human, animal and plant life: a critical review. Environ Int 99:107–119

American Cancer Society, https://www.cancer.org/cancer/cancer-causes/antiperspirants-and-breast-cancer-risk.html?sitearea=MED

Arnika, Arnika Association, Campaigns of the Toxics and Waste Programme. http://arnika.org/

ATSDR, Agency for Toxic Substances and Disease Registry, https://www.atsdr.cdc.gov/

Auriol M, Filali-Meknassi Y, Tyagi RD, Adams CD, Surampalli RY (2006) Endocrine disrupting compounds removal from wastewater, a new challenge. Process Biochem 41(3):525–539

Bester K, McArdell CHS, Wahlberg C, Bucheli TD (2010) Quantitative mass flows of selected xenobiotics in urban waters and wastewater treatment plants. Xenobiotics in the Urban Water Cycle. 509(23):3–26

Caliman FA, Gavrilescu M (2009) Pharmaceuticals, personal care products and endocrine disrupting agents in the environment—a review. Clean-Soil Air Water 37(4–5):277–303

Coleman HM, Chiang K, Amal R (2005) Effects of Ag and Pt on photocatalytic degradation of endocrine disrupting chemicals in water. Chem Eng J 113(1):65–72

EPA, U.S. Environmental Protection Agency. https://www.epa.gov/

Eriksson E, Baun A, Mikkelsen PS, Ledin A (2007a) Risk assessment of xenobiotics in stormwater discharged to Harrestrup Å, Denmark. Desalination 215:187–197

Eriksson E, Baun A, Scholes L (2007b) Selected stormwater priority pollutants—a European perspective. Sci Total Environ 383:41–51

European Commission, website of European Commission—https://ec.europa.eu

European Commission, Directive 2000/60/EC of the European Parliament and of the Council Establishing a Framework for the Community Action in the Field of Water Policy, 23 Oct 2000

FATE, http://fate.jrc.ec.europa.eu/rational/home.html

Fatta-Kassinos D, Bester K, Kümmerer K (2010) Xenobiotics in the urban water cycle mass flows, environmental processes, mitigation and treatment strategies. Environ Pollut 16(1):507

Gasperi J, Garnaud S, Rocher V, Moilleron R (2009) Priority pollutants in surface waters and settleable particles within a densely urbanised area: case study of Paris (France). Sci Total Environ (8):2900–2908

Gong Y, Han XD (2006) Nonylphenol-induced oxidative stress and cytotoxicity intesticular Sertoli cells. Reprod Toxicol 22(4):623–630

Harrigan JA, Vezina CM, Mcgarrile BP, Ersing N, Box HC, Maccubin AE, Olson JR (2004) DNA adduct formation in precision-cut rat liver and lung slices exposed to benzo a pyrene. Toxicol Sci 77(2):307–314

Heberer T (2002) Occurrence, fate, and removal of pharmaceutical residues in the aquatic environment: a review of recent research data. Toxicol Lett 131:5–17

Holoubek I et al (2000) Polychlorinated Biphenyls (PCBs)—World-Wide Contaminated Sites. TOCOEN Report No 173, Brno, May 2000

Hughes CL (1988) Phytochemical mimicry of reproductive hormones and modulation of herbivore fertility by phytoestrogens. Environ Health Perspect 78:171–175

IARC, International Agency for Research on Cancer, https://www.iarc.fr/en/about/index.php

Ingerslev F, Vaclavik E, Halling-Sorensen B (2003) Pharmaceuticals and personal care products: a source of endocrine disruption in the environment? Pure Appl Chem 75(11–12):1881–1893

IRZ, Integrated Pollution Register of the Czech Republic—https://www.irz.cz/

Jobling S, Tyler CR (2003) Endocrine disruption in wild freshwater fish. Pure Appl Chem 75(11–12):2219–2234

JRC—Joint Research Centre—Institute for Environment and Sustainability—http://ies-webarchive-ext.jrc.it

Kim I, Yamashita N, Tanaka H (2009) Performance of UV and UV/H2O2 processes for the removal of pharmaceuticals detected in secondary effluent of a sewage treatment plant in Japan. J Hazard Mater 166(2–3):1134–1140

Kopponen P, Sinkkonen S, Poso A et al (1994) Sulfur analogs of polychlorinated dibenzo-p-dioxins, dibenzofurans and diphenyl ethers as inducers of cyp1a1 in mouse hepatoma-cell culture and structure-activity-relationships. Environ Toxicol Chem 68:1543–1548

Kučerová P, Macková M, Macek T (1999) Perspectives of phytoremediation in decontamination of organic pollutants and xenobiotics. Chem Listy 93(8):19–26

Kvanli DM, Marisetty S, Anderson TA, Jackson WA, Morse AN (2008) Monitoring estrogen compounds in wastewater recycling systems. Water Air Soil Pollut 188(1–4):31–40

Lear L (1997) Rachel carson: witness for nature. Henry Holt, New York, ISBN 0-8050- 3428-5

Lemonick M (2007) Heroes of environmentalist: Theo Colborn. Magazine Time U.S., 17 Oct 2007, 45(1):27

Liu ZH, Kanjo Y, Mizutani S (2009) Removal mechanisms for endocrine disrupting compounds (Xenobiotika) in wastewater treatment—physical means, biodegradation, and chemical advanced oxidation: a review. Sci Total Environ 407(2):731–748

Marco-Urrea E, Perez-Trujillo M, Vicent T, Caminal G (2009) Ability of white-rot fungi to remove selected pharmaceuticals and identification of degradation products of ibuprofen by Trametes versicolor. Chemosphere 74(6):765–772

Maxa M et al (2002) Nebezpečné látky v odpadních vodách z chemického průmyslu České republiky (Odvětvová situační studie). TECHEM, Praha

MESAUC—Malta_EU Steering and Action Committee—meusac.gov.mt

Mestre AS, Pires J, Nogueira MF, Parra JB, Carvalho AP, Ania CO (2009) Waste-derived activated carbons for removal of ibuprofen from solution: Role of surface chemistry and pore structure. Biores Technol 100(5):1720–1726

Nebert DW, Roe AL, Dieter MZ, Sois WA, Yang YAND, Daton TP (2000) Role of the aromatic hydrocarbon receptor and Ah gene battery in the oxidative stress response, cell cycle control, and apoptosis. Biochem Pharmacol 59(1):65–85

Nghiem LD, Coleman PJ (2008) NF/RO filtration of the hydrophobic ionogenic compound triclosan: transport mechanisms and the influence of membrane fouling. Sep Purif Technol 62(3):709–716

Ni Y, Zhang Z, Zhang Q, Chen J, Wu Y, Liang X (2005) Distribution patterns of PCDD/Fs in chlorinated chemicals. Chemosphere 60(6):779–784

Pulkrabová J, Hrádková P, Hajšlová J, Poustka J, Nápravníková M, Poláček V (2009) Brominated flame retardants and other organochlorine pollutants in human adipose tissue samples from the Czech Republic. Environ Int 35:63–68

Randák T (2004) Informace o výsledcích výzkumu cizorodých látek a zdravotního stavu ryb v řece Labi v roce 2003. Jihočeská univerzita v Českých Budějovicích, Výzkumný ústav rybářský a hydrobiologický ve Vodňanech, 24 Apr 2004

REACH is the European Regulation on Registration, Evaluation, Authorisation and Restriction of Chemicals. https://ec.europa.eu/growth/sectors/chemicals/reach_cs

SCF (2002) Polycyclic Aromatic Hydrocarbons—Occurrence in foods, dietary exposure and health effects, SCF/CS/CNTM/PAH/29 ADD1 Final, prosinec 2002. (http://europa.eu.int/comm/food/fs/sc/scf/out154_en.pdf)

Scholes L, Revitt DM, Ellis JB (2005) Predicting the pollutant removal potentials of sustainable urban drainage systems. In: Proceedings of the 3rd national conference on sustainable drainage. Coventry, UK, pp 199–210

Snyder SA, Villeneuve DL, Snyder EM, Giesy JP (2001) Identification and quantification of estrogen receptor agonists in wastewater effluents. Environ Sci Technol 35(18):3620–3625

Snyder SA, Westerhoff P, Yoon Y, Sedlak DL (2003) Pharmaceuticals, personal care products, and endocrine disruptors in water: implications for the water industry. Environ Eng Sci 20(5):449–469

Suarez S, Carballa M, Omil F, Lema JM (2008) How are pharmaceutical and personal care products (PPCPs) removed from urban wastewaters? Rev Environ Sci Biotechnol 7(2):125–138

The Stockholm Convention on Persistent Organic Pollutants, http://chm.pops.int/TheConvention/Overview/TextoftheConvention/tabid/2232/Default.aspx

The Water Framework Directive 2000/60/EC

The Water Framework Directive: Tap into it! (2002) Luxembourg: Office for Official Publi-cations of the European Communities, p 12, ISBN 92-894-1946-6

Urban Waste Water Treatment Directive (UWWTD) site for Europe, http://uwwtd.oieau.fr/

Vaněk T, Soudek P, Tykva R, Kališová I (2002) Možnosti využití fytoremediace pro odstranění kontaminace způsobené toxickými kovy a radionuklidy Hornická Příbram ve vědě a technice, Příbram, CD ROM, 15–17 October 2002

Vrcek V (2017) Pharmacoecology—the environmental fate of pharmaceuticals. Kemija U Industriji-Journal of Chemists and Chemical Engineers 66(3–4):134–144

WFD CIRCABC—The Information Exchange Platform. http://ec.europa.eu/environment/water/water-framework/iep/index_en.htm

WHO (1991) Chlorobenzenes other than hexachlorobenzene. Health and safety guide no 128. IPCS International Programme on Chemical Safety, Geneva

WISE (2008), Water Information System for Europe, Water note 1, European Commission (DG Environment), ISBN 978-92-79-09282-4, online: http://ec.europa.eu/environment/water/participation/pdf/waternotes/water_note1_joining_forces.pdf, (downloaded: September 16, 2010)

Zhou H, Smith DW (2001) Advanced technologies in water and wastewater treatment. Can J Civ Eng 28:49–66

Chapter 12
Water and Aquatic Fauna on Drugs: What are the Impacts of Pharmaceutical Pollution?

Piotr Klimaszyk and Piotr Rzymski

Abstract Pharmaceutical pollution is becoming an unavoidable environmental issue of emerging concern. As forecasted, the consumption of medicinal drugs and their use in veterinary practice is expected to systematically increase over coming years, resulting in their increased discharge. The most commonly used pharmaceuticals include non-steroidal anti-inflammatory drugs (e.g., diclofenac, naproxen, ibuprofen), cardiovascular drugs (e.g., beta-blockers, diuretics, calcium channel blockers, lipid-regulating agents), antibiotics, oral contraceptives, anti-depressants, immunosuppressive drugs and cytostatics. Active pharmaceutical ingredients (APIs) are known to partially survive the conventional process of wastewater treatment. In freshwaters, they may undergo photodegradation, biodegradation, sorption to sediments and uptake by organisms. The latter results in metabolism or bioaccumulation, and potential toxicological effects and physiological responses. The magnitude of effects is largely modulated by the concentration of APIs, time of exposure and some environmental factors such as light and nutrient availability. The response to APIs in closely taxonomically related species may be significantly different. The concomitant presence of different APIs usually evokes potentiation of adverse effects. The most serious effects of pharmaceutical pollution evidenced so far for freshwaters include increase in antibiotic-resistant microorganisms, feminization, behavioral changes, and immunosuppression in fish. Beyond any doubt, it is imperative to support systematic research on API detection methods, to monitor the great number of APIs in wastewater, surface and groundwater, and tap water, and to assess the ecological risks arising from their increased presence in the freshwater environment.

Keywords Pharmaceutical pollution · Freshwater · Bioaccumulation · Fish physiology · Toxic effects · Environmental fate

---

P. Klimaszyk
Department of Water Protection, Faculty of Biology, Adam Mickiewicz University, Umultowska 89, 61-614 Poznań, Poland
e-mail: pklim@amu.edu.pl

P. Rzymski (✉)
Department of Environmental Medicine, Poznan University of Medical Sciences, Poznań, Poland
e-mail: rzymskipiotr@ump.edu.pl

M. Zelenakova (ed.), *Water Management and the Environment: Case Studies*, Water Science and Technology Library 86,
https://doi.org/10.1007/978-3-319-79014-5_12

12.1 Introduction

The problem of pharmaceutical discharge, known as pharmaceutical pollution, is considered unavoidable if one considers that the global pharmacological market and subsequent use of drugs are on the systematic rise (Boxall et al. 2012). Several hundred thousand tons of active pharmaceutical ingredients (APIs) are estimated to be consumed annually. It is forecasted that by 2020, the global use of medicines will reach 4.5 trillion doses worth a total of 1.4 trillion USD. The most often used pharmaceuticals include non-steroidal anti-inflammatory drugs (e.g., diclofenac, naproxen, ibuprofen), cardiological drugs (e.g., beta-blockers, diuretics, calcium channel blockers, lipid-regulating agents), antibiotics, oral contraceptives, anti-depressants, immunosuppressive drugs and cytostatics (IMS Institute for Healthcare Informatics 2015). These groups of APIs are the most often identified, at varying concentrations, in surface and groundwater, including sources of drinking water (Boxall et al. 2012). Public water treatment systems may not fully degrade certain APIs or by-products may be generated, e.g., via chlorination or ozonation (Khetan and Collins 2007). As a result, bioactive pharmaceuticals enter tap water mostly at ppb concentration ranges which causes public concern, although the World Health Organization (WHO) concluded in 2012 that at current levels, appreciable adverse impacts on human health are unlikely (WHO 2011).

Acute effects of pharmaceuticals to aquatic organisms are also unlikely, except for spills. The long-term ecological impacts of pharmaceutical pollution are, however, unknown, difficult to predict, and require extensive interdisciplinary and multifaceted research (Fent et al. 2006). It is obvious that potential effects depend on a combination of factors such as load of discharged APIs, their environmental concentrations, physicochemical features, number of co-occurring APIs and their potential additive, synergistic, or antagonistic interactions. Environmental effects, less or more subtile, can be expected for a number of reasons:

- some surface waters systematically receive treated wastewater which can contain high amounts of different pharmaceuticals at varying concentrations (ng L^{-1} to mg L^{-1} in extreme cases), and this can eventually lead to the occurrence of bioactive levels of APIs;
- continuous discharge may cause the loading levels to exceed degradation rates, even if some APIs do not possess a long half-life (and on the basis of this some pharmaceuticals are considered as pseudo-persistent contaminants);
- xenobiotic character of some APIs, their persistence and toxic action in non-target organisms;
- bioaccumulation of APIs in the food chain;
- the similar mechanism and pathways of action of some APIs in aquatic animals, mostly fish, as in humans.

There is some experimental evidence that pharmaceutical pollution can alter the growth of aquatic organisms on different organization levels, such as cyanobacteria, algae (Du et al. 2015; Bácsi et al. 2016), zooplankton (Flaherty and Dodson 2005;

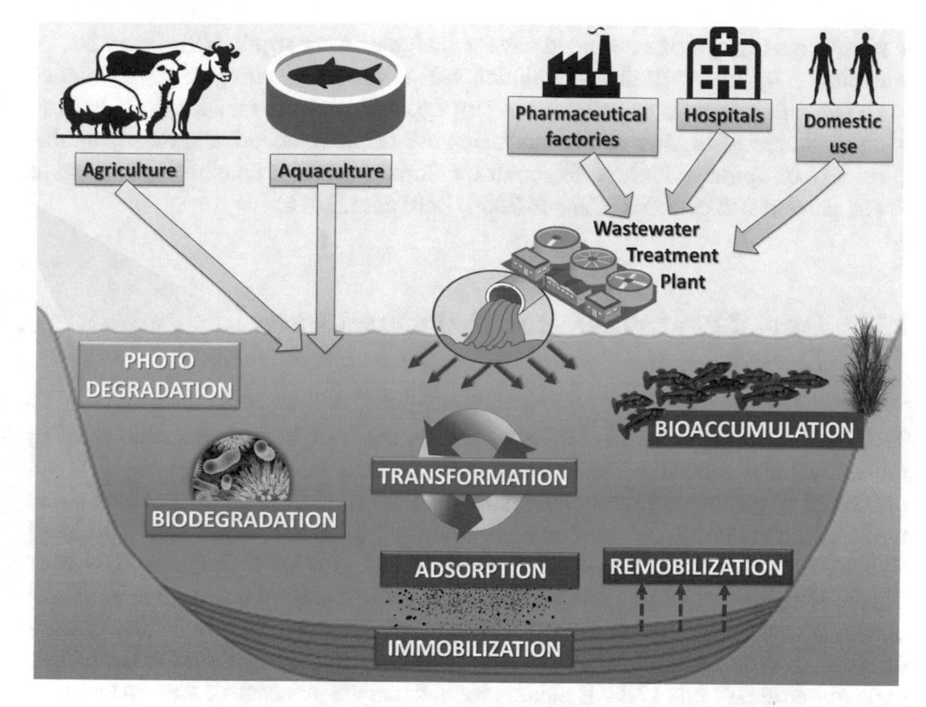

Fig. 12.1 Sources and fate of pharmaceuticals in the aquatic environment. *Source* Author

Richards et al. 2008; Sarma et al. 2014), or macroinvertebrates (Bringolf et al. 2010; López-Doval et al. 2012; Bundschuh et al. 2016). However, the most profound effects, some occurring at environmentally relevant API concentrations, have been found for fish, and depending on the pharmaceutical group have included endocrine effects, developmental alterations, and behavioral changes (Corcoran et al. 2010; Brodin et al. 2014).

12.2 Sources of Pharmaceuticals in the Environment

It is well established that the aquatic environment is being polluted with pharmaceuticals due to their extensive use in livestock and as human medicines (Fig. 12.1). APIs in the environment are known to originate from diverse sources, namely:

- hospital sewage (Mendoza et al. 2015);
- domestic wastewater due to the predominantly urinary excretion of unchanged forms or active metabolites (Barra Caracciolo et al. 2015);
- treated wastewater as traditional wastewater treatment plants is not able to completely remove excreted pharmaceuticals (Kot-Wasik et al. 2016);
- effluents from pharmaceutical manufactures (Cardoso et al. 2014);

- runoff from agriculture if treated sewage sludge is being applied (Sim et al. 2011);
- excretion by livestock due to common use of veterinary drugs (e.g., antibiotics and growth promotion agents) in modern agriculture (Łukaszewicz et al. 2016);
- aquaculture wastewater due to increasing use of pharmaceuticals, mostly antibiotics (e.g., tetracyclines, sulfonamides, chloramphenicol, ampicillin) to combat and prevent fish diseases (Cabello 2006; Bôto et al. 2016).

12.3 Fate of Pharmaceuticals in the Freshwater Environment

The increased concentrations of APIs are usually observed in environments receiving raw or treated wastewater. Their concentrations mostly fall into the ppt (ng L^{-1}) and ppb range (μg L^{-1}) and have predominantly been studied in rivers and streams as these are mostly the receivers of treated wastewater in various locations (Kunkel and Radke 2012). Examples of concentrations of various APIs identified in main rivers of Poland are summarized in Table 12.1. The highest levels were noted for some cardiological drugs (e.g., valsartan, sotalol, metoprolol), non-steroidal anti-inflammatory drugs (e.g., diclofenac), analgesics (e.g., tramadol), and seizure medications (e.g., carbamazepine) (Table 12.1). It should be noted that the presence of APIs in freshwaters is not subject to systematic monitoring and the available information is a result of scientific investigations and the development of detection methods. The presence of APIs in lakes is, in turn, less known and requires further studies (Blair et al. 2013).

Once discharged to the aquatic environment pharmaceuticals may undergo various fates (Fig. 12.2). Some may potentially be transformed during the wastewater treatment process, resulting in inactive compounds, those of lower or higher toxicity and persistence (Celiz et al. 2009). In some cases, the concentration of transformation products may exceed those of the parent compound which complicates a holistic assessment of pharmaceutical pollution in the aquatic environment. Such a phenomenon has been observed, for example, for atorvastatin, carbamazepine, and diclofenac (Langford and Thomas 2011, Jakimska et al. 2014).

The United Nations Environmental Programme (UNEP) sets half-life at 60 days to define a chemical as persistent in the aquatic environment. It is obvious that the persistence of APIs depends on a combination of inherited physicochemical properties of the compound and various environmental conditions. Experimental and in-field studies yield contradictory results on the persistence of APIs. For example, the half-life for carbamazepine during laboratory and in-field research has been reported to vary from 3.5 to over 230 days (Yamamoto et al. 2009) and from 63 to 1200 days, respectively (Tixier et al. 2003; Zou et al. 2015). On the other hand, all studies have noted that the half-life of diclofenac (which is one of the most often reported APIs in aquatic environment) is below the persistence threshold of 60 days (Bu et al. 2016) The half-life of another non-steroidal anti-inflammatory drug, ibuprofen, has been reported to range from 19 to 413 days, depending on experimental

Table 12.1 List of detected active pharmaceutical ingredients in main Polish rivers

Pharmaceutical group	Active pharmaceutical ingredient	Studied river	Maximum detected concentration [ng L^{-1}]	References
Cardiovascular drugs: calcium channel blockers	Amlodipine	Vistula	19	Giebułtowicz et al. 2016
	Diltiazem		24	
	Nifedipine		0.5	
Cardiovascular drugs angiotensin-converting enzyme inhibitors	Quinalapril		155	
	Ramipril		73	
Cardiovascular drugs: angiotensin II receptor antagonist	Losartan		610	
	Telmisartan		1130	
	Valsartan		5260	
		Warta	133	Kasprzyk-Hordern et al. 2007
Cardiovascular drugs: diuretics	Furosemide	Vistula	2670	Giebułtowicz et al. 2016
	Hydrochlorothiazide		1270	
Cardiovascular drugs: beta-blockers	Acebutolol		643	Giebułtowicz et al. 2016
	Atenolol		205	
	Bisoprolol		1470	
	Labetalol		3.3	
	Propranolol		69	
	Sotalol		2120	
	Metoprolol		2190	
		Warta	155	Kasprzyk-Hordern et al. 2007
Cardiovascular drugs: antiarrhythmics	Propafenone	Vistula	87	Giebułtowicz et al. 2016
Cardiovascular drug: lipid-regulating agents	Atorvastatin		114	Giebułtowicz et al. 2016; Giebułtowicz and Nałęcz-Jawecki 2016
	Bezafibrate		4.5	
		Warta	8.0	

(continued)

Table 12.1 (continued)

Pharmaceutical group	Active pharmaceutical ingredient	Studied river	Maximum detected concentration [ng L^{-1}]	References
	Ciprofibrate	Vistula	60	
	Clofibric acid		130	
	Fenofibrate		0.9	
	Gemfibrozil		3.0	
Immunosuppressive drugs	Mycophenolic acid		180	
Non-steroidal anti-inflammatory drugs	Ketoprofen	Warta	47	Kasprzyk-Hordern et al. 2007; Baranowska and Kowalski 2012
	Ibuprofen		76	
	Diclofenac		486	
		Oder	470	
		Vistula	140	
		Klodnica	70	
	Naproxen	Warta	130	
		Oder	140	
		Vistula	300	
	Aspirin	Oder	730	Baranowska and Kowalski 2012
		Vistula	400	
Analgesic	Paracetamol	Warta	90	
	Metamizole	Oder	900	
	Tramadol	Warta	2108	Kasprzyk-Hordern et al. 2007
	Codeine		15	
Seizure	Carbamazepine	Warta	794	
	Gabapentin		75	
Anti-depressants	Citalopram	Vistula	17.0	Giebułtowicz and Nałęcz-Jawecki 2014

(continued)

Table 12.1 (continued)

Pharmaceutical group	Active pharmaceutical ingredient	Studied river	Maximum detected concentration [ng L^{-1}]	References
	Doxepin		1.9	
	Fluoxetine		3.2	
	Mianserin		9.0	
	Mirtazepin		5.0	
	Moclobemid		28	
	Tianeptin		1.8	
	Trazodon		0.9	
	Venlaflaxin		250	
Antibiotics	Sulfamethoxazole	Warta	60	Kasprzyk-Hordern et al. 2007
	Sulfapyridine		31	
	Trimethoprim		27	

Source Author

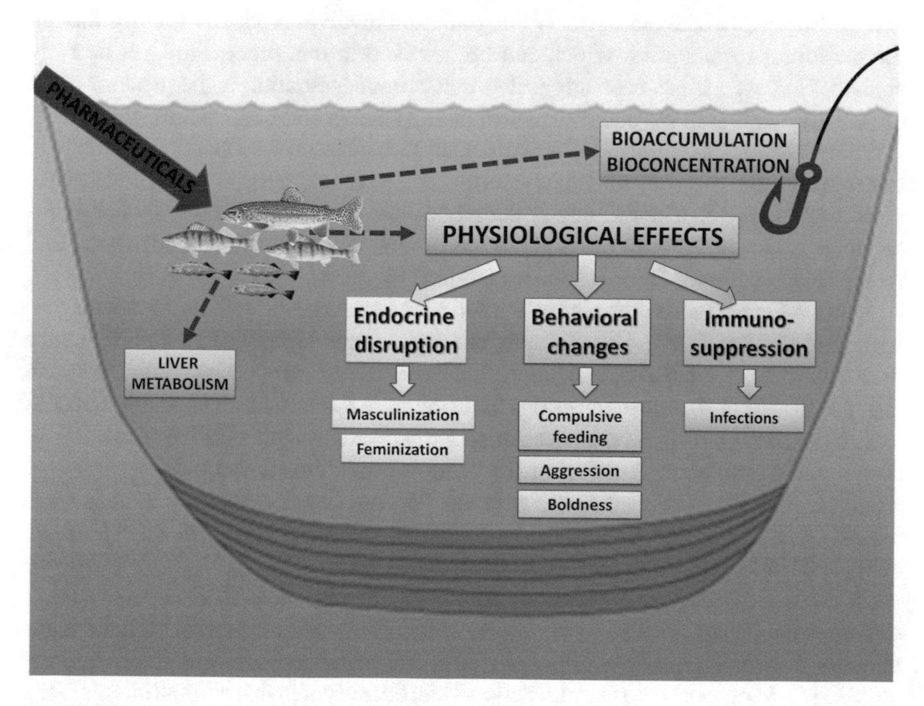

Fig. 12.2 Potential effects of pharmaceuticals on fish. *Source* Author

conditions (Yamamoto et al. 2009). Persistence of antibiotics is also very diversified with some, such as ampicillin, doxycycline, and oxytetracycline degrading well, and others, e.g., josamycin, proving resistant to degradation. As shown, some antibiotics may accumulate in sediments and may not biodegrade, particularly under anaerobic conditions (Maki et al. 2006; Kümmerer 2009).

Although most APIs may not persist in the environment, their continuous release to aquatic environments from wastewater treatment plants may result in loading levels exceeding degradation rates. For this reason, APIs are considered as a group of pseudo-persistent contaminants (Daughton et al. 2003). The degradation of APIs depends on a number of factors which, besides the inherent physicochemical properties of pharmaceutical include:

- water temperature;
- pH;
- availability of sunlight and oxygen;
- the composition and activity of the microbial community.

APIs may undergo photodegradation as various compounds of this group contain heteroatoms, aromatic rings, and other structures that can absorb light (direct photolysis) or react with photogenerated transient species (indirect photolysis) (Khetan and Collins 2007). The other route of degradation involves biodegradation, although the microbial communities which can be involved in this process are yet be fully studied. However, it has been suggested that photodegradation is the primary route of API elimination in the aquatic environment (Khetan and Collins 2007).

Pharmaceuticals can also be adsorbed on the surface of suspended matter and undergo sorption to sediments (Scheytt et al. 2005). Nevertheless, some of the initially sorbed APIs may also be eventually desorbed. The exact conditions responsible for re-mobilization of APIs in the aquatic environments are not fully understood and require further studies (Martínez-Hernández et al. 2014).

Some APIs may also bioaccumulate and bioconcentrate in various aquatic organisms (Brodin et al. 2014). For example, the popular non-steroidal anti-inflammatory drugs, diclofenac, naproxen, and ibuprofen, have all been shown to accumulate in fish (Lahti et al. 2011). The concentrations of the latter in the fish plasma and bile samples were 100–1000-fold higher than in corresponding water samples (Jeffries et al. 2015). Some antibiotics, particularly from the group of quinolones, can bioaccumulate at relatively high levels in bivalves (Li et al. 2012). Anti-depressants were evidenced to bioaccumulate in benthic invertebrates, periphyton, bivalves, snails, and fish (Brooks et al. 2005; Bringolf et al. 2010; Du et al. 2015; Grabicova et al. 2015). Although some APIs, e.g., macrolide antibiotic roxithromycin and beta-blocker propranolol, can enter food chains and be easily transferred through the trophic route, their accumulation in fish is rather low due to efficient liver metabolism (Liu et al. 2014; Ding et al. 2015). In other cases, pharmaceuticals can be present in fish compartments. This particularly concerns individuals inhabiting effluent-dominated streams. The bioaccumulation of APIs depends, inter alia, on their lipophilicity—a parameter which can significantly increase with pH as pharmaceuticals usually contain at least one ionizable group (Ingram et al. 2011). Fish can primarily uptake APIs from respiratory exchange (inhalation) and from diet (due to the potential of APIs to accumulate in the food web) (Du et al. 2014). The latter has been considered less often as it requires more complicated experimental design (Brodin et al. 2014). Despite the fact that accumulation of pharmaceuticals in aquatic biota has been extensively studied over the last few years, there remains a need to fully identify all transfer routes and evaluate environmental as well as human risks, analyze the content of APIs in waterborne foodstuffs and their potential contribution to unwanted exposure to pharmaceuticals (Puckowski et al. 2016). Pharmaceutical pollution could eventually result in the presence of APIs in the drinking water supply and tap water (Table 12.2).

12.4 The Effect of Pharmaceuticals on Aquatic Biota

To evaluate the potential effects which pharmaceuticals can exert on biocoenoses, it is important to consider their mechanisms of action and to elucidate whether these mechanisms may be similar in some non-target organisms (e.g., due to the presence of similar receptors and metabolism as in the target organism). Evolutionary similarities in receptors and molecular targets of pharmaceuticals are known to exist in fish while data for invertebrates is still scarce. Side effects which can occur in humans may provide some information on the possible adverse events that pharmaceuticals could exert on lower animals (Fent 2008).

The magnitude of effects of a pharmaceutical on aquatic organisms is largely modulated by its concentration and time of exposure. At extremely high concentrations of APIs (mg L^{-1}) which exceed by several or even a dozen orders of magnitude levels being observed in natural aquatic environments, deleterious effects, namely high mortality or significant growth impairment have been observed for most studied taxa (Crane et al. 2006; Brain et al. 2006; Nunes et al. 2014). At concentrations from ng L^{-1} to μg L^{-1} which were noted in freshwaters, the responses may not be as severe. However, exposure to pharmaceuticals may still represent a key factor in modifying the functioning of individuals as well as the evolution and survival of whole populations. It is rather expected that these effects are more likely to occur in rivers and streams, as in lakes, particularly in large reservoirs, a high dilution factor may decrease environmental risks (Blair et al. 2013).

Pharmaceutical pollution can lead to various changes ranging from biochemical modifications on a cellular level through behavioral alterations to modifications on a population or even ecosystem level (Vasseur and Cossu-Leguille 2006; Lynn et al. 2007; Nunes et al. 2014). The effects of APIs on organisms may be stimulated or hampered by some environmental factors (e.g., nutrients and light availability). Moreover, the response to APIs in closely taxonomically related species may be significantly different. For example, Nunes et al. (2014) found an approximately 14-fold difference in half the maximal effective concentration (EC_{50}) values for acute toxicity of paracetamol between *Daphnia longispina* and *Daphnia magna*, species which are phylogenetically very similar to each other. Moreover, variation in the

Table 12.2 List of pharmaceuticals detected at concentrations above 1.0 ng L^{-1} in tap water distributed in Poland

Pharmaceutical group	Compound	Maximum concentration [ng L^{-1}]	Detected locations	References
Cardiological	Acebutolol	4.0	Warsaw	Giebułtowicz et al. 2016
	Amlodipine	3.5		
	Bisoprolol	17.0		
	Clofibric acid	1.3		
	Diltiazem	1.4		
	Fenofibrate	1.1		
	Furosemide	29		
	Losartan	5.0		
	Metoprolol	14		
	Propafenone	4.0		
	Propranolol	7.0		
	Quinalapril	1.8		
	Sotalol	16		
	Telmisartan	23		
	Valsartan	27		
	Trimetazidine	4.2	Gdańsk	Kot-Wasik et al. 2016
	Nafronyl	3.8		
	Ramipril	2.8	Gdańsk, Warsaw	Kot-Wasik et al. 2016; Giebułtowicz et al. 2016
	Hydrochlorothiazide	26		
	Atenolol	1.5		
Anti-depressants	Citalopram	1.5	Warsaw	Giebułtowicz and Nałęcz-Jawecki 2014
	Venlaflaxin	1.9		
Anti-diabetic	Metformin	8.0	Gdańsk	Kot-Wasik et al. 2016
Gastric	Ranitidine	5.6		
Analgesic	Paracetamol	118.9		Caban et al. 2015
Antibiotics	Erythromycin	6.0		Kot-Wasik et al. 2016
Seizure	Carbamazepine	6.0		
Contraceptives	Levonorgestrel	46.4		
Hormone replacement	Progesteron	4.8		
Non-steroidal anti-inflammatory drugs	Ketoprofen	166.9		
	Ibuprofen	223.6		
	Diclofenac	114.3		

Source Author

response of the same test species was noted in different studies; the reasons behind these discrepancies are yet to be elucidated (Henschel 1997; Kim et al. 2007). It should also be noted that data from laboratory studies does not fully reflect the real impact the APIs may have on aquatic biocenoses. The majority of experimental investigations have tested the effects of only single APIs at a time, while in the aquatic environment APIs tend to occur as a mixture of various chemical compounds (potentially interacting with each other), often in the presence of other organic waste contaminants (Kolpin et al. 2002). Such a concomitant presence of different APIs usually evokes potentiation of adverse effects, and these effects may be triggered at lower concentrations than those identified in experimental tests involving a single API (Fent 2008; Quinn et al. 2009).

The following sub-chapters review the effect of selected groups of APIs, namely antibacterial pharmaceuticals, non-steroidal anti-inflammatory drugs, anti-depressants, anti-epileptic drugs, estrogenic endocrine disrupting pharmaceuticals and cardiological medicines on aquatic biota. These groups are the most commonly identified in freshwaters.

12.4.1 Anti-microbials

Discharge of antiseptic pharmaceuticals to freshwater environments may potentially affect naturally occurring microorganisms. However, antibiotics as well as other compounds of anti-microbial activity (e.g., sulfonamides) have not been found to have a directly negative impact on microorganisms at environmentally relevant concentrations. Only in a few cases has the reduction of microbial assemblages in water ecosystems been observed as a result of significant and long-lasting discharge of antiseptic drugs. For example, Hansen et al. (1992) found reduced microbial densities and activity in sediments below fish farm cages—the result of persistent veterinary application of anti-bacterial agents.

The greatest concern over the effects of APIs on microorganisms is the development of drug resistance and the potential consequences this phenomenon could have on aquatic biota and human health (Crane et al 2006). As observed, common use of antibiotics has led to the appearance of antibiotic resistant phenotypes among bacteria (Borg et al. 2010; Shryock and Richwine 2010). Antibiotic resistance in freshwaters was identified in the late twentieth century (Ohlsen et al. 1998; Boon and Cattanach 1999), and currently, it is commonly found in inland surface waters. It is suggested that drug resistance has been mostly caused by the release of resistant strains with urban and rural wastewater, and possible gene transfer rather than only by acquisition of resistance by the naturally living bacteria due to exposure to discharged antibacterial compounds (Crane et al. 2006). Boon and Cattanach (1999) found a relation between human pressure and antibiotic resistance and proposed using resistance as a bacteriological parameter of water quality assessment. It was found that the discharge of effluent from urban wastewater treatment plants contributes to increased resistance to antibiotics in *Escherichia coli* (Koczura et al. 2012) and all heterotrophic bacteria (Koczura et al. 2016). The elevated frequency and density of antibiotic resistant bacteria is usually observed in rivers downstream of the discharge of wastewater treatment plants. Pharmaceutical pollution and treatment of wastewater also contribute to an increase in the frequency of sulfonamide resistant bacteria in freshwaters (Munir et al. 2011; Makowska et al. 2016). As stated by Crane et al. (2006), the specific implications of anti-microbial resistance for aquatic ecosystem functioning remain unknown. However, increasing use of anti-microbial drugs is likely to effect the aquatic environment and may have consequences for human health.

12.4.2 Non-steroidal Anti-inflammatory Drugs

Non-steroidal anti-inflammatory drugs (NSAIDs) are a class of pharmaceuticals that exhibit anti-pyretic and analgesic activity, and induce anti-inflammatory effects at higher doses. NSAIDs inhibit COX_1 and COX_2 enzymes and consequently decrease the synthesis of prostaglandin. The COX enzymes and prostaglandins are involved in a number of physiological processes such as inflammation and pain, blood flow in kidney and coagulation processes (Fent 2008). The side effects of NSAIDs (renal, gastric failure, and damage) are related to the physiological function of prostaglandins and are generally connected with long-term exposure. Inducible COX homologues have been found in different species of fish (Zou et al. 1999; Roberts et al. 2000; Ishikawa and Herschmann 2007) and invertebrates (Pedibhotla et al. 1995). In fish, prostaglandins play a key role in reproduction, namely stimulation of ovulation, oestradiol production, and promotion of sex-specific behavior (Corcoran et al. 2010). NSAIDs can hamper the production of cortisol and this can significantly affect osmoregulation function in fish (Gravel et al. 2009). NSAIDs have also been connected with cardiac abnormalities and lower heart rate, depletion of glycogen in the liver and fish embryo teratogenicity. Cleuvers (2008) found that NSAIDs are responsible for contraction or atony of muscles. Muscle contraction was suspected to decrease the activity of the crustacean *Gammarus* sp. after exposure to 10 ng L^{-1} of ibuprofen (De Lange et al. 2006). Short-term acute toxicity tests demonstrated the highest sensitivity of phytoplankton to NSAIDs when compared to zooplankton and benthic macroinvertebrates (Ferrari et al. 2004; Fent 2008). It should be noted that these tests employed concentrations much above those observed in natural freshwaters. Chronic NSAIDs side effects were stated for all groups of aquatic animals. For example, in cladocerans and rotifers, they caused a decrease in reproduction rate (Ferrari et al. 2004; Marques et al. 2004). Changes in amphipod behavior induced by NSAIDs were observed by (De Lange et al. 2006). In fish, renal lesions, alteration of gills, and other subtle subcellular effects were observed at concentrations ranging below 5 $\mu g\ L^{-1}$ of diclofenac (Schwaiger et al. 2004). NSAIDs were also found to be responsible for disturbance of fish embryo development and hatching (Hallare et al. 2004).

12.4.3 Anti-depressants

Anti-depressants represent a group of pharmaceuticals commonly prescribed to treat major depressive disorders and other conditions such as obsessive compulsive disorder, dysthymia, and mood instability. Anti-depressants may have varying modes of action including selective serotonin reuptake inhibition, monoamine oxidase inhibition, or serotonin–norepinephrine reuptake inhibition.

The effect of anti-depressants on aquatic biota has mostly been studied for selective serotonin reuptake inhibitors (SSRI). Their action leads to an increase in serotonin concentration in the synapse space. As a neurotransmitter, serotonin and its receptors can be found in both vertebrates and invertebrates—the presence of SSRI in the environment can potentially exert various effects on biota. Serotonin is known to directly act on the immune system, stimulate hunger, influence behavior, and modulate sexual function (Fent 2008). However, various species are characterized by different sensitivity and reaction to anti-depressants; thus, there is no one universal effect of these compounds in aquatic biota (Sumpter et al. 2014). Induction of parturition and spawning was observed in freshwater mussels at low concentrations of fluoxetine (Fong 1998). Stimulation of reproduction was also observed for crustaceans *D. magna* and *Ceriodaphnia dubia* exposed to 36 and 56 $\mu g\ L^{-1}$, respectively (Flaherty et al. 2001). Some other studies found cladocerans to be unaffected by SSDI (Brooks et al. 2003). In fish, SSDI exposure has been shown to decrease territorial aggressive behavior in males (Semsar et al. 2004; Kania et al. 2012). Dzieweczynski and Hebert (2012) found decreased *Betta splendens* fish male–male aggressive reaction after short time exposure to SSDI while Kania et al. (2012) observed *B. splendens* male aggression to diminish only after 14 days of fish exposure to fluoxetine. Moreover, SSDI was also found to affect female-direct behavior (courtship) of *B. splendens* fish (Dzieweczynski and Hebert 2012) and the social hierarchy of *Salvelinus alpinus* (Winberg et al. 1991).

As demonstrated, SSDI exposure may also decrease fish ability to capture prey (Gaworecki and Klaine 2008), alter the feeding rate (Stanley et al. 2007), and impact the response to jeopardy (Barry 2013). All of these effects can have a significant impact on fish mortality rates. It is important to note that these adverse impacts were observed at very low concentrations falling in the 0.3–0.5 $\mu g\ L^{-1}$ range. Acute tests show the potentially high toxicity of SSRI to aquatic biota, particularly to phytoplankton as compared to other organisms (Brooks et al. 2003).

12.4.4 Anti-epileptic Drugs

Anti-epileptic drugs decrease neuronal activity. This effect is achieved either by blocking the voltage-dependent sodium channels of excitatory neurons (carbamazepine) or by increasing the inhibitory effect of neurotransmitter gamma-aminobutyric acid (GABA) by binding on a specific site on the corresponding receptor (pharmaceuticals belonging to the benzodiazepine family) (Fent 2008). GABA receptors are known to occur in fish (Meissl and Ekstrom 1991) and aquatic invertebrates (Concas et al. 1998; Gallardo et al. 2000).

As found by Oetken et al. (2005), carbamazepine reveals low acute toxicity in benthic invertebrates. However, chronic exposure to carbamazepine-rich sediments resulted in the significant inhibition of pupation and emergence in the nonbiting midge *Chironomus riparius.* Ferrari et al. (2003) reported chronic toxicity of carbamazepine in cladocerans—*C. dubia* and in the rotifers *Brachionus calyciflorus* (no observed effect concentration $NOEC(_{7d}) = 25\ \mu g\ L^{-1}$ and $NOEC(_{2d}) = 377\ \mu g\ L^{-1}$, respectively), and in early life stages of the zebrafish *Danio rerio* ($NOEC(10d) = 25\ mg\ L^{-1}$). Thaker (2005), in turn, observed sub-lethal effects of carbamazepine in *Daphnia* at 92 mg L^{-1} and lethal effect in zebrafish at 43 mg L^{-1}. Pascoe et al. (2003) found that diazepam at 10 $\mu g\ L^{-1}$ can hamper polyp regeneration of the cnidarian *Hydra vulgaris.* Concas et al. (1998) revealed that 100 μM of diazepam can potentiate the effects of GABA in *H. vulgaris.*

As antiepileptic drugs may affect the activity of the GABA neurotransmitter, they may also trigger behavioral effects in chronically exposed individuals. As found, these pharmaceuticals can disturb day–night activity patterns, increase boldness, disrupt, or modify specimen response for predator occurrence and others, both in invertebrates (Hong et al. 2016) and vertebrates (Nassef et al. 2010). This effect may significantly affect the survivability of individuals or populations exposed to these types of drugs. Carbamazepine was also found to modify behavior of some aquatic biota. For instance, Nassef et al. (2010) revealed that chronic exposures may result in decreased swimming speed and preying abilities of the Japanese medaka fish *Oryzias latipes*.

APIs belonging to the benzodiazepine family and carbamazepine can be classified as potentially harmful to aquatic organisms because most of the acute toxicity data are below 100 mg L^{-1} (Fent 2008).

12.4.5 Estrogenic Pharmaceuticals

The biological effects of the estrogenic endocrine disrupting pharmaceuticals (EEDP) on aquatic animals (especially fish) are better known than for any other API. Estrogenic endocrine-disrupting compounds are an abundant chemical group that consists of substances of natural and synthetic origin. EEDP such as 17-α-ethynylestradiol (EE), 17-β-oestradiol (E) and others have frequently been detected at low concentrations (ng or $\mu g\ L^{-1}$) in freshwaters (Kolpin et al. 2002) and coastal waters near wastewater treatment plant effluents or river mouths (Hutchinson et al. 1999). They have tendency to bioconcentrate in animal bodies up to extremely high rates—over 10,000-fold, even after a short time of exposure (Corcoran et al. 2010). Significantly high levels of EEDP have been identified in the gonads of aquatic animals, highlighting that these APIs may pose an important ecological threat (Gibson et al. 2005). Physiological effects of EEDP, commonly used in humans as part of contraception or as hormone replacement therapy, may be very broad and diversified—depending on the specific compound and organism phylum. A number of animal tissues express estrogen receptors, namely those that can be found in the immune and cardiovascular system, brain, reproductive system, mammary glands, liver, and kidneys (Muller 2004, Campbell et al. 2006). Invertebrates tend to be less prone to EEDP compared to aquatic vertebrates (Hutchinson et al. 1999, Crane et al. 2006). Nevertheless, many studies show a biological effect of EEDP on aquatic invertebrates even at low (ng L^{-1}), environmentally relevant concentrations. Acute or chronic exposure effects have been observed for copepods (Forget-Leray et al. 2005) daphnids (Crane et al. 2006), amphipods (Vandenburgh et al. 2003), and snails (Belfroid and Leonards 1996). Increased mortality of aquatic invertebrates was mostly observed only after exposure to very high concentrations of EEDP, exceeding those found in the environment. Vandenburgh et al. (2003) found that exposure of the crustacean *Hyalella azteca* to sub-lethal concentrations of EEDP results in altered sexual development. Repression of male secondary characteristics and changes in the morphology of the reproductive tracts in *H. azteca* was observed. Sex ratio tended to be in favor of female development. Hermaphroditism and disturbed maturation of germ cells and spermatogenesis in male individuals were observed. All these effects were amplified at low concentrations of EE (0.1–10 $\mu g\ L^{-1}$) implicating that they are receptor-mediated responses (Vandenburgh et al. 2003). On the other hand, studies by Hutchinson et al. (1999) and Breitholtz and Bengtsson (2001) showed no significant impact of EEDP on development and reproduction of aquatic copepods.

Importantly, EE is very reactive in fish. Despite that sensitivity to EE can vary across fish species, at environmentally-relevant concentrations it can cause feminization in males through induction of vitellogenin and development of female reproductive tract in testis and development of oocytes in testis (Seki et al. 2002; Chikae et al. 2003; Corcoran et al. 2010). It should also be noted that despite estrogenic effects in males, some APIs can cause masculinization in female fish. For example, Hattori et al. (2009) found that food contamination with cortisol triggered masculinization in *Odonthesthes bonariensis*, while Sanchez et al. (2011) observed it in wild gudgeons (*Gobio gobio*) inhabiting streams affected by EEDP.

As demonstrated, EEDP can affect mitochondrial function (through gene elicitation), energy metabolism, and cell-cycle control in fish (Filby et al. 2007; Corcoran et al. 2010). The occurrence of EEDP in aquatic environments may represent a threat to fish and decrease their survival. Nash et al. (2004) and Kidd et al. (2007) found that EE at a concentration of 5 ng L^{-1} ceased reproductive functions in *D. rerio* and *Pimephales promelas* fish. Importantly, the experiment that employed *P. promelas* was conducted on an ecosystem scale. Some available data demonstrate that EE is responsible for fertility reduction in amphibians (Gyllenhammar et al. 2009) and are suspected to be one of the key factors that drive the global decline in amphibian abundance (Orton and Tyler 2015).

12.4.6 Cardiolovascular Drugs

This group consists of drugs to treat conditions of the heart or the circulatory or vascular system, and it encompasses pharmaceuticals belonging to different classes, namely calcium channel blockers (e.g., amlodypine, verapamil), angiotensin converting enzyme (ACE) inhibitors (e.g., cilazapril, enalapril), angiotensin II receptor antagonists (e.g., valsartan, telmisartan), diuretics (e.g., furosemid, hydrochlorothiazade), beta-blockers (e.g., bisoprolol, propranolol), antiarrhythmics (e.g., flecainide, amiodarone), and lipid-regulating agents (e.g., atorvastatin, clofibrate, simvastatin). As found, various cardiovascular drugs may undergo abiotic degradation or biodegradation after being discharged but newly formed compounds may reveal greater toxicity or retain the properties of the parent API (Stankiewicz et al. 2015). Despite the fact that cardiovascular pharmaceuticals are one of the most often prescribed medicines and that they have been identified in wastewater, surface water, groundwater, and tap water (Tables 12.1 and 12.2), only little is known as to their potential effects on aquatic biota. Most of the available data concerns beta-blockers and lipid regulating agents. Further studies on the potential impacts cardiovascular drugs may have on non-target organisms are urgently needed.

It is known that fish have beta-adrenergic signaling mechanisms analogous to human cardiovascular receptors. Thus, it is hypothetically plausible that exposure to beta-blockers could cause cardiovascular dysfunction, and impact fish survival (Owen et al. 2007). Propranolol, which is one of the most often prescribed beta-blockers, was found to be actively uptaken by fish from water, although it adversely affected their growth only at very high concentration (10 mg L^{-1}), largely exceeding those observed in freshwaters (Owen et al. 2009). Moreover, life cycle exposure to propranolol did not decrease egg quality, fertility, and hatching in fish (Parrott and Balakrishnan 2016). The existing data suggest that it is unlikely for beta-blockers to pose a relevant risk to fish at levels currently found in the aquatic environment. However, one should note that these compounds are general (propranolol) or selective (metoprol) blockers of adrenergic beta receptors and can cause additive (or even potentially hyperadditive) effects if they co-occur as a mixture (Stankiewicz et al. 2015). As demonstrated, zooplankton may be more sensitive to calcium channel blockers and beta-blockers. For example, verapamil and metoprol have the ability to significantly affect heart rates in *D. magna*. Interestingly, high concentrations caused lowered heart rate while at low concentrations acceleration was observed (Villegas-Navarro et al. 2003). The EC_{50} for propranolol in acute 48 h toxicity tests was 0.8 mg L^{-1} for *C. dubia* but 1.6 mg L^{-1} for *D. magna*, indicating that sensitivity to beta-blockers may be highly species specific (Huggett 2002; Ferrari et al. 2004).

Lipid-regulating agents generally can be divided into statins and fibrates. The former decrease cholesterol levels in plasma, the latter activate the lipoprotein lipase enzyme which is responsible for the conversion of very low-density lipoprotein (VLDL) to high-density lipoproteins (HDL). As found, simvastatin revealed greater toxicity to zebrafish embryos, compared to diclofenac, sertraline and propranolol, caused abnormalities at lower concentrations and death at higher levels (Ribeiro et al. 2015). Simvastatin and atorvastatin were also toxic to primary hepatocytes of *Oncorhynchus mykiss* in a dose-dependent manner (Ellesat et al. 2010). Simvastatin was able to adversely affect reproduction and population growth of the amphipod *Gammarus locusta* at the ng L^{-1} range (Neuparth et al. 2014). Fibrates act by binding to peroxisome proliferator-activated receptors (PPARs). In isolated fish, hepatocytes clofibrate caused induction of mRNA of PPARα and PPARγ at the 0.5–2 mM range (Ibabe et al. 2005). PPARs are commonly found in different fish species (Fent 2008). Chronic exposure of *D. magna* to gemfibrozil did not affect longevity but some reproductive parameters including number of young per female or brood size were affected. Gemfibrozil was also found to significantly decrease cholesterol levels in daphnids (Salesa et al. 2017).

One of the most commonly found diuretics in freshwater is furosemide (Tables 12.1 and 12.2). As demonstrated, it caused growth inhibition in *C. dubia* and *B. calyciflorus* with the lowest observed effect concentration (LOEC) of 0.3 and 1.25 mg L^{-1}, respectively. Importantly, the products of furosemide photodegradation had a decidedly lower LOEC—0.02 mg L^{-1} in *C. dubia* and 0.31 mg L^{-1} in *B. calyciflorus* (Isidori et al. 2006).

12.5 Conclusions

Pharmaceutical pollution is becoming a serious environmental issue, although the ecological and ecotoxicological consequences are still mostly unknown. The most obvious impacts were evidenced in fish. The observed responses depend on the type of API and include feminization in response to synthetic hormones and behavioral alterations in response to anti-depressants. The released pharmaceuticals can survive traditional wastewater and drinking water treatment and can be present at very low (ppt) concentrations in drinking water leading to unintended human exposures. It is unknown whether this can lead to any long-term adverse effects on health. It appears that pharmaceutical pollution is rather unavoidable, if one considers the forecast increase in use of medicinal drugs. It is imperative to monitor APIs in aquatic ecosystems and sources of drinking water, investigate the potential health effects of their presence in the environment, and develop methods that will increase the efficiency of API removal from wastewater and water used for human purposes.

References

Bácsi I, B-Béres V, Kókai Z et al (2016) Effects of non-steroidal anti-inflammatory drugs on cyanobacteria and algae in laboratory strains and in natural algal assemblages. Environ Pollut 212:508–518

Baranowska I, Kowalski B (2012) A rapid UHPLC method for the simultaneous determination of drugs from different therapeutic groups in surface water and wastewater. Bull Environ Contam Toxicol 89:8–14

Barra Caracciolo A, Topp E, Grenni P (2015) Pharmaceuticals in the environment: biodegradation and effects on natural microbial communities, a review. J Pharm Biomed Anal 15:25–36
Barry MJ (2013) Effect of fluoxetine on swimming and behavioural responses of the Arabian killfish. Ecotoxicol 22:425–432
Belfroid A, Leonards P (1996) Effect of ethinyl oestradiol on the development of snails and amphibians. SETAC 17th Annual Meeting, Washington DC, USA
Blair BD, Crago JP, Hedman CJ, Klaper RD (2013) Pharmaceuticals and personal care products found in the Great lakes above concentrations of environmental concern. Chemosphere 93:2116–2123
Boon PI, Cattanach M (1999) Antibiotic resistance of native and faecal bacteria isolated from rivers, reservoirs and sewage treatment facilities in Victoria, south-eastern Australia. Lett App Microbiol 28:164–168
Borg MA, Zarb P, Scicluna EA, Rasslan O, Gür D, Ben Redjeb S, Elnasser Z, Daoud Z (2010) Antibiotic consumption as a driver for resistance in Staphylococcus aureus and *Escherichia coli* within a developing region. Am J Infect Control 38:212–226
Bôto M, Almeida CMR, Mucha AP (2016) Potential of constructed wetlands for removal of antibiotics from saline aquaculture effluents. Water 8(10):465
Boxall AB, Rudd MA, Brooks BW et al (2012) Pharmaceuticals and personal care products in the environment: what are the big questions? Environ Health Perspect 120:1221–1229
Brain RA, Sanderson H, Sibley PK, Solomon KR (2006) Probabilistic ecological hazard assessment: evaluating pharmaceutical effects on aquatic higher plants as an example. Ecotoxicol Environ Saf 64:128–135
Breitholtz M, Bengtsson BE (2001) Oestrogens have no hormonal effect on the development and reproduction of the harpacticoid copepod Nitocra spinepes. Mar Pollut Bull 42:879–886
Bringolf RB, Heltsley RM, Newton TJ et al (2010) Environmental occurrence and reproductive effects of the pharmaceutical fluoxetine in native freshwater mussels. Environ Toxicol Chem 29:1311–1318
Brodin T, Piovano S, Fick J et al (2014) Ecological effects of pharmaceuticals in aquatic systems-impacts through behavioural alterations. Philos Trans R Soc Lond B Biol Sci 369:20130580
Brooks BW, Chambliss CK, Stanley JK et al (2005) Determination of select antidepressants in fish from an effluent-dominated stream. Environ Toxicol Chem 24:464–469
Brooks BW, Foran CM, Richards SM et al (2003) Aquatic ecotoxicology of fluoxetine. Toxicol Lett 142:169–183
Bu Q, Shi X, Yu G, Huang J, Wang B (2016) Assessing the persistence of pharmaceuticals in the aquatic environment: Challenges and needs, Emerg Contam 2(3):145–147
Bundschuh M, Hahn T, Ehrlich B et al (2016) Acute toxicity and environmental risks of five veterinary pharmaceuticals for aquatic macroinvertebrates. Bull Environ Contam Toxicol 96:139–143
Caban M, Lis E, Kumirska J, Stepnowski P (2015) Determination of pharmaceutical residues in drinking water in Poland using a new SPE-GC-MS(SIM) method based on Speedisk extraction disks and DIMETRIS derivatization. Sci Total Environ 15:402–411
Cabello FC (2006) Heavy use of prophylactic antibiotics in aquaculture: a growing problem for human and animal health and for the environment. Environ Microbiol 8:1137–1144
Campbell CG, Borglin SE, Green B et al (2006) Biologically directed environmental monitoring, fate, and transport of estrogenic endocrine disrupting compounds in water: A review. Chemosphere 65(8):1265–1280
Cardoso O, Porcher JM, Sanchez W (2014) Factory-discharged pharmaceuticals could be a relevant source of aquatic environment contamination: review of evidence and need for knowledge. Chemosphere 115:20–30
Celiz MD, Tso J, Aga DS (2009) Pharmaceutical metabolites in the environment: analytical challenges and ecological risks. Environ Toxicol Chem 28:2473–2484
Chikae M, Ikeda R, Hasan Q et al (2003) Effect of alkylphenols on adult male medaka: Plasma vitellogenin goes up to the level of estrous female. Environ Tox Pharma 15(1):33–36

Cleuver M (2008) Chronic mixture toxicity of pharmaceuticals to *daphnia*—the example of non-steroidal anti-inflammatory drugs. In: Pharmaceuticals in environment, Springer, Berlin pp. 227–284

Concas A, Pierobon P, Mostallino MC et al (1998) Modulation of aminobutyricacic (GABA) receptors and the feeding response by neurosteroids in *Hydra vulgaris*. Neuroscience 85:979–988

Corcoran J, Winter MJ, Tyler CR (2010) Pharmaceuticals in the aquatic environment: a critical review of the evidence for health effects in fish. Crit Rev Toxicol 40:287–304

Crane M, Watts C, Boucard T (2006) Chronic aquatic environmental risk from exposure to human pharmaceuticals. Sci Tot Environ 367:23–41

Daughton CG (2003) Cradle-to-cradle stewardship of drugs for minimizing their environmental disposition while promoting human health. II. Drug disposal, waste reduction, and future directions. Environ Health Perspect 111:775–785

De Lange HJ, Noordoven W, Murk AJ et al (2006) Behavioural responses of *Gammarus pulex* (Crustacea, Amphipoda) to low concentrations of pharmaceuticals. Aquat Toxicol 78:209–216

Ding J, Lu G, Li S et al (2015) Biological fate and effects of propranolol in an experimental aquatic food chain. Sci Total Environ 1:31–39

Du B, Haddad SP, Luek A et al (2014) Bioaccumulation and trophic dilution of human pharmaceuticals across trophic positions of an effluent-dependent wadeable stream. Philos Trans R Soc Lond B Biol Sci 369:1656

Du B, Haddad SP, Scott WC et al (2015) Pharmaceutical bioaccumulation by periphyton and snails in an effluent-dependent stream during an extreme drought. Chemosphere 119:927–934

Dzieweczynski TL, Hebert OL (2012) Fluoxetine alters behavioral consistency of aggression and courtship in male Siamense fightingfish, *Betta splendens*. Physiol Behav 20:92–97

Ellesat KS, Tollefsen KE, Asberg A et al (2010) Cytotoxicity of atorvastatin and simvastatin on primary rainbow trout (*Oncorhynchus mykiss*) hepatocytes. Toxicol In Vitro 24:1610–1618

Fent K (2008) Effects of pharmaceuticals on aquatic organisms. In: Pharmaceuticals in environment, Springer, Berlin, pp. 175–203

Fent K, Weston AA, Caminada D (2006) Ecotoxicology of human pharmaceuticals. Aquat Toxicol 76:122–159

Ferrari B, Mons R, Vollat B et al (2004) Environmental risk assessment of six human pharmaceuticals: are the current environmental risk assessment procedures sufficient for the protection of the aquatic environment? Environ Toxicol Chem 23:1344–1354

Ferrari B, Paxeus N, Lo Giudice R, Pollio A, Garric J (2003) Ecotoxicological impact of pharmaceuticals found in treated wastewaters: study of carbamazepine, clofibric acid, and diclofenac. Ecotox Environ Safety 55:359–370

Filby AL, Thorpe KL, Maack G, Tyler CR (2007) Gene expression profiles revealing the mechanisms of antiandrogen- and estrogen-induced feminization in fish. Aquat Toxicol 81(2):219–231

Flaherty CM, Dodson SI (2005) Effects of pharmaceuticals on *Daphnia* survival, growth, and reproduction. Chemosphere 61:200–207

Flaherty CM, Kashian DR, Dodson SI (2001) Ecological impacts of pharmaceuticals on zooplankton: the effects of three medications on *Daphnia magna*. In: Proceedings of the annual meeting of the society of environmental toxicology and chemistry, Baltimore

Fong PP (1998) Zebra mussel spawning is induced in low concentrations of putative serotonin reuptake inhibitors. Biol Bull 194:143–149

Forget-Leray J, Landriau I, Minier C, Leboulenger F (2005) Impact of endocrine toxicants on survival, development, and reproduction of the estuarine copepod Eurytemora affinis (Poppe). Ecotox Environ Saf 60(3):288–294

Gallardo WG, Hagiwara A, Hara K et al (2000) GABA, 5-HT and amino acids in the rotifers Brachionus plicatilis and Brachionus rotundiformis. Comp Biochem Phys A 127(3):301–307

Gaworecki KM, Klaine SJ (2008) Behavioural and biochemical responses of hybrid striped bass during and after fluoxetine exposure. Aquat Toxicol 88:207–213

Gibson R, Smith MD, Spary CJ et al (2005) Mixtures of estrogenic contaminants in bile of fish exposed to wastewater treatment works effluents. Environ Sci Technol 39:246–271

Giebułtowicz J, Nałęcz-Jawecki G (2014) Occurrence of antidepressant residues in the sewage-impacted Vistula and Utrata rivers and in tap water in Warsaw (Poland). Ecotoxicol Environ Saf 104:103–109

Giebułtowicz J, Nałęcz-Jawecki G (2016) Occurrence of immunosuppressive drugs and their metabolites in the sewage-impacted Vistula and Utrata Rivers and in tap water from the Warsaw region (Poland). Chemosphere 148:137–147

Giebułtowicz J, Stankiewicz A, Wroczyński P, Nałęcz-Jawecki G (2016) Occurrence of cardiovascular drugs in the sewage-impacted Vistula River and in tap water in the Warsaw region (Poland). Environ Sci Pollut Res Int 23:24337–24349

Grabicova K, Grabic R, Blaha M et al (2015) Presence of pharmaceuticals in benthic fauna living in a small stream affected by effluent from a municipal sewage treatment plant. Water Res Apr 1:145–153

Gravel A, Wilson JM, Pedro DFN, Vijayan FM (2009) Non-steroidal anti-inflammatory drugs disturb in osmoregulatory, metabolic and cortisol responses associated with seawater exposure in rainbow trout. Comp Biochem Phys C 149:481–490

Gyllenhammar I, Holm L, Eklund R, Berg C (2009) Reproductive toxicity in Xenopus tropicalis after developmental exposure to environmental concentrations of ethynylestradiol. Aquat Toxicol 91(2):171–178

Hallare AV, Kohler HR, Triebskorn R (2004) Developmental toxicity and stress protein responses in zebrafish embryos after exposure to diclofenac and its solvent, DMSO. Chemosphere 56:659–666

Hansen PK, Lunestad BT, Samuelsen OB (1992) Effects of oxytetracycline, oxolinic acid, and flumequine on bacteria in an artificial marine fish farm sediment. Can J Microb 38(12):1307–1312

Hattori RS, Fernandino JI, Kishii A et al (2009) Cortisol-induced masculinization: does thermal stress affect gonadal fate in pejerrey, a teleost fish with temperature-dependent sex determination? PLoS ONE 4(8):e6548

Henschel KP, Wenzel A, Diedrich M, Fliedner A (1997) Environmental hazard assessment of pharmaceuticals. Regul Toxicol Pharm 25(3):220–225

Hong K-B, Yooheon Park Y, Hyung Joo Suh HJ (2016) Sleep-promoting effects of a GABA/5-HTP mixture: behavioral changes and neuromodulation in an invertebrate model. Life Sci 150:42–47

Huggett DB, Brooks BW, Peterson B et al (2002) Toxicity of selected beta adrenergic receptor-blocking pharmaceuticals (b-blockers) on aquatic organisms. Arch Environ Contam Toxicol 43:229–235

Hutchinson TH, Pounds NA, Hampel M, Williams TD (1999) Impact of natural and synthetic steroids on the survival, development and reproduction of marine copepods (Tisbe battagliai). Sci Total Environ 233:167–179

Ibabe A, Herrero A, Cajaraville MP (2005) Modulation of peroxisome proliferator-activated receptors (PPARs) by PPAR[alpha]- and PPAR[gamma]-specific ligands and by 17[beta]-estradiol in isolated zebrafish hepatocytes. Toxicol In Vitro 19:725–735

IMS Institute for Healthcare Informatics (2015) Global use of medicines in 2020

Ingram T, Richter U, Mehling T, Smirnova I (2011) Modelling of pH dependent noctanol/water partition coefficients of ionisable pharmaceuticals. Fluid Phase Equilib 305:197–203

Ishikawa TO, Herschman HR (2007) Two inducible, functional cyclooxygenase-2 genes are present in the rainbow trout genome. J Cell Biochem 102:1486–1492

Isidori M, Nardelli A, Parrella A et al (2006) A multispecies study to assess the toxic and genotoxic effect of pharmaceuticals: furosemide and its photoproduct. Chemosphere 63:785–793

Jakimska A, Śliwka-Kaszyńska M, Reszczyńska J et al (2014) Elucidation of transformation pathway of ketoprofen, ibuprofen, and furosemide in surface water and their occurrence in the aqueous environment using UHPLC-QTOF-MS. Anal Bioanal Chem 406:3667–3680

Jeffries KM, Brander SM, Britton MT et al (2015) Chronic exposure to low and high concentration of ibuprofen elicit different gene response patterns in a euryhaline fish. Environ Sci Pollut Res 22:17397–17413

Kania BF, Gralak MA, Wielgosz M (2012) Four-week fluoxetine exposure diminish aggressive behavior of male Siamense figtingfish (*Betta splendens*). J Behav Brain Sci 2:185–190

Kasprzyk-Hordern B, Dinsdale RM, Guwy AJ (2007) Multi-residue method for the determination of basic/neutral pharmaceuticals and illicit drugs in surface water by solid-phase extraction and ultra performance liquid chromatography-positive electrospray ionisation tandem mass spectrometry. J Chromatogr A 1161:132–145

Khetan SK, Collins TJ (2007) Human pharmaceuticals in the aquatic environment: a challenge to green chemistry. Chem Rev 107:2319–2364

Kidd KA, Blanchfield PJ, Mills KH et al (2007) Collapse of a fish population after exposure to a synthetic estrogen. P Natl Aca Sci USA 104(21):8897–8901

Kim Y, Choi K, Jung J et al (2007) Aquatic toxicity of acetaminophen, carbamazepine, cimetidine, diltiazem and six major sulfonamides, and their potential ecological risks in Korea. Environ Int 33:370–375

Koczura R, Mokracka J, Jabłońska L et al (2012) Antimicrobial resistance of integronharboring ‘ isolates from clinical samples wastewater treatment plant and river water. Sci Tot Environ 414:680–685

Koczura R, Mokracka J, Taraszewska A, Łopacinska N (2016) Abundance of Class 1 integron-integrase and sulfonamide resistance genes in river water and sediment is affected by anthropogenic pressure and environmental factors. Microbial Ecol 72:909–916

Kolpin DK, Furlong ET, Meyer MT (2002) Pharmaceuticals, hormones, and other organic wastewater contaminants in U.S. streams, 1999–2000: a national reconnaissance. Environ Sci Technol 36:1202–1211

Kot-Wasik A, Jakimska A, Śliwka-Kaszyńska M (2016) Occurrence and seasonal variations of 25 pharmaceutical residues in wastewater and drinking water treatment plants. Environ Monit Assess 188:188–661

Kümmerer K (2009) Antibiotics in the aquatic environment—a review-part I. Chemosphere 75:417–434

Kunkel U, Radke M (2012) Fate of pharmaceuticals in rivers: deriving a benchmark dataset at favorable attenuation conditions. Water Res 46:5551–5565

Lahti M, Brozinski JM Jylhä A et al (2011) Uptake from water, biotransformation, and biliary excretion of pharmaceuticals by rainbow trout. Environ Toxicol Chem 30:1403–1411

Langford K, Thomas KV (2011) Input of selected human pharmaceutical metabolites into the Norwegian aquatic environment. J Environ Monitor 13:416–421

Li W, Shi Y, Gao L et al (2012) Investigation of antibiotics in mollusks from coastal waters in the Bohai Sea of China. Environ Pollut Mar 162:56–62

Liu J, Lu G, Wang Y et al (2014) Bioconcentration, metabolism, and biomarker responses in freshwater fish Carassius auratus exposed to roxithromycin. Chemosphere 99:102–108

López-Doval JC, Kukkonen JV, Rodrigo P, Muñoz I (2012) Effects of indomethacin and propranolol on *Chironomus riparius* and Physella (Costatella) acuta. Ecotoxicol Environ Saf 78:110–115

Łukaszewicz P, Maszkowska J, Mulkiewicz E (2016) Impact of veterinary pharmaceuticals on the agricultural environment: a re-inspection. Rev Environ Contam Toxicol. https://doi.org/10.1007/398_2016_16

Lynn SE, Egar JM, Walker BG et al (2007) Fish on Prozac: a simple, noninvasive physiology laboratory investigating the mechanisms of aggressive behavior in *Betta splendens*. Adv Physiol Educ 31(4):358–363

Maki T, Hasegawa H, Kitami H et al (2006) Bacterial degradation of antibiotic residues in marine fish farm sediments of Uranouchi Bay and phylogenetic analysis of antibiotic-degrading bacteria using 16S rDNA sequences. Fisheries Sci 72:811–820

Makowska N, Koczura R, Mokracka J (2016) Class 1 integrase, sulfonamide and tetracycline resistance genes in wastewater treatment plant and surface water. Chemosphere 144:1665–1673

Marques CR, Abrantes N, Goncalves F (2004) Life-history traits of standard and autochthonous cladocerans: I. Acute and chronic effects of acetylsalicylic acid. Environ Toxicol 19:518–526

Martínez-Hernández V, Meffe R, Herrera S et al (2014) Sorption/desorption of non-hydrophobic and ionisable pharmaceutical and personal care products from reclaimed water onto/from a natural sediment. Sci Total Environ 472:273–281

Meissl H, Ekstrom P (1991) Action of gamma-aminobutyric-acid (GABA) in the isolated photosensory pineal organ. Brain Res 562(1):71–78

Mendoza A, Aceña J, Pérez S et al (2015) Pharmaceuticals and iodinated contrast media in a hospital wastewater: a case study to analyse their presence and characterise their environmental risk and hazard. Environ Res 140:225–241

Muller SO (2004) Xenoestrogens: mechanisms of action and detection methods. Anal Bioanal Chem 378:582–587

Munir M, Wong K, Xagoraraki I (2011) Release of antibiotic resistant bacteria and genes in the effluent and biosolids in five wastewater utilities in Michigan. Water Res 45:681–693

Nash JP, Kime DE, Van der Ven LTM et al (2004) Long-Term Exposure to Environmental Concentrations of the Pharmaceutical Ethynylestradiol Causes Reproductive Failure in Fish. Environ Health Persp 112(17):1725–1733

Nassef M, Matsumoto S, Seki M et al (2010) Acute effects of triclosan, diclofenac and carbamazepine on feeding performance of Japanese medaka fish (*Oryzias latipes*). Chemosphere 80:1095–1100

Neuparth T, Martins C, Santos CB et al (2014) Hypocholesterolaemic pharmaceutical simvastatin disrupts reproduction and population growth of the amphipod *Gammarus locusta* at the ng/L range. Aquat Toxicol 155:337–347

Nunes B, Antunes SC, Santos J et al (2014) Toxic potential of paracetamol to freshwater organisms: a headache to environmental regulators? Ecotox Environ Safe 107:178–185

Oetken M, Nentwig G, Loffler D et al (2005) Effect of pharmaceuticals on aquatic invertebrates. Part I. The antiepileptic drug carbamazepine. Archiv Environ Con Tox 49:353–361

Ohlsen K, Ziebuhr W, Koller K et al (1998) Effect of subinhibitory concentrations of antibiotics on alphatoxon (hla) gene expression on methicillin-sensitive and methicillin-resistant Staphylococus aureus isolates. Antimicrob Agents Chem 42:2817–2823

Orton F, Tyler CR (2015) Do hormone-modulating chemicals impact on reproduction and development of wild amphibians? Biol Rec Camb Philosoph Soc 90(4):1100–1117

Owen SF, Giltrow E, Huggett DB et al (2007) Comparative physiology, pharmacology and toxicology of beta-blockers: mammals versus fish. Aquat Toxicol 82:145–162

Owen SF, Huggett DB, Hutchinson TH et al (2009) Uptake of propranolol, a cardiovascular pharmaceutical, from water into fish plasma and its effects on growth and organ biometry. Aquat Toxicol 93:217–224

Parrott JL, Balakrishnan VK (2016) Life-cycle exposure of fathead minnows to environmentally relevant concentrations of the β-blocker drug propranolol. Toxicol Chem, Environ. https://doi.org/10.1002/etc.3703

Pascoe D, Karntanut W, Müller CT (2003) Do pharmaceuticals affect freshwater invertebrates? a study with *Hydra vulgaris*. Chemosphere 51:521–528

Pedibhotla VK, Sarath G, Sauer JR, Stanleysamuelson DW (1995) Prostaglandin biosynthesis and subcellularlocalization of prostaglandin-H synthase activity in the lone star tick, Amblyommaamericanum. Insect Biochem Molec 25:1027–1039

Puckowski A, Mioduszewska K, Łukaszewicz P et al (2016) Bioaccumulation and analytics of pharmaceutical residues in the environment: a review. J Pharm Biomed Anal 5:232–255

Quinn B, Gagne F, Blaise C (2009) Evaluation of the acute, chronic and teratogenic effects of mixture of eleven pharmaceuticals on the cnidarian, Hydra attenuate. Sci Tot Environ 407:1072–1079

Ribeiro S, Torres T, Martins R, Santos MM (2015) Toxicity screening of diclofenac, propranolol, sertraline and simvastatin using *Danio rerio* and Paracentrotus lividus embryo bioassays. Ecotoxicol Environ Saf 114:67–74

Richards SM, Kelly SE, Hanson ML (2008) Zooplankton chitobiase activity as an endpoint of pharmaceutical effect. Arch Environ Contam Toxicol 54:637–644

Roberts SB, Langenau DM, Goetz FW (2000) Cloning and characterization of prostaglandin endoperoxide synthase-1 and -2 from the brook trout ovary. Mol Cell Endocrin 160:89–97

Salesa B, Ferrando MD, Villarroel MJ, Sancho E (2017) Effect of the lipid regulator Gemfibrozil in the Cladocera *Daphnia magna* at different temperatures. J Environ Sci Health A Tox Hazard Subst Environ Eng 52:228–234

Sanchez W, Sremski W, Piccini B et al (2011) Adverse effects in wild fish living downstream from pharmaceutical manufacture discharges. Environ Int 37(8):1342–1348

Sarma SS, González-Pérez BK, Moreno-Gutiérrez RM, Nandini S (2014) Effect of paracetamol and diclofenac on population growth of Plationus patulus and Moina macrocopa. J Environ Biol 35:119–126

Seki M, Yokota H, Matsubara H, et al (2002) Effect of ethinylestradiol on the reproduction and induction of vitellogenin and testis-ova in medaka (Oryzias latipes). Environ Toxicol Chem 21:1692–1698

Scheytt T, Mersmann P, Lindstädt R, Heberer T (2005) Determination of sorption coefficients of pharmaceutically active substances carbamazepine, diclofenac, and ibuprofen, in sandy sediments. Chemosphere 60:245–253

Schwaiger J, Ferling H, Mallow U (2004) Toxic effects of the non-steroidal anti-inflammatory drug diclofenac. Part I: histopathological alterations and bioaccumulation in rainbow trout. Aquat Toxicol 68:141–150

Semsar K, Perreault HAN, Godwin J (2004) Fluoxetine treated male wrasses exhibit low AVT expression. Brain Res 1029:141–147

Shryock TR, Richwine A (2010) The interface between veterinary and human antibiotic use. Ann NY Acad Sci 1213:92–105

Sim WJ, Lee JW, Lee ES et al (2011) Occurrence and distribution of pharmaceuticals in wastewater from households, livestock farms, hospitals and pharmaceutical manufactures. Chemosphere 82:179–186

Stankiewicz A, Giebułtowicz J, Stankiewicz U et al (2015) Determination of selected cardiovascular active compounds in environmental aquatic samples-methods and results, a review of global publications from the last 10 years. Chemosphere 138:642–656

Stanley JK, Ramirez AJ, Chambliss CK, Brooks BW (2007) Enantiospecific sublethal effects of the antidepressant fluoxetine to a model aquatic vertebrate and invertebrate. Chemosphere 69:9–16

Sumpter JP, Donnachie RL, Johnson AC (2014) The apparently very variable potency of antidepressant fluoxetine. Aquat Toxicol 151:57–60

Thaker PD (2005) Pharmaceutical data elude researchers. Environ Sci Technol 139:193A–194A

Tixier C, Singer HP, Oellers S, Müller SR (2003) Occurrence and fate of carbamazepine, clofibric acid, diclofenac, ibuprofen, ketoprofen, and naproxen in surface waters. Environ Sci Technol 37:1061–1068

Vandenburgh GF, Adriaens D, Verslycke T, Janssen CR (2003) Effects of 17α-ethinyloestradiol on sexual development of the amphipod Hyalella azteca. Ecotoxicol Environ Saf 54:216–222

Vasseur P, Cossu-Leguille C (2006) Linking molecular interactions to consequent effects of persistent organic pollutants (POPs) upon populations. Chemosphere 62(7):1033–1042

Villegas-Navarro A, Rosas-L E, Reyes JL (2003) The heart of *Daphnia magna*: effects of four cardioactive drugs. Comp Biochem Phys C 136:127–134

WHO (World Health Organization) (2011) Pharmaceuticals in drinking water. WHO/HSE/WSH/11.05. WHO, Geneva

Winberg S, Nilsson GE, Olsén KH (1991) Social rank and brain levels of monoamines and monoamine metabolites in arctic char *Salvelinus alpinus* (L). J Comp Physiol A 168:241–246

Yamamoto H, Nakamura Y, Moriguchi S et al (2009) Persistence and partitioning of eight selected pharmaceuticals in the aquatic environment: laboratory photolysis, biodegradation, and sorption experiments. Water Res 43:351–362

Zou J, Neuman NF, Holland JW et al (1999) Fish macrophages express a cyclo-oxygenase-2 homologue after activation. Biochem J 340:153–159

Zou H, Radke M, Kierkegaard A et al (2015a) Using chemical benchmarking to determine the persistence of chemicals in a Swedish lake. Environ Sci Technol 49:1646–1653
Zou H, Radke M, Kierkegaard A et al (2015b) Using chemical benchmarking to determine the persistence of chemicals in a Swedish lake. Environ Sci Technol 49:1646–1653

Chapter 13
Constructed Wetlands and Groundwater Infiltration Treating Industrial Wastewater, Treatment Efficiency, and Pollution Tracing

Ketil Haarstad

Abstract Three treatment systems for wastewater from two landfills, one active and one closed, and an industrial location including a quarry have been monitored continuously for over a decade. The wastewater from the active landfill is infiltrated through an extensive unsaturated zone into groundwater and subsequently into a large river system. The wastewater from the closed landfill is treated in a constructed wetland (CW) and the industrial low-grade wastewater in filter dams. The treatment systems operate well with the specific wastewaters, high-concentration leachate from waste in infiltration systems, low-concentration leachate in constructed wetlands, and wastewater from inert waste in filter dams. The landfilling of organic waste was restricted to low limit values for more than a decade ago, but it is hard to see any changes in leachate due to changes in waste landfilling regulations. The heavy carbon stable isotope 13C is useful in tracing landfill leachate and to evaluate dilution into other water bodies. The adding of P to the aeration pond treating low-concentration leachate did not help in the removal of N; on the contrary, the concentration of ammonia was sharply decreased when the adding of P was discontinued.

Keywords Industrial sites · Tracers · Constructed wetland treatment

13.1 Introduction

How do wetlands compare to soil infiltration when it comes to treating industrial wastewater? This chapter looks at two types of industrial wastewater, (a) from a sanitary waste landfill for ordinary waste and (b) wastewater from low-risk industrial activities such as quarries and inert waste landfilling. The wastewater from these types of locations is very different. Wastewater or leachate from waste landfills is, depending on their age, characterized by high concentrations, dominated by organic

K. Haarstad (✉)
Research Professor, Division of Environment and Natural Resources, NIBIO—Norwegian Institute of Bioeconomy Research, No-1431 Aas, 115, Oslo, Norway
e-mail: ketil.haarstad@nibio.no

M. Zelenakova (ed.), *Water Management and the Environment: Case Studies*, Water Science and Technology Library 86,
https://doi.org/10.1007/978-3-319-79014-5_13

matter, nitrogen, and selected heavy metals, but potentially also a number of highly toxic organic pollutants, most typically aliphatic hydrocarbons, aromatics, polyaromatics, and others. The recipients are often evaluated based on health risk classes, from 1 (background levels) to 5 (extremely polluted/hazardous waste). Changes of classes due to emissions are not beneficial, although a change from class 1 to class 2 is not critical, while from class 4 to class 5 means a full risk assessment has to be made (SFT 2003). For compounds without any acceptance limit values, the concept of pollution index can be used. This says that if the concentration of a wastewater is 10 times higher than the corresponding concentration in a particular recipient a risk evaluation will have to be carried out.

For groundwater infiltration, an area of influence can be established, in such a way that the industrial area including a zone 200 m downstream may be acceptable for increased concentrations due to diffuse emissions and assuming that the groundwater further downstream is unaffected and that the internal aquifer is of no significant interest (SFT 2003).

Wetlands can remove a large range of pollutants, also from Norwegian locations (Haarstad 2008; Haarstad and Borch 2008; Haarstad and Mæhlum 2008, 2009; Haarstad et al. 2012), due to a versatility in removal processes and materials. They are, however, also sensitive to climate and large variations in wastewater volume. For example, a review of the removal by CWs the highest pesticide removal by Vymazal and Brezinova (2015) was achieved for pesticides of the organochlorine, strobilurin/strobin, organophosphate, and pyrethroid groups, while the lowest removals were observed for pesticides of the triazinone, aryloxyalkanoic acid, and urea groups. The removal of pesticides generally increases with increasing value of KOC, but the relationship is not strong. However, at the location "closed landfill" the pesticide mecoprop, belonging to the aryloxyalkanoic group, was successfully removed in the CW, probably due to relatively high concentrations (Haarstad and Mæhlum 2008).

Non-hazardous industrial wastewaters are different from leachate, usually carrying suspended solids and, depending on the source of soil landfilling, other pollutants such as heavy metals. The example here has emission targets saying that 90% of a weekly mixed sample shall not exceed 50 mg/l suspended solids (SS), 10 mg/l hydrocarbons, and a pH between 6 and 9 (EPA, 9.12.2014).

For waste operation that includes organic waste, the leachate will usually have increased level of the heavy stable isotope 13C (Haarstad and Mæhlum 2012), mainly due to the sequencing by methanogenic bacteria. This is used to trace waters polluted by landfill leachate. The content of the isotope is given as per mille relative ratio between the sample and a reference value.

The purpose of this chapter is to study the difference of the treatment of industrial wastewater in processes based on groundwater infiltration, constructed wetlands, and ponds/dams by looking the typical landfill leachate parameters and the emission permit parameters. In addition, the usefulness of using the stable isotope heavy carbon as a tracer is studied.

13.2 Methods—Locations

13.2.1 Active Landfill

The wastewater from the active landfill has a volume of 21,000 m^3/yr. (last 5 years average). It is treated in an aeration pond before infiltration to groundwater (Fig. 13.1). The unsaturated zone is approximately 30 m thick. The distance to the river recipient is ca 300 m. Two groundwater recharge points are sampled just before the groundwater reaches the river. Infiltration systems depend to a large degree on the biological activity of the top soil zone and the particle size and mineralogy of the unsaturated zone. Beyond the soil layer, there is not biological activity.

Treated leachate samples are collected in a well where the groundwater exits the area (B4). The unsaturated zone at this point is 22 m.

13.2.2 Closed Landfill

The wastewater from the closed landfill has a volume of 27,000 m^3/yr. (last 5 years average). It is treated in an aeration pond, a biodam with emergent plants, and finally in a horizontal flow constructed wetland (CW) (Fig. 13.2).

Constructed wetlands can be designed with a multitude of treatment elements, and for horizontal or vertical flow. The flow, however, is difficult to control and is often a major treatment limitation.

A downstream groundwater well is also sampled in addition to upstream and downstream creek working as a recipient. The sampling is carried out according to a simplified program compared to active landfills, but with a more comprehensive program every fifth year.

13.2.3 Quarry

The quarry is located in Oslo and is receiving inert waste such as non-polluted soil and construction materials. The wastewater from the quarry has a volume of 40,000 m^3/yr. It is treated in three constructed dams with barriers that are constructed to remove particles by filtration (CW) (Fig. 13.3). The wastewater is led to a local creek with sensors both upstream and downstream measuring turbidity, suspended solids, electrical conductivity, and temperature and water level. The quarry is licensed with limited values for the wastewater; 90% of weekly logging values are below 50 mg/l suspended solids (SS), 10 mg/l hydrocarbons, and a pH value between 6 and 9 (Fylkesmannen, 9.12.2014). Filter dams are easy to construct but are subject to flow problems due to clogging or a lack of effect due to bad design.

Fig. 13.1 Active landfill with leachate infiltration to groundwater. *Photograph* Haarstad

13.3 Sampling

The sampling is part of ongoing monitoring programs at each site. The waste treatment is regulated by law and supervised by the local EPAs. The samples are analyzed according to a recommended program, with more comprehensive sampling every

Fig. 13.2 Closed landfill with leachate treatment in wetland. *Photograph* Haarstad

fifth year (SFT 2003). There is a separate program for industrial activities handling assumed un-polluted soil and waste. These mostly focus on the emission of particles. Sampling programs for landfills include both grab and time proportional mixed samples, taken at a minimum of four times a year and fortnightly, respectively. The quarry is also sampled quarterly but is in addition monitored by continuous loggers in the creek upstream and downstream.

Fig. 13.3 Quarry with gravel production, wastewater treatment in ponds. *Photograph* Haarstad

13.4 Results

13.4.1 Active Landfill

Figure 13.4 shows the concentrations of selected parameters in raw and treated leachate. Nitrogen and COD dominate the leachate. As a consequence of adopting the EU waste directive, the content of organic matter in landfilled waste was regulated in 2004 (FOR 2004), limiting the TOC to 10% or the glow loss to 20% DM. This cannot be seen as a reduction in COD over the period 2004 to 2016 (Fig. 13.4), which shows a stable and high, for Norwegian conditions, level, and indeed not in the concentrations of BOD. In addition, there are relatively high concentrations of hydrocarbons (oljeforbindelser) here represented by the sum of aliphatic hydrocarbons with C10–C40 reaching 40 mg/L. The source of the hydrocarbons has been shown to be predominately biosolids wastewater from a biogas reactor on the location.

It is obvious from the figure that the infiltration system is able to remove most of the pollutants except for PAH (Fig. 13.4). There are large and systematic variations in the concentrations of COD and NH_4–N, probably due to hydraulic conditions. The concentration of hydrocarbon shows high variation and concentration over a shorter period. The source of this increase is under investigation.

The mean value of the stable isotope 13C in the leachate (sigevann) is 15.8 and in the treated groundwater (B4) 10.3 (Fig. 13.5), indicating a dilution factor of 1.53. The values in the unaffected groundwater (B8, K1, and K2) and the river (elv) are all negative, with the upstream well (B8) showing the lowest values. The groundwater at the point of discharging into the river, and the river itself, shows no sign of pollution from the leachate.

The removal is far the greatest for COD with ca. 265 tonnes per year from the leachate (Table 13.1), followed by nitrogen (45 tonnes), mostly as ammonia. The removal is good for all parameters in Table 13.1 except for iron (Fe), which is mobilized from the soil due to emissions of large amount of oxygen-depleting compounds creating anaerobic conditions in parts of the aquifer.

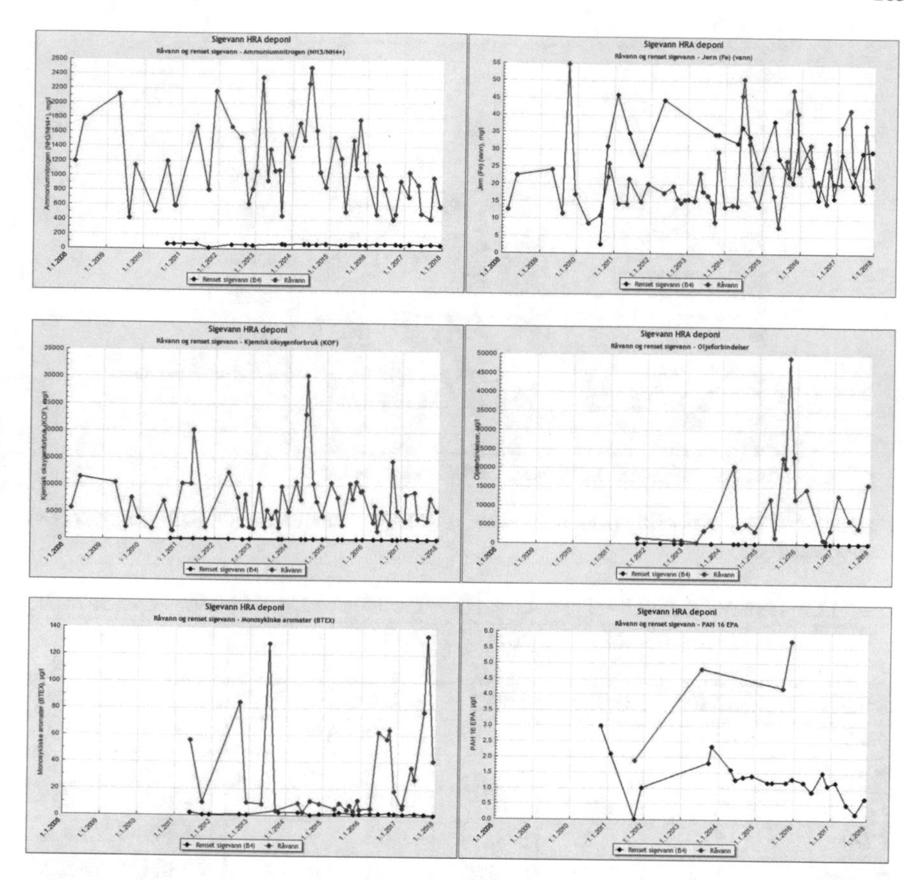

Fig. 13.4 Concentrations from 10 years of grab sampling of ammonia (NH_4–N), iron (Fe), chemical oxygen demand (KOF), hydrocarbons (oljeforbindelser), aromatics (BTEX), and PAH in raw and treated leachate from the active landfill. *Source* own data

13.4.2 Closed Landfill

Figure 13.6 shows the concentrations of selected parameters in raw and treated leachate at the closed landfill. This leachate is in the same way as the leachate from the active landfill, dominated by nitrogen and COD, but at much lower concentrations. The system is able to remove most of the pollutants. Pre-2004, the leachate was showing very high concentrations of pesticides, but they disappeared probably because a container was emptied. A striking effect is the disappearance of ammonium in the treated wastewater after 2013, corresponding with the end of a routine of adding phosphorus to the aeration pond. There is no corresponding reduction in the concentration of COD or a systematic change in the content of the stable isotope 13C in the treated wastewater (Fig. 13.7).

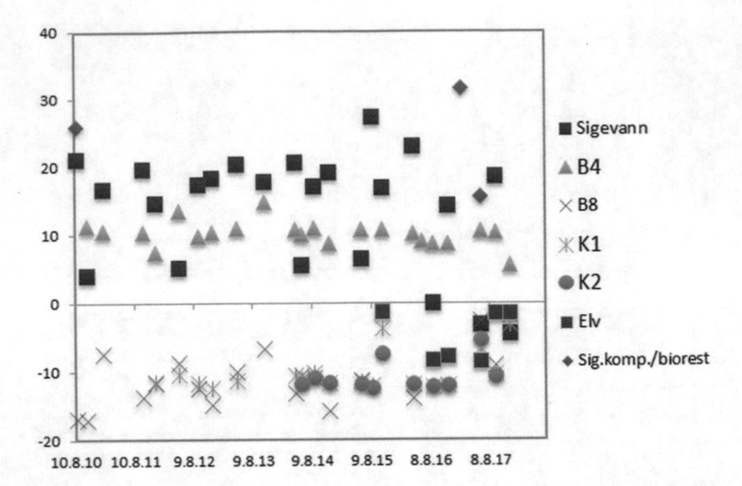

Fig. 13.5 ^{13}C (%) in leachate (sigevann), wells (B4, B8), groundwater (K1, K2), and river (elv). *Source* own data

Table 13.1 Removal (kg/year and percentage) from the infiltrated leachate from an active landfill

	2016	2015	2014	2013	%	Average	Non-diluted
NH_4–N	19,400	38,956	41,025	11,061	95	24,103	36,878
BOD	–	–	–	12,974	99	12,974	19,850
P, total	–	–	–	179	99	179	274
Fe	−7	−121	−180	−159	−61	−146	−223
COD	153,819	275,357	328,594	47,813	99	173,401	265,304
N, total	24,236	45,132	53,933	11,131	95	29,102	44,525
Suspended solids	–	–	–	93	–	13,456	20,588

Source own data

The mean 13C in the leachate is 6.1 and in the treated groundwater −4.3, indicating a dilution factor of 1.70 (Fig. 13.7).

The removal in the CW at the closed landfill shows that the plant operates well although the removal percentage for COD is low (Table 13.2). The COD from old landfill is usually very inert and hard to remove, but also shows low toxicity and constitutes minor problems. The annual removals including dilution are much lower than for the active landfill (Table 13.2).

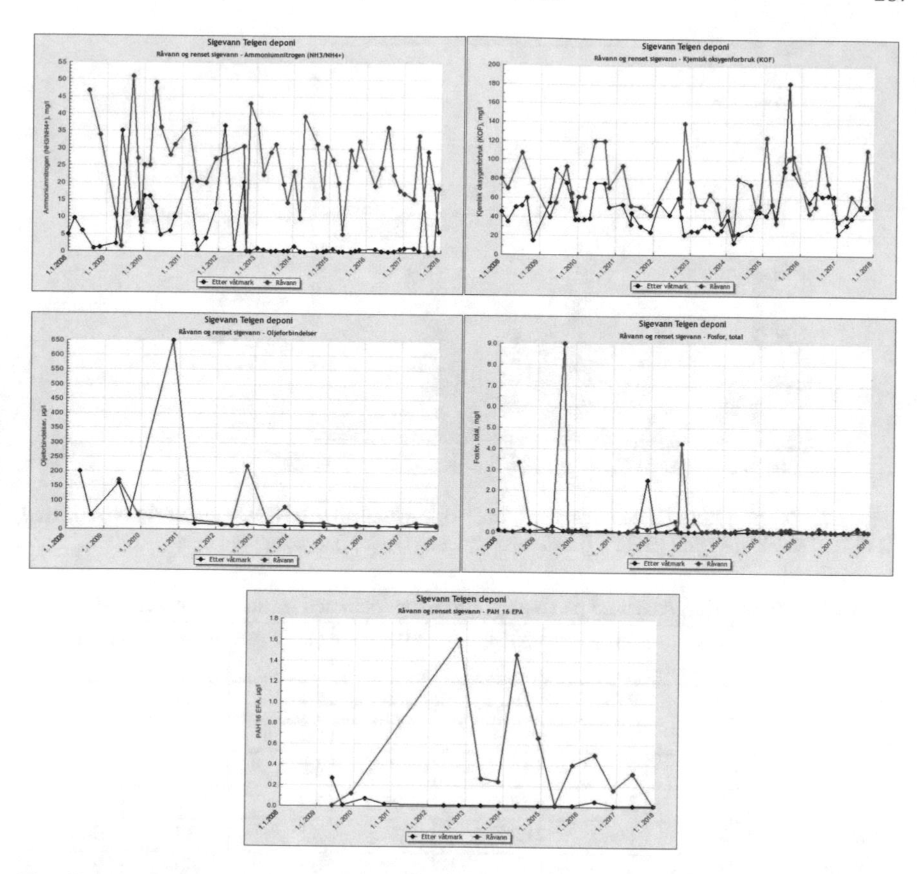

Fig. 13.6 Concentrations from 10 years of grab sampling of ammonia (NH_4–N), chemical oxygen demand (KOF), hydrocarbons (oljeforbindelser), phosphorus (F), and PAH in raw and treated leachate from the active landfill. *Source* own data

13.4.3 Quarry

The treatment at the quarry is designed to remove particles measured as suspended solids. The mean size of the dust from the area is about 100 m. According to filtration theory, a filter can remove from water particles that are up to 4 times finer than the mean grain size of the filter, indicating that the ultimate filter should be around 400 m. We had no means of separating the fractions according to this, so the finest filter used as core in the filter was made up of the fraction 1–2 mm and an outer layer of 2–4 mm gravel. As shown in Fig. 13.3, it seems to function well producing a visibly transparent water in the final dam before emission.

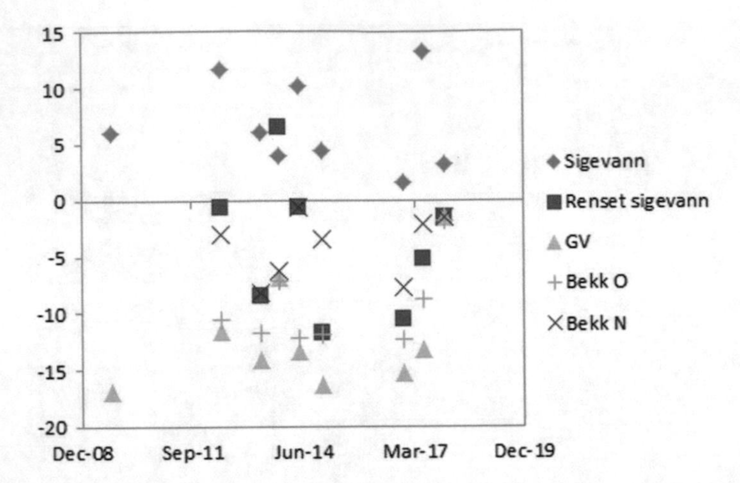

Fig. 13.7 ^{13}C (‰) in leachate (sigevann), treated leachate after wetland (renset sigevann), well (GV), and creek upstream (Bekk O) and downstream (Bekk N). *Source* own data

Table 13.2 Removal (kg/year and percentage) from the infiltrated leachate from a closed landfill.

	2016	2015	2014	2013	%	Average	Non-diluted
NH_4-N	625	813	769	540	89	604	1026
BOD				11	98	11	18
P, total	1	2	1	3	61	8	13
Fe	32	380	599	206	80	563	958
COD	154	108	686	626	32	535	909
N, total	722	–	–	557	92	640	1087
Suspended solids	1546	2390	1780	1490	97	2125	3613

Source own data

The concentrations of selected parameters in grab samples from the creek upstream and downstream the quarry are the same, except for a slightly higher value for electrical conductivity downstream (Fig. 13.8). The logger unit downstream, however, shows a different picture with repeated but infrequent violations of the limited value of 50 mg/l SS. Seen as weekly samples, the limit of 90% of the values does not exceed the limited value, showing good performance of the filtration dams.

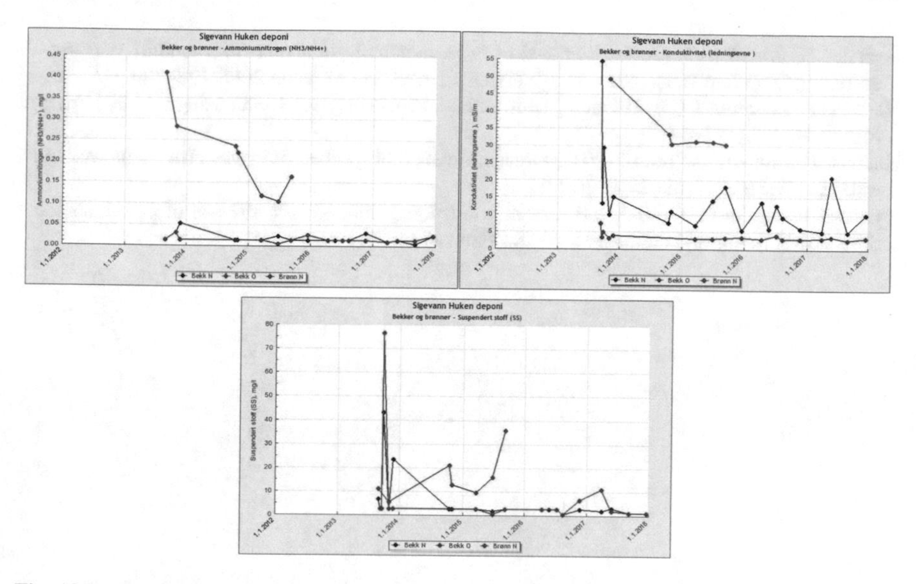

Fig. 13.8 Concentrations from 3 years of grab sampling of ammonia (NH_4–N), electrical conductivity and suspended solids in the creek upstream (Bekk O) and downstream (Bekk N) the quarry. Also included samples from abandoned well (Brønn N). *Source* own data

13.5 Conclusions

- The treatment systems operate well with the specific wastewaters, high-concentration leachate from waste in infiltration systems, low-concentration leachate in constructed wetlands, and wastewater from inert waste in filter dams.
- It cannot be expected to see changes in leachate due to changes in waste landfilling regulations.
- The heavy carbon stable isotope 13C is useful in tracing landfill leachate and to evaluate dilution into other water bodies.
- The adding of P to the aeration pond treating low-concentration leachate did not help in the removal of N; on the contrary, the concentration of ammonia was sharply decreased when the adding of P was discontinued.

References

For (2004) Regulations on recirculation and treatment of waste. Miljødepartementet, Oslo

Haarstad K (2008) Longterm leakage of DDT and other pesticides from a tree nursery landfill. Ground Water Monit Rem 28(4):107–111

Haarstad K, Borch H (2008) Halogenated compounds, PCB and pesticides in landfill leachate, downstream lake sediments and fish. J Environ Sci Health, Part A 43:1346–1352

Haarstad K, Mæhlum T (2008) Pesticides in Norwegian landfill leachate. Open Environ Biol Monit J 1:8–15

Haarstad K, Mæhlum T (2009) Removal of environmental pollutants from landfill leachate—a screening from four treatment plants. Norwegain Engl Summ. Vann 2(44):178–186

Haarstad K, Mæhlum T (2012) Tracing solid waste leachate in groundwater using ^{13}C. Isot Environ Health Stud 48(3):1–14

Haarstad K, Bavor J, Mæhlum T (2012) Organic and metallic pollutants in water treatment wetlands: a review. Water Sci Tech 65(1):76–99

Vymazal J, Brezinova T (2015) The use of constructed wetlands for removal of pesticides from agricultural runoff and drainage: a review. Environ Int 75:11–20

Printed in the United States
By Bookmasters